T0296997

CAMBRIDGE LIBRARY COLLECTION

Books of enduring scholarly value

Mathematical Sciences

From its pre-historic roots in simple counting to the algorithms powering modern desktop computers, from the genius of Archimedes to the genius of Einstein, advances in mathematical understanding and numerical techniques have been directly responsible for creating the modern world as we know it. This series will provide a library of the most influential publications and writers on mathematics in its broadest sense. As such, it will show not only the deep roots from which modern science and technology have grown, but also the astonishing breadth of application of mathematical techniques in the humanities and social sciences, and in everyday life.

Dynamics

Sir Horace Lamb (1849–1934) the British mathematician, wrote a number of influential works in classical physics. A pupil of Stokes and Clerk Maxwell, he taught for ten years as the first professor of mathematics at the University of Adelaide before returning to Britain to take up the post of professor of physics at the Victoria University of Manchester (where he had first studied mathematics at Owens College). As a teacher and writer his stated aim was clarity: 'somehow to make these dry bones live'. His Dynamics was first published in 1914 and the second edition, offered here, in 1923: it remained in print until the 1960s. It was intended as a sequel to his Statics (also reissued in this series), and like its predecessor is a textbook with examples.

Cambridge University Press has long been a pioneer in the reissuing of out-of-print titles from its own backlist, producing digital reprints of books that are still sought after by scholars and students but could not be reprinted economically using traditional technology. The Cambridge Library Collection extends this activity to a wider range of books which are still of importance to researchers and professionals, either for the source material they contain, or as landmarks in the history of their academic discipline.

Drawing from the world-renowned collections in the Cambridge University Library, and guided by the advice of experts in each subject area, Cambridge University Press is using state-of-the-art scanning machines in its own Printing House to capture the content of each book selected for inclusion. The files are processed to give a consistently clear, crisp image, and the books finished to the high quality standard for which the Press is recognised around the world. The latest print-on-demand technology ensures that the books will remain available indefinitely, and that orders for single or multiple copies can quickly be supplied.

The Cambridge Library Collection will bring back to life books of enduring scholarly value across a wide range of disciplines in the humanities and social sciences and in science and technology.

Dynamics

Horace Lamb

CAMBRIDGE UNIVERSITY PRESS

Cambridge New York Melbourne Madrid Cape Town Singapore São Paolo Delhi

Published in the United States of America by Cambridge University Press, New York

www.cambridge.org
Information on this title: www.cambridge.org/9781108005333

This edition first published 1961
This digitally printed version 2009

ISBN 978-1-108-00533-3

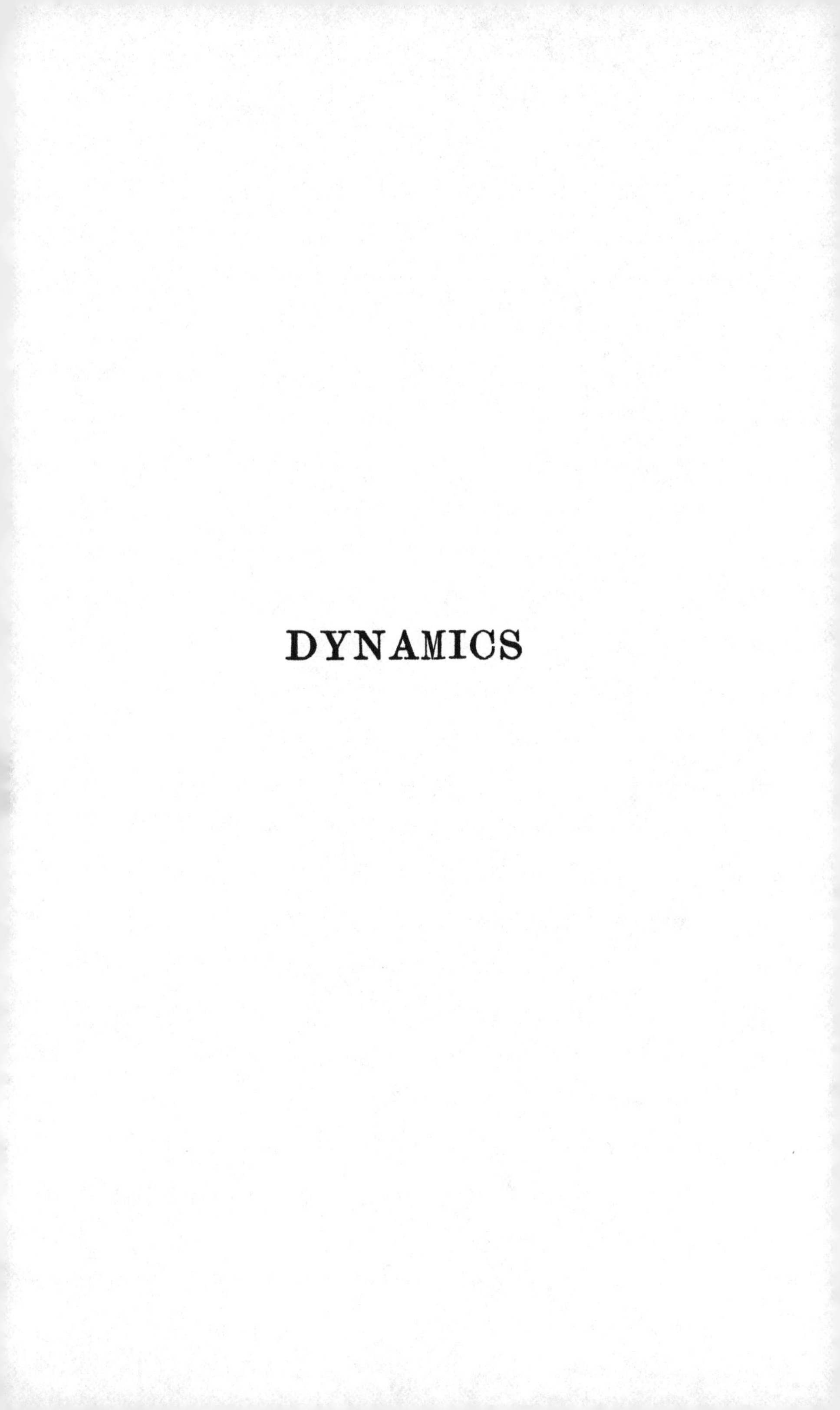

DYNAMICS

DYNAMICS

By

HORACE LAMB

CAMBRIDGE
AT THE UNIVERSITY PRESS
1961

PUBLISHED BY
THE SYNDICS OF THE CAMBRIDGE UNIVERSITY PRESS
Bentley House, 200 Euston Road, London, N.W.1
American Branch: 32 East 57th Street, New York 22, N.Y.
West African Office: P.O. Box 33, Ibadan, Nigeria

First edition 1914
Reprinted 1920
Second edition 1923
Reprinted 1926
1929
1942
1945
1946
1947
1951
1960
1961

Printed in Great Britain at the University Press, Cambridge
(Brooke Crutchley, University Printer)

PREFACE

THIS book is a sequel to a treatise on Statics published a little more than a year ago, and has a similar scope. To avoid repetitions, numerous references to the former volume are made.

A writer who undertakes to explain the elements of Dynamics has the choice, either to follow one or other of the traditional methods which, however effectual from a practical point of view, are open to criticism on logical grounds, or else to adopt a treatment so abstract that it is likely to bewilder rather than to assist the student who looks to learn something about the behaviour of actual bodies which he can see and handle. There is no doubt as to which is the proper course in a work like the present; and I have not hesitated to follow the method adopted by Maxwell, in his *Matter and Motion*, which forms, I think, the best elementary introduction to the 'absolute' system of Dynamics. Some account of the more abstract, if more logical, way of looking at dynamical questions is, however, given in its proper place, which is at the end, rather than at the beginning of the book.

There is some latitude of judgment as to the order in which the different parts of the subject should be taken. To many students it is more important that they should gain, as soon as possible, some power of dealing with the simpler questions of 'rigid' Dynamics, than that they should master the more intricate problems of 'central forces,' or of motion under various laws of resistance. This consideration has dictated the arrangement here adopted, but as the later chapters are largely independent of one another, they may be read in a different order without inconvenience.

Some pains have been taken in the matter of examples for practice. The standard collections, and the text-books of several generations, supply at first sight abundant material for appropriation, but they do not always reward the search for problems which are really exercises on dynamical theory, and not merely algebraical or trigonometrical puzzles in disguise. In the present treatise, preference has been given to examples which are simple rather than elaborate from the analytical point of view. Most of those which are in any degree original have been framed with this intention.

I am again greatly indebted to Prof. F. S. Carey, and Mr J. H. C. Searle, for their kindness in reading the proofs, and for various useful suggestions. The latter has moreover verified most of the examples. Miss Mary Taylor has also given kind assistance in the later stages of the passage through the press.

H. L.

THE UNIVERSITY,
MANCHESTER.
January 1914.

In this issue a few pages have been re-written, and a number of errors have been corrected, chiefly in the examples. An additional set of miscellaneous exercises has also been inserted near the end. It will be understood that the 'Appendix,' which remains unaltered, is intended merely as a summary of the 'Newtonian' standpoint, from which an elementary text-book is necessarily written.

H. L.

CAMBRIDGE.
February, 1923.

CONTENTS

CHAPTER I.

KINEMATICS OF RECTILINEAR MOTION.

CHAPTER II.

DYNAMICS OF RECTILINEAR MOTION.

CHAPTER III.

TWO-DIMENSIONAL KINEMATICS.

CHAPTER IV.

DYNAMICS OF A PARTICLE IN TWO DIMENSIONS. CARTESIAN COORDINATES.

CHAPTER V.

TANGENTIAL AND NORMAL ACCELERATIONS. CONSTRAINED MOTION.

CHAPTER VI.

MOTION OF A PAIR OF PARTICLES.

CHAPTER VII.

DYNAMICS OF A SYSTEM OF PARTICLES.

CHAPTER VIII.

DYNAMICS OF RIGID BODIES.

ROTATION ABOUT A FIXED AXIS.

CHAPTER IX.

DYNAMICS OF RIGID BODIES (CONTINUED). MOTION IN TWO DIMENSIONS.

CHAPTER X.

LAW OF GRAVITATION.

CHAPTER XI.

CENTRAL FORCES.

CHAPTER XII.

DISSIPATIVE FORCES.

CHAPTER XIII.

SYSTEMS OF TWO DEGREES OF FREEDOM.

APPENDIX.

CHAPTER I

KINEMATICS OF RECTILINEAR MOTION

1. Velocity.

We begin with the elementary kinematical notions relating to motion in a straight line.

The position P of a moving point at any given instant t, i.e. at the instant when t units of time have elapsed from some particular epoch which is taken as the zero of reckoning, is specified by its distance x from some fixed point O on the line, this distance being reckoned positive or negative according to the side of O on which P lies. In any given case of motion x is then a definite and continuous function of t. When the form of this

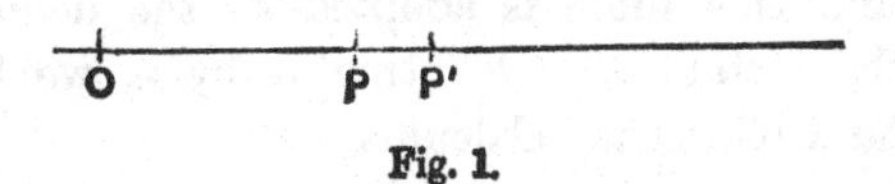

Fig. 1.

function is known it is often convenient to represent it graphically, in an auxiliary diagram, by means of a curve constructed with t as abscissa and x as ordinate. This may be called the 'space-time curve*.'

If equal spaces are described, in the same sense, in any two equal intervals of time the moving point is said to have a 'constant

* There are various classical experiments, especially in Acoustics, where such curves are produced by mechanical or optical contrivances, as e.g. in studying the nature of the vibration of a tuning-fork, or of a point of a piano-wire. On a different scale we have the graphical records of the oscillations of the barometer, &c.

velocity*,' the magnitude of this velocity being specified by the space described in the unit time. This space must of course have the proper sign attributed to it. Hence if the position change from x_0 to x in the time t, the velocity is $(x - x_0)/t$. Denoting this by u, we have

$$x = x_0 + ut. \quad \text{.........................(1)}$$

The space-time curve is therefore in this case a straight line.

If the velocity is not constant, the space described in any interval of time, divided by that interval, gives a result which may be called the 'mean velocity' in the interval. That is to say, a point having a constant velocity equal to this would describe an equal space in the same interval. Thus if, in Fig. 1, the points P, P' denote the positions at the instants t, t', respectively, the mean velocity in the interval $t' - t$ is

$$PP'/(t' - t), \text{ or } (x' - x)/(t' - t),$$

where x, x' are the abscissae of P, P', respectively. If we write $x' = x + \delta x$, $t' = t + \delta t$, so that δx, δt denote corresponding increments of x and t, the mean velocity is denoted by

$$\frac{\delta x}{\delta t}.$$

In all cases which it is necessary to consider†, this fraction has a definite limiting value when the interval δt is indefinitely diminished; and this limit is adopted as the definition of the 'velocity at the instant t.' Denoting it by u, we have, in the notation of the Differential Calculus,

$$u = \frac{dx}{dt}. \quad \text{.........................(2)}$$

In the space-time curve, above referred to, the velocity at any instant is represented by the 'gradient' of the curve at the

* The phrase 'uniform velocity' is often used; but it seems preferable to use the word 'constant' when invariability in *time* is meant, the term 'uniform' being reserved to express invariability in *space*. Thus a constant field of force would be one which does not alter with the time; whilst a uniform field would be one which has the same properties at every point. Cf. Maxwell, *Matter and Motion*, London, 1876, pp. 24, 25.

† Except in the conventional treatment of problems of impact, where we may have different limits according as δt is positive or negative. See Chap. VI.

corresponding point, i.e. by the trigonometrical tangent of the angle which the tangent line to the curve, drawn in the direction of t increasing, makes with the positive direction of the axis of t, the differential coefficient dx/dt corresponding exactly to the dy/dx ordinarily employed in the Calculus.

The velocity u is in general a definite and continuous function of t, and may be represented graphically by a curve, called the 'velocity-time' curve, constructed with t as abscissa and u as ordinate. Since, by integration of (2), we have

$$x = \int u\,dt, \quad \text{.........................(3)}$$

it appears that the area swept over by the ordinate of this curve in any interval of time gives the space described in that interval. The integral in (3) corresponds in fact to the ordinary $\int y\,dx$ of the Calculus.

If P_1, P_2 be two moving points, and x_1, x_2 their coordinates, then putting

$$\xi = P_1P_2 = x_2 - x_1, \quad \text{....................(4)}$$

we have

$$\frac{d\xi}{dt} = \frac{dx_2}{dt} - \frac{dx_1}{dt}, \quad \text{.......................(5)}$$

i.e. the velocity of P_2 relative to P_1 is the difference of the velocities of P_1 and P_2.

2. Acceleration.

When the velocity increases by equal amounts (of the same sign) in any two equal intervals of time, the motion is said to be 'uniformly accelerated,' or (preferably) the moving point is said to have a 'constant acceleration'; and the amount of this acceleration is specified by the velocity gained per unit time*. Hence if the velocity change from u_0 to u in the time t, the acceleration is $(u - u_0)/t$. Denoting it by α, we have

$$u = u_0 + \alpha t. \quad \text{.........................(1)}$$

The velocity-time curve is in this case a straight line.

* A negative acceleration is sometimes described as a 'retardation.'

In the general case, the increment of the velocity in any interval of time, divided by that interval, gives a result which may be called the 'mean rate of acceleration,' or briefly the 'mean acceleration' in that interval. That is to say, a point having a constant acceleration equal to this would have its velocity changed by the same amount in the same interval. Hence if u, u' denote the velocities at the instants t, t', respectively, the mean acceleration in the interval $t'-t$ is $(u'-u)/(t'-t)$, or

$$\frac{\delta u}{\delta t},$$

if we write $u' = u + \delta u$, $t' = t + \delta t$.

In all important cases, except those of impact, this fraction has a definite limiting value when δt is indefinitely diminished, and this limit is adopted as the definition of the 'acceleration at the instant t.' Hence, denoting the acceleration by α, we have

$$\alpha = \frac{du}{dt}. \quad \text{..........................(2)}$$

It appears that the acceleration is represented by the gradient, in the velocity-time curve.

Since $u = dx/dt$, we have

$$\alpha = \frac{d^2x}{dt^2}. \quad \text{..........................(3)}$$

It is often convenient to use the 'fluxional' notation, in which differentiations with respect to the time are denoted by dots placed over the symbol of the dependent variable. Thus the velocity may be denoted by $\dot{x}$, and the acceleration by $\dot{u}$ or $\ddot{x}$.

Another very important expression for the acceleration is obtained if we regard the velocity (u) as a function of the position (x). It is to be noted that, since the moving point may pass through a given position more than once, there may be more than one value of u corresponding to a given value of x. But if we fix our attention on one of these we have

$$\alpha = \frac{du}{dt} = \frac{du}{dx}\frac{dx}{dt} = u\frac{du}{dx}. \quad \text{.................(4)}$$

If x_1, x_2 be the coordinates of two moving points P_1, P_2 and if

$$\xi = P_1P_2 = x_2 - x_1, \quad \text{.....................(5)}$$

we have
$$\frac{d^2\xi}{dt^2} = \frac{d^2x_2}{dt^2} - \frac{d^2x_1}{dt^2}; \quad \text{.....................(6)}$$

i.e. the acceleration of P_2 relative to P_1 is the difference of the accelerations of these points. In particular, if the velocity of P_1 be constant, so that $d^2x_1/dt^2 = 0$, we have

$$\frac{d^2\xi}{dt^2} = \frac{d^2x_2}{dt^2};$$

i.e. the acceleration of a moving point is the same whether it be referred to a fixed origin, or to an origin which is in motion with constant velocity.

Ex. 1. If x be a quadratic function of t, say

$$x = \tfrac{1}{2}(\alpha t^2 + 2\beta t + \gamma), \quad \text{.............................(7)}$$

we have
$$\dot{x} = \alpha t + \beta, \quad \ddot{x} = \alpha. \quad \text{.............................(8)}$$

The acceleration is therefore constant.

Ex. 2. If
$$x = a\cos(nt + \epsilon), \quad \text{.............................(9)}$$

we have
$$\dot{x} = -na\sin(nt + \epsilon), \quad \text{.........................(10)}$$

$$\ddot{x} = -n^2a\cos(nt + \epsilon) = -n^2x. \quad \text{.......................(11)}$$

The space-time curve is a curve of sines; and the velocity-time curve is a similar curve whose zero points synchronize with the maxima and minima of the former. See Fig. 4, p. 26.

Ex. 3. If
$$x = Ae^{nt} + Be^{-nt}, \quad \text{.............................(12)}$$

we have
$$\dot{x} = nAe^{nt} - nBe^{-nt}, \quad \text{.............................(13)}$$

$$\ddot{x} = n^2Ae^{nt} + n^2Be^{-nt} = n^2x. \quad \text{.....................(14)}$$

Ex. 4. If u^2 is a quadratic function of x, say

$$u^2 = Ax^2 + 2Bx + C, \quad \text{.............................(15)}$$

we have
$$u\frac{du}{dx} = Ax + B. \quad \text{.............................(16)}$$

The acceleration therefore varies as the distance from the point $x = -B/A$, unless $A = 0$, in which case it is constant.

3. Units and Dimensions.

The unit of *length* is generally defined by some material standard, or as some convenient multiple or submultiple thereof.

Thus in the metric system we have the metre, with its subdivisions the decimetre, centimetre, etc., and its multiple the kilometre. The standard metre was originally intended to represent the ten-millionth part of a quadrant of the earth's meridian as closely as possible. The agreement, though since found not to be exact*, is very close; but the practical and legal definition of the metre is of course by reference to the material standard, and not to the earth's dimensions. The reason for this particular choice of the standard was that on the decimal division of the quadrant a minute of latitude on the earth's surface corresponds to a kilometre.

In the British system of measurement we have the standard yard, with its subdivisions of foot and inch, and its multiple the mile. There is here no simple relation to the earth's dimensions, and the sea-mile, which corresponds to a (sexagesimal) minute of latitude, differs considerably from the statute mile of 1760 yards.

The relations between the two systems of length measurement are shewn by the following table†, which gives the factors required to reduce the various British units to centimetres, with their reciprocals, to four significant figures.

	Cm.	Reciprocals
Inch	2·540	·3937
Foot	30·48	·03281
Yard	91·44	·01094
Mile	$1{\cdot}609 \times 10^5$	$6{\cdot}214 \times 10^{-6}$
Sea-mile	$1{\cdot}852 \times 10^5$	$5{\cdot}398 \times 10^{-6}$

Owing to the decimal basis of the metric system the relations between other units can be read off at once. Thus a mile is 1609 metres; and a kilometre ($=10^5$ cm.) is 1094 yards.

* The most authoritative value for the length of the earth-quadrant is 10,001,869 metres (Clarke, *Geodesy*, London, 1880).

† Taken from Everett's *Units and Physical Constants*.

In this book we shall employ mainly the foot or the centimetre, the latter being the unit now generally adopted in scientific measurements.

For the measurement of *time* some system based on the earth's rotation is universally adopted, all clocks and watches being regulated ultimately by reference to this. From a purely scientific standpoint the simplest standard would be the sidereal day, i.e. the period of a complete rotation of the earth relatively to the fixed stars; but this would have the serious inconvenience that ordinary time-keepers would have to be discarded for scientific purposes. The units commonly employed are based on the 'mean solar day,' i.e. the average interval between two successive transits of the sun across any given meridian. This bears to the sidereal day the ratio 1·00274. In scientific measurements the unit is generally the mean solar second, i.e. the $\frac{1}{86400}$th part of the mean solar day, whilst for practical purposes the hour, or the day, or year, are of course often employed.

The units of length and time are in the first instance arbitrary and independent. Those of velocity and acceleration depend upon them, and are therefore classed as 'derived' units. The unit velocity, i.e. the velocity which is represented by the number 1 according to the definition of Art. 1, is such that a unit of length is described in the unit time. Its magnitude therefore varies directly as that of the unit length, and inversely as that of the unit time. It is therefore said to be of one 'dimension' in length, and *minus* one dimension in time. If we introduce symbols L and T to represent the magnitudes of the units of length and time respectively, this may be expressed concisely by saying that the unit velocity is L/T, or LT^{-1}. The number which expresses any actual velocity will of course vary *inversely* as the magnitude of this unit. When it is necessary to specify the particular unit adopted we may do this by the addition of words such as 'feet per second,' or 'miles per hour,' or more briefly 'ft./sec.,' or 'mile/hr.'

The unit acceleration is such that a unit of velocity is acquired in the unit time. Its dimensions are therefore indicated by $\mathsf{LT}^{-1}/\mathsf{T}$, or L/T^2, or LT^{-2}. An actual acceleration may be specified by a number followed by the words 'feet per second per second,' or 'miles per hour per hour,' &c., as the case may be. These indications are conveniently abbreviated into 'ft./sec.^{2},' or 'mile/hr.^{2},' &c. The *double* reference to time in the specification of an acceleration is insisted upon sufficiently in elementary works on Mechanics; but the student may notice that it is indicated again by the form of the differential coefficient d^2x/dt^2.

Ex. To translate from the mile and hour to the foot and second as fundamental units we may write

$$L' = 5280\ L, \quad T' = 3600\ T.$$

Hence
$$L'/T' = \tfrac{22}{15}\ L/T.$$

The units of velocity on the two systems are therefore as 22 to 15, and the numerical values of any given velocity as 15 to 22. Thus a speed of 60 miles an hour is equivalent to 88 feet per second.

Again
$$L'/T'^2 = \tfrac{11}{27000}\ L/T^2,$$

so that the units of acceleration are as 11 to 27000.

4. The Acceleration of Gravity.

It may be taken as a result of experiment, although the best experimental evidence is indirect (Art. 11), that a particle falling freely at any given place near the earth's surface has a definite acceleration g, the same for all bodies.

The precise value of g varies however with the locality, increasing from the equator towards the poles, and diminishing slightly with altitude above the sea-level. There are also local irregularities of comparatively small amount. According to recent investigations*, the value of g at sea-level is represented to a high degree of accuracy by the formula

$$g = 978{\cdot}03\ (1 + {\cdot}0053 \sin^2 \phi), \quad \text{..............(1)}$$

where ϕ is the latitude, the units being the centimetre and the second. In terms of the foot and the second this makes

$$g = 32{\cdot}088\ (1 + {\cdot}0053 \sin^2 \phi). \quad \text{..................(2)}$$

The total variation from equator to pole is therefore a little more than one-half per cent. In latitude 45° we have $g = 980{\cdot}62$, or 32·173, according to the units chosen.

* F. R. Helmert, *Encycl. d. math. Wiss.*, Bd. vi.

The variation with altitude is given by the formula

$$g' = g(1 - \cdot 0000003h), \quad \text{.....................(3)}$$

where g is the value at the sea-level, and g' that at a height h (in metres) above this level. This variation is accordingly for most purposes quite unimportant.

For illustrative purposes it is in general sufficiently accurate to assume $g = 980$ cm./sec.^{2}, or $= 32$ ft./sec.^{2}, the latter number being specially convenient for mental calculations, on account of its divisibility.

6. Differential Equations.

The formulæ (3) and (4) of Art. 2 enable us to find at once expressions for the acceleration when the position (x) is given as a function of the time, or the velocity as a function of the position. But in dynamical questions we have more usually to deal with the inverse problem, where the acceleration is given as a function of the time or the position, or both, or possibly of the velocity as well, and it is required to find the velocity and the position at any assigned instant. We notice here one or two of the more important types of differential equation which thus present themselves, and the corresponding methods of solution.

1. The acceleration may be given as a function of the time; thus

$$\frac{d^2x}{dt^2} = f(t). \quad \text{..........................(1)}$$

This can be integrated at once with respect to t. We have

$$\frac{dx}{dt} = \int f(t)\,dt + A = f_1(t) + A, \quad \text{..............(2)}$$

say, where $f_1(t)$ stands for *any* indefinite integral of $f(t)$, and the additive constant A is arbitrary. Integrating again we have

$$x = \int f_1(t)\,dt + At + B, \quad \text{.....................(3)}$$

where B is a second arbitrary constant.

The reason why two arbitrary constants appear in this solution is that a point may be supposed to start at a given instant from any arbitrary position with any arbitrary velocity, and to be governed as to its subsequent motion by the law expressed in (1). A solution, to be general, must therefore be capable of adjustment to these arbitrary initial conditions. See Ex. 1 below. The reason why the arbitrary element in the solution occurs in the particular form $At+B$ is that the superposition of any constant velocity does not affect the acceleration.

2. The acceleration may be given as a function of the position; thus

$$\frac{d^2x}{dt^2}=f(x). \quad \text{.........................(4)}$$

If we multiply both sides of this equation by dx/dt it becomes integrable with respect to t; thus

$$\frac{dx}{dt}\frac{d^2x}{dt^2}=f(x)\frac{dx}{dt},$$

$$\frac{1}{2}\left(\frac{dx}{dt}\right)^2=\int f(x)\frac{dx}{dt}\,dt+A$$

$$=\int f(x)\,dx+A, \quad \text{....................(5)}$$

by the ordinary formula for change of variable in an indefinite integral.

This process will occur over and over again in our subject, and the result (5) has usually an important interpretation as the 'equation of energy.' It may be obtained in a slightly different manner as follows. Taking x as the independent variable we have, in place of (4),

$$u\frac{du}{dx}=f(x), \quad \text{............................(6)}$$

which is integrable with respect to x; thus

$$\tfrac{1}{2}u^2=\int f(x)\,dx+A. \quad \text{...................(7)}$$

In either way we obtain u^2 or $\dot{x}^2$ as a function of x, say

$$\left(\frac{dx}{dt}\right)^2=F(x), \quad \text{......................(8)}$$

whence

$$\frac{dx}{dt}=\pm\sqrt{\{F(x)\}}. \quad \text{......................(9)}$$

The two signs relate to the two directions in which the moving point may pass through the position x. In particular problems both cases may require to be taken into consideration.

For the further integration we may write (9) in the form

$$\frac{dt}{dx} = \pm \frac{1}{\sqrt{\{F(x)\}}}, \quad \text{.....................(10)}$$

whence, integrating with respect to x,

$$t = \pm \int \frac{dx}{\sqrt{\{F(x)\}}} + B. \quad \text{.....................(11)}$$

This solution really contains *two* arbitrary constants, since one is already involved in the value of $F(x)$. That one of the arbitrary constants would consist in an addition to t might have been foreseen from the form of the differential equation (4), which is unaltered in form if the origin of t be shifted.

Such further types of differential equation as present themselves will be most conveniently dealt with as they arise.

Ex. 1. In the case of a particle moving vertically under gravity, we have, if the positive direction of x be upwards,

$$\frac{d^2x}{dt^2} = -g, \quad \text{.....................................(12)}$$

$$\frac{dx}{dt} = -gt + A, \quad x = -\tfrac{1}{2}gt^2 + At + B. \quad \text{..................(13)}$$

If the initial conditions are that $x = x_0$, $\dot{x} = u_0$, for $t = 0$, we have $u_0 = A$, $x_0 = B$, whence

$$\dot{x} = u_0 - gt, \quad x = x_0 + u_0 t - \tfrac{1}{2}gt^2. \quad \text{.....................(14)}$$

The space-time curve is therefore a parabola, and the velocity-time curve a straight line, as shewn in Fig. 2 (p. 12), which relates to the case of $x_0 = 12$, $u_0 = 48$, in foot-second units.

Treating the question by the second method we have

$$u\frac{du}{dx} = -g, \quad \text{.................................(15)}$$

$$\tfrac{1}{2}u^2 = -gx + C, \quad \text{.................................(16)}$$

whence

$$\tfrac{1}{2}u^2 + gx = \tfrac{1}{2}u_0^2 + gx_0, \quad \text{..............................(17)}$$

a result which may also be deduced from (14).

If x_1, x_2 and u_1, u_2 be the positions and velocities at the instants t_1, t_2, respectively, we have from (14)

$$\frac{x_2-x_1}{t_2-t_1}=u_0-\tfrac{1}{2}g(t_1+t_2)=\tfrac{1}{2}(u_1+u_2). \quad \text{.................(18)}$$

The mean velocity in any interval of time is therefore equal to the velocity at the middle instant of the interval, and also to the arithmetic mean of the initial and final velocities. These are, moreover, obvious corollaries from the fact that the velocity curve is straight.

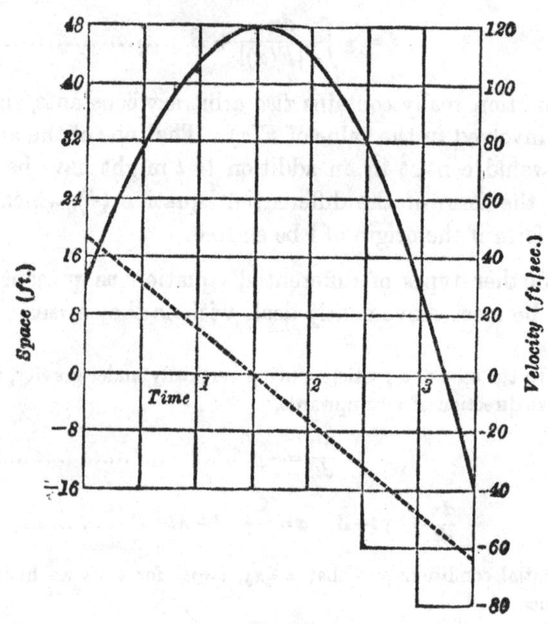

Fig. 2.

Ex. 2. If the acceleration be a circular function of the time, say

$$\frac{d^2x}{dt^2}=f\cos nt, \quad \text{............................(19)}$$

we find
$$\frac{dx}{dt}=\frac{f}{n}\sin nt+A, \quad x=-\frac{f}{n^2}\cos nt+At+B. \quad \text{...............(20)}$$

If $x=x_0$, $\dot{x}=u_0$, for $t=0$, we have $u_0=A$, $x_0=-f/n^2+B$,

whence
$$\dot{x}=u_0+\frac{f}{n}\sin nt, \quad x=x_0+u_0t+\frac{f}{n^2}(1-\cos nt). \quad \text{...............(21)}$$

Ex. 3. If the acceleration be directed always towards a fixed point (the origin) in the line of motion, and vary as the distance from this point, we have

$$\frac{d^2x}{dt^2} = -n^2 x, \quad \text{.................................(22)}$$

where n^2 is a given positive constant. The *minus* sign is required by the fact that $\ddot{x}$ is negative when x is positive, and *vice versâ*. An equivalent form is

$$u\frac{du}{dx} = -n^2 x. \quad \text{.................................(23)}$$

From this, or from (22), we have, by the methods above explained,

$$u^2 = C - n^2 x^2. \quad \text{.................................(24)}$$

Since u^2 is essentially positive, the arbitrary constant C must be positive, so that we may write $C = n^2 a^2$, where a is arbitrary. Hence

$$\left(\frac{dx}{dt}\right)^2 = n^2 (a^2 - x^2). \quad \text{...........................(25)}$$

This shews that, whatever the initial conditions, the value of x must be confined between certain limits $\pm a$. We are therefore at liberty to write

$$x = a \cos \theta, \quad \text{.................................(26)}$$

where θ is a new variable. Substituting in (25) we find

$$\left(\frac{d\theta}{dt}\right)^2 = n^2, \quad \frac{d\theta}{dt} = \pm n. \quad \text{.........................(27)}$$

Hence $$\theta = \pm(nt + \epsilon), \quad \text{.................................(28)}$$

where ϵ is a second arbitrary constant. The general solution of (22) is therefore

$$x = a \cos (nt + \epsilon), \quad \text{.................................(29)}$$

involving the two arbitrary constants a and ϵ.

If we put

$$a \cos \epsilon = A, \quad a \sin \epsilon = -B, \quad \text{.........................(30)}$$

we have the alternative form

$$x = A \cos nt + B \sin nt. \quad \text{............................(31)}$$

Conversely we can pass from (31), where A and B are arbitrary, to the form (29), the values of a and ϵ being derived from (30).

The differential equation (22) occurs so often in dynamical problems that it is well to remember, once for all, that the general solution has the form (29) or (31). It is indeed evident (cf. Art. 2, Ex. 2) that the formula (29) or (31) does in fact satisfy (22), and that since there are two arbitrary constants at our disposal, it can be made to fit any prescribed initial conditions. For instance, if $x = x_0$, $\dot{x} = u_0$, for $t = 0$, we have, in (31), $x_0 = A$, $u_0 = nB$, whence

$$x = x_0 \cos nt + \frac{u_0}{n} \sin nt. \quad \text{..........................(32)}$$

Ex. 4. If the acceleration be always *from* the origin, and proportional to the distance, the differential equation has the form

$$\frac{d^2x}{dt^2}=n^2x. \quad \text{.................................(33)}$$

It has been seen (Art. 2, Ex. 3) that this is satisfied by

$$x=Ae^{nt}+Be^{-nt}, \quad \text{.................................(34)}$$

and since there are two arbitrary constants the solution is complete. Thus if $x=x_0$, $\dot{x}=u_0$, for $t=0$, we have $x_0=A+B$, $u_0=n(A-B)$, whence

$$x=x_0\cosh nt+\frac{u_0}{n}\sinh nt, \quad \text{.............................(35)}$$

in the notation of hyperbolic functions. This result may be compared with (32).

EXAMPLES. I.

1. A steamer takes a time t_1 to travel a distance a up a river, and a time t_2 to return. Prove that the speed of the steamer relative to the water is

$$\frac{a(t_1+t_2)}{2t_1t_2}.$$

Shew that this is greater than the speed calculated from the arithmetic mean of the times.

2. Assuming that the acceleration of gravity at the distance of the moon is $\frac{1}{3600}$ of its value at the earth's surface, express it in terms of the kilometre and hour as units. [35·3.]

3. If the greatest admissible acceleration or retardation of a train be 3 ft./sec.², find the least time from one station to another at a distance of 10 miles, the maximum speed being 60 miles per hour. [10 m. 30 s.]

4. A particle is projected vertically upwards, and is at a height h after t_1 secs., and again after t_2 secs. Prove that

$$h=\tfrac{1}{2}gt_1t_2,$$

and that the initial velocity was

$$\tfrac{1}{2}g(t_1+t_2).$$

5. The speed of a train increases at a constant rate α from 0 to v, then remains constant for an interval, and finally decreases to 0 at a constant rate β. If l be the total distance described, prove that the total time occupied is

$$\frac{l}{v}+\frac{1}{2}v\left(\frac{1}{\alpha}+\frac{1}{\beta}\right).$$

For what value of v is the time least?

6. A bullet describes two consecutive spaces of 150 ft. in ·754 sec. and ·764 sec., respectively. Find its retardation, and its velocity when half-way.

[3·43 ft./sec.^{2}; 198 ft./sec.]

7. A bullet travelling horizontally, pierces in succession three thin screens placed at equal distances a apart. If the time from the first to the second be t_1, and from the second to the third t_2, prove that the retardation (assumed to be constant) is

$$\frac{2a(t_2-t_1)}{t_1t_2(t_1+t_2)},$$

and that the velocity at the middle screen is

$$\frac{a(t_1^2+t_2^2)}{t_1t_2(t_1+t_2)}.$$

8. If the coordinates of a point moving with constant acceleration be x_1, x_2, x_3 at the instants t_1, t_2, t_3, respectively, prove that the acceleration is

$$\frac{2\{(x_2-x_3)t_1+(x_3-x_1)t_2+(x_1-x_2)t_3\}}{(t_2-t_3)(t_3-t_1)(t_1-t_2)}.$$

9. If a point move with constant acceleration, the *space*-average of the velocity over any distance is

$$\frac{2}{3}\cdot\frac{u_1^2+u_1u_2+u_2^2}{u_1+u_2},$$

where u_1, u_2 are the initial and final velocities.

Is this greater or less than the *time*-average?

10. Shew, graphically or otherwise, that the following three quantities, viz. (1) the mean velocity in a given interval of time, (2) the arithmetic mean of the initial and final velocities, and (3) the velocity at the middle instant of the interval, are in general distinct.

Shew that no two of these quantities can be always equal unless the acceleration be constant.

11. Prove that if a curve be constructed with the space described by a moving point as abscissa, and the velocity as ordinate, the acceleration will be represented by the subnormal.

Illustrate this by the case of constant acceleration.

12. Prove that if the time (t) be regarded as a function of the position (x) the retardation is

$$u^3\frac{d^2t}{dx^2},$$

where u is the velocity.

13. If t is a quadratic function of x, the acceleration varies inversely as the cube of the distance from a fixed point.

14. If x^2 is a quadratic function of t, the acceleration varies as $1/x^3$, except in a particular case.

15. If a point describes a parabola with constant velocity, the foot of the ordinate has an acceleration varying inversely as the square of the distance from the directrix.

16. A point P describes a straight line with constant velocity, and Q is the projection of P on a fixed straight line from a fixed centre O. Prove that Q describes its straight line with an acceleration varying as the cube of the distance from a fixed point on it, except in a particular case.

How is the fixed point in question determined geometrically?

17. Prove that a point cannot move so that its velocity shall vary as the distance it has travelled from rest.

Can it move so that its velocity varies as the square root of the distance?

18. Prove that if a point moves with a velocity varying as its distance from a fixed point which it is approaching, it will not quite reach that point in any finite time.

19. The velocity of an airship moving horizontally with the engines shut off was observed at successive instants of time. When the reciprocal of the velocity was plotted against the time the graph was found to be a straight line. Prove that the retardation varied as the square of the velocity.

What is the corresponding graph of the relation between velocity and space described?

20. A crank OQ revolves about O with constant angular velocity ω, and a connecting rod QP is hinged to it at Q, whilst P is constrained to move in a straight line through O [S. 15]. If $OQ=a$, $QP=l$, and θ is the angle QOP, prove that the acceleration of P is

$$\omega^2 a \cos\theta + \frac{\omega^2 a^2}{l} \cos 2\theta,$$

approximately, if a/l be small.

CHAPTER II

DYNAMICS OF RECTILINEAR MOTION

6. Dynamical Principles. Gravitational Units.

The object of the science of Dynamics is to investigate the motion of bodies as affected by the forces which act upon them. Some system of physical assumptions, to be justified ultimately by comparison with experience, is therefore necessary as a basis. For the present we consider specially cases where the motion and the forces are in one straight line.

The subject may be approached from different points of view, and the fundamental assumptions may consequently be framed in various ways, but the differences must of course be mainly formal, and must lead to the same results when applied to any actual dynamical problem. In the present Article, and the following one, two distinct systems are explained. Both systems start from the idea of 'force' as a primary notion, but they differ as to the principles on which different forces are compared.

The first assumption which we make is in each case that embodied in Newton's 'First Law,' to the effect that a material particle* persists in its state of rest, or of motion in a straight line with constant velocity, except in so far as it is compelled to change that state by the action of force upon it†. In other words, acceleration is the result of force, and ceases with the force. This is sometimes called the 'law of inertia.'

* For the sense in which the word 'particle' is used see *Statics*, Art. 6.

† "Corpus omne perseverare in statu suo quiescendi vel movendi uniformiter in directum, nisi quatenus illud a viribus impressis cogitur statum suum mutare." (*Philosophiae naturalis principia mathematica*, London, 1687.) The law of inertia dates from Galileo (1638).

The first of the two systems to be explained proceeds from a purely terrestrial and local standpoint, and adopts the system of force-measurement, in terms of gravity, with which we are familiar in Statics. The physical assumption now introduced is that the acceleration produced in a given body by the action of any force is proportional to the magnitude of that force. It can therefore be found by comparison with the known acceleration (g) produced in the same body when falling freely under its own gravity. It will be noted that the assumption here made is the simplest that we can frame consistent with the law of inertia, although its validity must rest of course not on its simplicity but on its conformity with experience.

It is convenient here to distinguish between the 'weight' and the 'gravity' of a body. When we say that a body has a 'weight W,' as determined by the balance, we mean that the downward pressure which it exerts on its supports, when at rest, is in the ratio W to the pressure exerted under like circumstances, and therefore in the same locality, by the standard pound or kilogramme, or whatever the unit is*. The 'weight' of a body is therefore, on this definition, a numerical constant attached to it; it is the same at all places, since a variation in the intensity of gravity would affect the body and the standard alike.

By the 'gravity' of a body, on the other hand, is meant the downward pull of the earth upon it. This is, on statical principles, *equal to* the pressure which the body exerts on its supports when at rest, but is of course to be distinguished from it. It is known to vary somewhat with the latitude, and with altitude above sea-level.

It follows from the principle above laid down that if α be the acceleration produced in a body of weight W by a force P (i.e. by a force equal to the gravity of a weight P) we have

$$\frac{\alpha}{g} = \frac{P}{W}. \qquad (1)$$

This is, from the present point of view, the fundamental equation of Dynamics.

* The sense in which the word 'weight' is here temporarily used is of course only one of the various meanings which it bears in ordinary language. See Art. 8.

The force required to produce an acceleration α in a body whose weight is W is therefore

$$P = \frac{W}{g} \cdot \alpha. \qquad \text{.........................(2)}$$

The factor W/g accordingly measures the 'inertia' of the body, i.e. the degree of sluggishness with which it yields to the action of force. It is now usual to designate inertia, when regarded as a measurable quality, by the term 'mass'; the mass of the body on the present reckoning is therefore W/g.

The above procedure is simple and straightforward, and perfectly accurate if the meanings of the symbols are carefully observed; but it is open to the reproach that the unit of force implied, viz. the attraction of the earth on the standard body (pound, kilogramme, &c.) varies somewhat from place to place. Hence if measurements made at different places are to be compared with accuracy, a correction on this account has to be applied. The numerical value of the 'mass' of a body on the present system is also variable on account of the variation of g.

The total variation in the intensity of gravity over the earth's surface is, however, only one-half per cent., and this degree of vagueness is for many practical purposes quite unimportant. The numerical data on which an engineer, for example, has to rely, such as strengths of materials, coefficients of friction, &c., are as a rule affected by much greater uncertainty. For this reason the gravitational system of force-measurement is retained by engineers without inconvenience, even in dynamical questions where gravity is not directly concerned.

But when, as in many scientific measurements, greater precision is desired and is possible, it becomes necessary either to express the results in terms of gravity at some particular station on the earth's surface which is taken as a standard, or to have recourse to some less arbitrary dynamical system, independent of terrestrial or other gravity. The latter procedure is clearly preferable, and in the application to questions of Astronomy almost essential. We proceed accordingly, in the next Article, to give another statement of fundamental dynamical principles, and to explain the 'absolute' system of force-measurement to which it leads.

7. The Absolute System of Dynamics.

For purposes of explanation it is convenient to appeal to a series of ideal experiments. We imagine that we have some means, independent of gravity, of applying a constant force, and of verifying its constancy, e.g. by a spring-dynamometer which is stretched or deformed to a constant extent; but we do not presuppose any graduated scale by which different forces can be compared numerically*.

The first experimental result which we may suppose to be established in this way is that a constant force acting on a body produces a constant acceleration, i.e. the velocity changes by equal amounts in equal times.

It is observed, again, that the same force applied in succession to different bodies produces in general different degrees of acceleration. This is described as due to differences in the 'inertia,' or 'mass,' of the respective bodies. Two bodies which acquire equal velocities in equal times, when acted upon by the same force, are regarded as dynamically equivalent, and their masses are said to be 'equal.' The standard, or unit, of mass must therefore be that of some particular piece of matter, chosen in the first instance arbitrarily, e.g. the standard pound or kilogramme. A body is said to have the mass m (where m is an integer) when it can be divided into m portions each of which is dynamically equivalent to the unit. Similarly, if a unit mass be subdivided into n dynamically equivalent pieces, the mass of each of these is said to be $1/n$. It is evident that on these lines a complete scale of mass can be constructed, and that the mass of any body whatever can be indicated by a numerical quantity.

The next statement is that the accelerations produced in different bodies by the same force are inversely proportional to the respective masses as above defined. If we introduce the term 'momentum' to designate the product of the mass of a body into its velocity, we may say that a given force generates always the same momentum in the same time, whatever the body on which it acts.

* The course of the exposition is substantially that adopted by Maxwell in his *Matter and Motion*. The sequence of ideas is somewhat different from that of Newton, but this does not affect the final results.

The momentum generated per unit time is therefore an invariable characteristic of a force, and expresses all that is necessary to be known about the force, from the dynamical point of view. It is therefore conveniently taken as the *measure* of the force. The unit force on this reckoning is accordingly one which generates unit momentum in unit time, or unit acceleration in unit mass. This is called the 'absolute' unit, since it is the same in all places and at all times.

Hence if F be the absolute measure of a force, the acceleration a which it produces in a mass m is given by the equation

$$ma = F. \qquad\qquad (1)$$

It will be noticed that in this exposition there is no mention of gravity. The ideal experiments referred to may be supposed carried out in some remote region of space where gravity is insensible; or, to adopt an illustration used by Lord Kelvin, we may imagine them to be performed in a central spherical cavity in the interior of the earth. Whatever the other inconveniences attending research in such a central institution, the theory of Attractions assures us that gravitation would not intervene to mar the simplicity of the experiments.

It is hardly necessary to say that the real evidence for the correctness of our fundamental principles is indirect. The various statements which have been made cannot be tested singly, but only as a whole. The most striking verification is afforded by the agreement with observation of the predictions of physical Astronomy, which are based solely on the Laws of Motion and Newton's Law of Gravitation.

8. Application to Gravity.

If we apply the preceding principles to the case of ordinary terrestrial gravity, we learn that the attraction of the earth on a mass m is expressed by mg in absolute measure, this being the momentum generated per second in the body when falling freely. Since the value of g at a given place is known to be the same for all bodies, it follows that the attractions of the earth on different bodies are proportional to the respective masses. Hence bodies which have equal weights, as tested by the balance, may be asserted to have equal masses in the dynamical sense. This gives practically the most convenient method of comparing masses.

It follows also that the two systems of force-measurement, viz. the absolute and the gravitational, are consistent, in the sense that the *ratio* of any two given forces has the same numerical value on either system. We may remark, indeed, that the fundamental equations of the two systems, viz.

$$ma = F, \quad \text{.............................(1)}$$

and

$$Wa = Pg \quad \text{.............................(2)}$$

are really equivalent, although the units implied are different. For if they are applied to the same problem, m is numerically equal to W, whilst F is the absolute measure of the gravity of a body whose mass is numerically equal to P, and is accordingly equal to Pg.

In the rest of this treatise we follow the absolute system, as by far the most convenient for general application, and the equations of motion which we shall use will therefore be on the model of (1). If we adopt the foot, pound, and second as units of length, mass, and time, the absolute unit of force will be such that acting for one second on a mass of one pound it generates a velocity of one foot per second. This unit is sometimes called the 'poundal,' but the name is falling into disuse. Since the year 1875 the centimetre, gramme, and second have been generally adopted by physicists as fundamental units. The absolute unit of force on this system* is called the 'dyne.'

In terms of ordinary gravity the poundal is equal to the gravity of about 1/32·2 of a pound, or (roughly) half an ounce; whilst the dyne is equal to the gravity of about 1/981 of a gramme, or (roughly) a milligramme.

A book on Dynamics can hardly evade all notice of verbal questions. It is scarcely necessary to insist on the special sense which the word 'force' has come to bear in Mechanics. It is perhaps unfortunate that some more technical term was not introduced instead of a word which in popular language has so many

* Usually referred to, for brevity, as the 'C.G.S.' system.

different meanings*. The usage is however long established, and must be accepted.

There is not the same agreement, although there has been much controversy, as to the use of the word 'weight.' In ordinary language this is employed in a great variety of senses. Thus it may mean the actual statical *pressure* which a body exerts on whatever is supporting it, as when we speak of the 'weight' of a burden; it may also mean (as in Art. 6) the *ratio* which this pressure bears to that exerted by a pound or a ton; it is often used virtually in the sense of *mass*, as when we refer to the 'weight' of a projectile; when, again, we speak of the 'weight' of a blow the idea is (vaguely) that of *momentum*. The one sense in which the word is *never* used in popular language is that of the gravitational attraction on a body. This is of course equal to the statical pressure above referred to, but is not identical with it; it is a force exerted *on* a body, not *by* it. Unfortunately this new and alien sense is precisely that which some writers of eminence have sought to attach exclusively to the word 'weight' in Mechanics. In the author's opinion it is best not to attempt to specialize altogether the meaning of so familiar a word, but to use it freely in whatever sense may be convenient, whenever there is no risk of misunderstanding. When there is danger of confusion, some other term such as 'mass,' or 'gravity,' may be employed to indicate precisely the sense which it is wished to convey.

9. General Equation of Motion. Impulse.

Let us now suppose that a particle of mass m moving in a given straight line is subject to a force X, which may be constant or variable, also in that line. Since X denotes the momentum which would be produced in unit time if the force were to remain constant at its actual value, the momentum produced in the infinitesimal time δt is $X\,\delta t$. Hence if u be the velocity at the instant t, we have

$$\delta(mu) = X\,\delta t,$$

* The Latin equivalent (*vis*) is used by Newton quite generally, special meanings being indicated by the addition of qualifying adjectives. Thus we meet constantly with such combinations as *vis motrix*, *vis acceleratrix*, *vis inertiae*, of which the first alone is a 'force' in the modern specialized sense.

or $$m\frac{du}{dt} = X. \quad\text{.......................(1)}$$

If x be the abscissa of the particle measured from some fixed point in the line, we have $u = dx/dt$, and therefore

$$m\frac{d^2x}{dt^2} = X. \quad\text{.......................(2)}$$

The product $X\delta t$ is called the 'impulse' of the force in the time δt. The 'total' impulse in the interval from t_0 to t_1 is defined as the time-integral

$$\int_{t_0}^{t_1} X\,dt. \quad\text{.......................(3)}$$

This may be represented graphically by means of a curve constructed with t as abscissa and X as ordinate. The impulse is then represented by the area swept over by the ordinate.

If we integrate the equation (1) with respect to t between the limits t_0 and t_1 we obtain

$$mu_1 - mu_0 = \int_{t_0}^{t_1} X\,dt. \quad\text{.......................(4)}$$

The increment of the momentum in any interval of time is therefore equal to the impulse.

The theory of rectilinear motion under a constant force is sufficiently illustrated in elementary treatises. We proceed to the discussion of some important cases where the force X is variable.

10. Simple-Harmonic Motion.

The case of a particle attracted towards a fixed point O in the line of motion, with a force varying as the distance from that point, is important, as typical of the most general case of a dynamical system of one degree of freedom oscillating about a position of stable equilibrium.

If K denote the force at unit distance, the force at distance x will be $-Kx$, the sign being always opposite to that of x. The differential equation is accordingly

$$m\frac{d^2x}{dt^2} = -Kx. \quad\text{.......................(1)}$$

If we write

$$n^2 = K/m, \qquad\qquad (2)$$

the solution is, as in Art. 5, Ex. 3,

$$x = A \cos nt + B \sin nt, \qquad\qquad (3)$$

or

$$x = a \cos (nt + \epsilon), \qquad\qquad (4)$$

where the constants A, B, or a, ϵ, are arbitrary. The motion is therefore periodic, the values of x and dx/dt recurring whenever nt increases by 2π. The interval

$$T = \frac{2\pi}{n} = 2\pi \sqrt{\left(\frac{m}{K}\right)} \qquad\qquad (5)$$

is therefore called the 'period'*; it is the same whatever the initial conditions, and the oscillations are therefore said to be 'isochronous.'

The type of motion represented by the formula (4) is of great importance in Mechanics, and in various branches of Physics; it is called a 'simple-harmonic,' or (sometimes) a 'simple' vibration. Its character may be illustrated in various ways. For instance, if with O as centre we describe a circle of radius a, and imagine

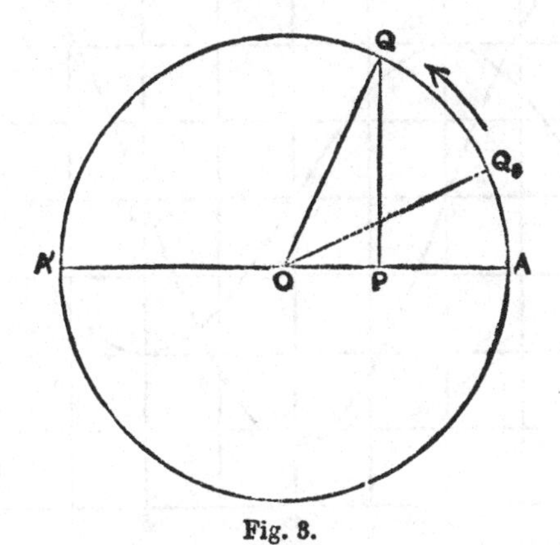

Fig. 8.

a point Q to describe this circle continually in the positive direction with the constant angular velocity n, the orthogonal projection (P)

* The reciprocal ($n/2\pi$) of the period gives the number of complete vibrations per unit time; this is called the 'frequency' in Acoustics.

of Q on a fixed diameter AOA' will execute a vibration of the above character. To correspond with (4) we must make the point Q start at the instant $t=0$ from a position Q_0 such that the angle $AOQ_0 = \epsilon$. The abscissa of P at time t is then

$$x = a \cos AOQ = a \cos (Q_0OQ + AOQ_0) = a \cos (nt + \epsilon). \quad \ldots(6)$$

The figure gives also a simple representation of the velocity; thus

$$\dot{x} = -na \sin (nt + \epsilon) = -n \, . \, PQ, \quad \ldots\ldots\ldots\ldots(7)$$

provided PQ be reckoned positive or negative according as it lies above or below the line AA'. This is otherwise evident on resolving the velocity (na) of Q parallel to OA.

The distance a of the extreme positions A, A' from O is called the 'amplitude' of the vibration, and the angular distance AOQ of the guiding point Q on the auxiliary circle from A is called the 'phase.' The angle AOQ_0, or ϵ, may therefore be called the 'initial phase*.'

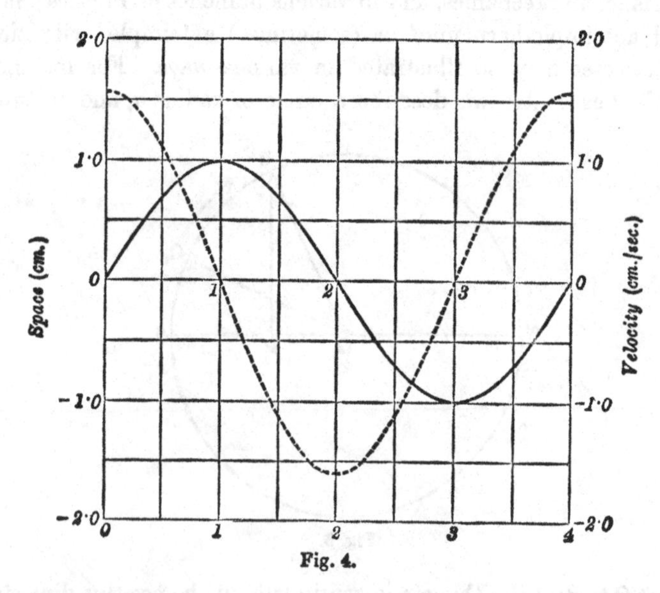

Fig. 4.

It is important also to notice the forms which the space-time and velocity-time curves of Art. 1 assume in the present case.

* It is sometimes called the 'epoch,' a technical term borrowed from Astronomy.

It appears from (4) and (7) that both are sine-curves, and that the zero points of one correspond to maxima or minima of the other. The annexed figure refers to a case where the amplitude is 1 cm. and the period 4 secs. The space-scale, which refers to the continuous curve, is shewn on the left, and the velocity-scale, corresponding to the dotted curve, on the right.

Ex. 1. The vertical oscillations of a weight which hangs from a fixed point by a helical spring come under this case, provided we assume that the spring obeys Hooke's Law of elasticity, and that the inertia of the spring itself may be neglected in comparison with that of the suspended body.

When the body is displaced vertically through a space x from the position of equilibrium, a restoring force $-Kx$ is called into play, where K is a constant measuring what we may call the 'stiffness' of the particular spring. The equation of motion is therefore identical with (1), and the period is accordingly

$$T = 2\pi\sqrt{(m/K)}. \qquad \text{.......................(8)}$$

This increases with the mass m, but is diminished by increase of the stiffness K. It is to be noticed that the period does not depend on the intensity of gravity, which merely affects the position of equilibrium.

Fig. 5.

The result is however conveniently expressed in terms of the statical increase of length which is produced by the weight when hanging in equilibrium. Denoting this increase by c, we have

$$Kc = mg, \qquad \text{....................................(9)}$$

so that the period is

$$T = 2\pi\sqrt{\left(\frac{c}{g}\right)}. \qquad \text{............................(10)}$$

It will be seen presently (Art. 11) that this is the same as the period of oscillation of a pendulum of length c. A similar mode of expression applies in many analogous problems.

Conversely, from observation of T and of the statical elongation c, we can infer the value of g, using the formula

$$g = \frac{4\pi^2 c}{T^2}, \qquad \text{...................................(11)}$$

but the method is not a good one, owing to the neglect of the inertia of the spring. If we assume, however, that this is equivalent to an addition m' to the suspended mass, we have from (8)

$$KT_1^2 = 4\pi^2(m_1 + m'), \quad KT_2^2 = 4\pi^2(m_2 + m'), \qquad \text{...........(12)}$$

whilst from (9),

$$Kc_1 = m_1 g, \quad Kc_2 = m_2 g, \quad \text{..........................(13)}$$

where the suffixes refer to experiments made with two different values of the suspended mass. These equations lead to

$$g = \frac{4\pi^2 (c_1 - c_2)}{T_1^2 - T_2^2}. \quad \text{..........................(14)}$$

Ex. 2. The vertical oscillations of a ship also come under the present theory, if we neglect the *inertia* of the water.

If ρ denote the density of the water, and V the volume displaced by the ship, its mass is ρV. If the ship be depressed through a small space x, the buoyancy is increased by $g\rho Ax$, in dynamical measure, where A denotes the area of the water-line section [*S.* 101]*. Hence

$$\rho V \frac{d^2x}{dt^2} = -g\rho Ax, \quad \text{..........................(15)}$$

or if $h = V/A$, i.e. h denotes the mean depth of immersion,

$$\frac{d^2x}{dt^2} + \frac{g}{h} x = 0. \quad \text{..........................(16)}$$

The period is therefore

$$T = 2\pi \surd(h/g), \quad \text{..........................(17)}$$

the same as for a simple pendulum of length h.

Ex. 3. A particle m, attached to the middle point of a wire which is tightly stretched between fixed points with a tension P, makes small lateral oscillations.

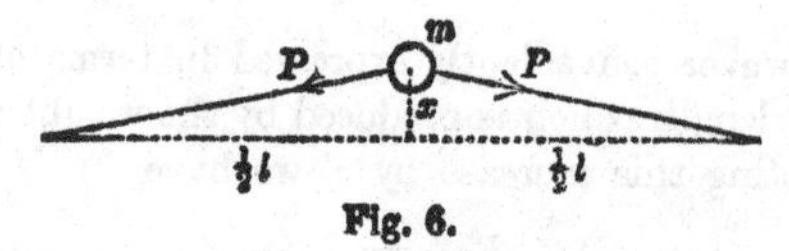

Fig. 6.

If we neglect the inertia of the wire, and assume the lateral displacement x to be so small compared with the length l that the change of tension is a negligible fraction of P, the restoring force is $-2Px/\frac{1}{2}l$, whence

$$m\frac{d^2x}{dt^2} = -4P\frac{x}{l}, \quad \text{..........................(18)}$$

approximately. The solution is as in (3), provided

$$n^2 = 4P/ml; \quad \text{..........................(19)}$$

and the period is therefore

$$T = \frac{2\pi}{n} = \pi \sqrt{\left(\frac{ml}{P}\right)}. \quad \text{..........................(20)}$$

For instance, if

$$l = 100 \text{ cm.}, \quad m = 5 \text{ gm.}, \quad P = 981 \times 1000 \text{ gm.cm./sec.}^2,$$

the period is ·071 sec., and the frequency accordingly 14·1.

* References in this form are to the author's *Statics*, Cambridge, 1912; thus '[*S.* 101]' means '*Statics*, Art. 101.'

11. The Pendulum.

The small oscillations of the bob of a pendulum are not strictly rectilinear, but by a slight anticipation of principles to be stated later they may be brought under the present treatment.

A particle of mass m, suspended from a fixed point O by a light string or rod of length l, is supposed to make small oscillations in a vertical plane about its position of equilibrium. We assume that the inclination θ of the string to the vertical is never more than a few degrees, so that the vertical displacement of the particle, viz. $l(1-\cos\theta)$, or $2l\sin^2\frac{1}{2}\theta$, is a small quantity of the second order, and may therefore be neglected. The vertical component $(P\cos\theta)$ of the tension P of the string may therefore be equated to the gravity mg of the particle, so that, to the same degree of approximation,

$$P = mg. \quad \text{.......................(1)}$$

Hence if x denote the horizontal displacement of the particle, we have

Fig. 7.

$$m\frac{d^2x}{dt^2} = -P\sin\theta = -mg\frac{x}{l}, \quad \text{...........(2)}$$

or

$$\frac{d^2x}{dt^2} = -n^2x, \quad \text{.........................(3)}$$

provided

$$n^2 = g/l. \quad \text{.........................(4)}$$

The horizontal motion is therefore simple-harmonic, and the period is

$$T = 2\pi\sqrt{\left(\frac{l}{g}\right)}. \quad \text{.........................(5)}$$

It is to be remembered that since it is only by an approximation that the differential equation was reduced to the form (3), it is essential for the validity of the simple-harmonic solution in any particular case that the initial conditions should be such as are consistent with the approximation. Thus if the initial displacement be x_0, and the initial velocity u_0, we have, as in Art. 5, Ex. 3, for the approximate solution,

$$x = x_0\cos nt + \frac{u_0}{n}\sin nt. \quad \text{..................(6)}$$

The ratios x_0/l and u_0/nl must therefore both be small. The latter condition requires, by (4), that the ratio $u_0/\sqrt{(gl)}$ should be small, i.e. the initial velocity u_0 must be small compared with that acquired by a particle in falling freely through a height equal to half the length of the pendulum.

The exact theory of the simple pendulum will be discussed in Art. 37.

The ideal simple pendulum, consisting of a mass concentrated in a point and suspended by a string devoid of gravity and inertia, cannot of course be realized, but by means of a metal ball suspended by a fine wire a good approximation to the value of g can be obtained, the formula being

$$g = 4\pi^2 l/T^2. \qquad\qquad (7)$$

The more accurate methods of determining g will be referred to later.

The theory of the pendulum is important as leading to the most precise means of verifying that the acceleration of gravity at any given place is the same for all bodies, and consequently that the gravity of a body varies as its mass. If, without making any assumption on this point, we denote the gravity of the particle by G, the equation (2) is replaced by

$$m\frac{d^2x}{dt^2} = -G\frac{x}{l}, \qquad\qquad (8)$$

and the period of a small oscillation is therefore

$$T = 2\pi\sqrt{\left(\frac{ml}{G}\right)}. \qquad\qquad (9)$$

The experimental fact, observed by Newton, and since abundantly confirmed by more refined appliances, that for pendulums of the same length the period is independent of the mass or material of the bob, shews that the ratio G/m is at the same place the same for all bodies.

12. Disturbed Simple-Harmonic Motion. Constant Disturbing Force.

When a particle is subject not only to an attractive force varying as the distance, as in Art. 10, but also to a given extraneous or 'disturbing' force whose accelerative effect is X, the equation of motion takes the form

$$\frac{d^2x}{dt^2} + n^2x = X. \qquad (1)$$

Thus, if the disturbing force be constant, producing (say) an acceleration f, we have

$$\frac{d^2x}{dt^2} + n^2x = f. \qquad (2)$$

We may write this in the form

$$\frac{d^2}{dt^2}\left(x - \frac{f}{n^2}\right) + n^2\left(x - \frac{f}{n^2}\right) = 0, \qquad (3)$$

the solution of which is

$$x - \frac{f}{n^2} = A\cos nt + B\sin nt. \qquad (4)$$

The interpretation is that the particle can execute a simple-harmonic oscillation of arbitrary amplitude and initial phase about the new equilibrium position ($x = f/n^2$). The period $2\pi/n$ of this oscillation is the same as in the undisturbed motion.

Ex. Suppose that the particle, initially at rest, is acted on by a constant disturbing force for one-sixth of a period, that the force then ceases for one-sixth of a period, and is afterwards applied and maintained constant.

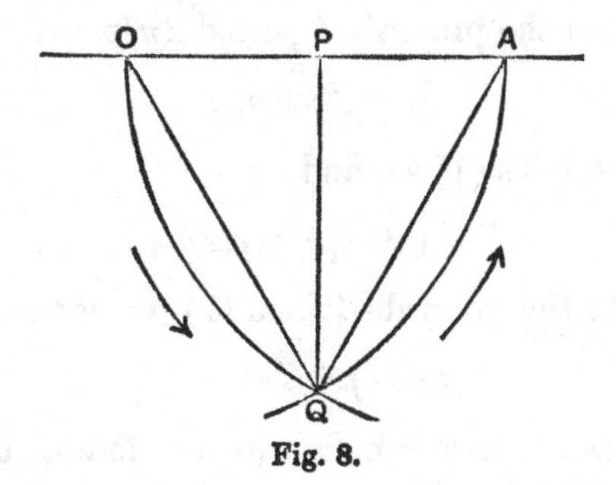

Fig. 8.

Let O be the original position of equilibrium, and make $OA = f/n^2$, where f is the constant acceleration due to the disturbing force. The particle will proceed to oscillate about A as a new position of equilibrium, with the amplitude OA, and the guiding point will therefore begin to describe a circle about A, starting from O. After one-sixth of a period it will be in the position Q, such that the angle $OAQ = \frac{1}{3}\pi$, and the vibrating particle will therefore be at P, the mid-point of OA. Since, by hypothesis, the disturbing force now ceases, the guiding point will proceed to describe a circle about O, starting at Q since the initial velocity is $n \,.\, PQ$. Since $AOQ = \frac{1}{3}\pi$, the particle will after the further lapse of one-sixth of a period arrive with zero velocity at A. Hence if the disturbing force be now applied again, the particle will remain at rest at A.

This example illustrates a rule devised by Gauss for applying the current in an undamped tangent galvanometer. It is assumed that the deflections are so small that the restoring couple on the needle can be taken as proportional to the deflections as indicated by the scale of the instrument. The needle being at rest, the current is applied (by pressing a key) until the deflection has one-half the estimated equilibrium value; it is then interrupted until the needle comes to rest, and finally applied again. If the estimated deflection were correct, and the operations exactly performed, the needle would remain at rest. In practice these conditions cannot be exactly fulfilled, and the result is that the needle finally oscillates about the new equilibrium position. But since the range is small, this position can be determined with great accuracy.

13. Periodic Disturbing Force.

Another very important case is where the given disturbing force is a simple-harmonic function of the time, say

$$X = f \cos pt, \qquad \text{(1)}$$

where the value of p is given. To solve the equation, we may inquire what extraneous force is necessary to maintain a simple-harmonic motion of the prescribed period $2\pi/p$, say

$$x = C \cos pt. \qquad \text{(2)}$$

Substituting in Art. 12 (1) we find

$$X = (n^2 - p^2)\, C \cos pt, \qquad \text{(3)}$$

which agrees with the prescribed form (1) provided

$$(n^2 - p^2)\, C = f. \qquad \text{(4)}$$

The given force can therefore maintain the 'forced' oscillation

$$x = \frac{f}{n^2 - p^2} \cos pt. \qquad \text{(5)}$$

To this value of x we may obviously add the terms

$$A \cos nt + B \sin nt,$$

since they do not affect the required value of X; they represent in fact the 'free' oscillations which can exist independently of X. We thus obtain the complete solution

$$x = \frac{f}{n^2 - p^2} \cos pt + A \cos nt + B \sin nt, \quad \text{............(6)}$$

containing two arbitrary constants.

The forced oscillation (5) has the same period $2\pi/p$ as the disturbing force. Its phase is the same with that of the force, or the opposite, according as $p \lessgtr n$, that is, according as the imposed period is longer or shorter than that of the free vibration. If the imposed period be infinitely long, we have $p = 0$, and the displacement is at each instant that which could be maintained by a steady force equal in magnitude to the instantaneous value of the actual force. Borrowing a term from the theory of the Tides, we may call this the statical or 'equilibrium' value of the displacement. Denoting it by $\bar{x}$ we have

$$\bar{x} = \frac{f}{n^2} . \cos pt. \quad \text{........................(7)}$$

Hence for any other period we have, in the forced oscillation,

$$x = \frac{\bar{x}}{1 - p^2/n^2}. \quad \text{..........................(8)}$$

As p increases from 0, the amplitude of the forced oscillation increases, until when p is nearly equal to n, i.e. when the imposed period is nearly equal to the natural period, it becomes very great. If the differential equation is merely an approximation to the actual conditions, as in the case of the pendulum, the solution (6) may before this cease to be applicable, as inconsistent with the fundamental assumption as to the smallness of x on which the approximation was based. It may be added that frictional forces, such as are always in operation to some extent, may also become important. This question will be considered later (Chap. XII), but we have already an indication that a vibration of abnormal amplitude may occur whenever there is approximate coincidence between the free and the forced periods. This phenomenon, which is incident to all vibrating systems, is known as 'resonance,' the term being borrowed from Acoustics, where illustrations of the principle are frequent and conspicuous.

In the case of exact coincidence between the two periods the preceding method fails, but the difficulty is resolved if we examine the limiting form of the solution (6) for $p \to n$, when the constants A, B have been adapted to prescribed initial conditions. If we have $x = x_0$, $\dot{x} = u_0$, for $t = 0$, we find

$$A + \frac{f}{n^2 - p^2} = x_0, \quad nB = u_0, \quad \text{..................(9)}$$

whence

$$x = x_0 \cos nt + \frac{u_0}{n} \sin nt + \frac{f}{n^2 - p^2} (\cos pt - \cos nt). \quad \text{...(10)}$$

The last term may be written in the form

$$\frac{ft}{n+p} \cdot \frac{\sin \frac{1}{2}(n-p)t}{\frac{1}{2}(n-p)t} \cdot \sin \frac{1}{2}(n+p)t,$$

where the second factor has the limiting value unity. Hence for $p \to n$ we have

$$x = x_0 \cos nt + \frac{u_0}{n} \sin nt + \frac{ft}{2n} \sin nt. \quad \text{...........(11)}$$

The last term may be described as representing a simple-harmonic vibration whose amplitude increases proportionally to t. For a reason already given this result is usually only valid, in practical cases, for the earlier stages of the motion.

If p be increased beyond the critical value, the phase of the forced oscillation is reversed, and its amplitude continually diminishes.

The preceding theory may be illustrated by the case of a pendulum whose point of suspension is moved horizontally to and fro in a given manner. It is assumed that the conditions are such that the inclination of the string to the vertical is always small.

If ξ denote the displacement of the point of suspension, and x that of the bob, from a fixed vertical line, we have

$$m \frac{d^2x}{dt^2} = -mg \cdot \frac{x - \xi}{l} \quad \text{..............(12)}$$

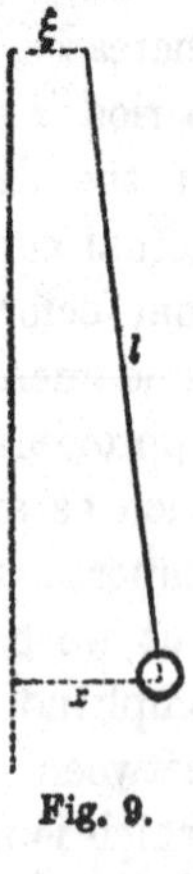

Fig. 9.

or, if $n^2 = g/l$, as in Art. 11,

$$\frac{d^2x}{dt^2} + n^2 x = n^2 \xi. \quad \text{.........................(13)}$$

This equation is the same as if the upper end of the string were fixed and the bob were acted on by a horizontal force whose accelerative effect is $n^2\xi$.

If the imposed motion be simple-harmonic, say

$$\xi = a \cos pt, \quad \text{..........................(14)}$$

the solution is as in (5), with $f = n^2a$. If we put $p^2 = g/l'$, i.e. the prescribed period is that of a pendulum of length l', the forced oscillation is

$$x = \frac{n^2a}{n^2 - p^2} \cos pt = \frac{l'}{l' - l} \xi. \quad \text{..............(15)}$$

This result is illustrated by the annexed Fig. 10 for the two cases where $l' > l$ and $l' < l$, respectively. The pendulum moves as if suspended from C as a fixed point, the distance CP being equal to l'. The solution (15) is in fact obvious from the figure*.

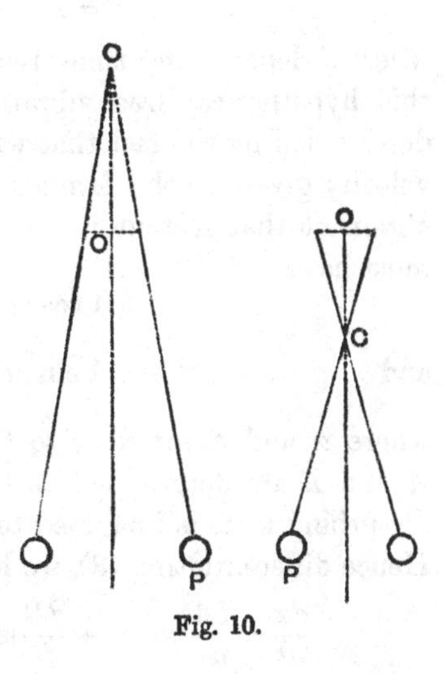

Fig. 10.

The displacement relative to the point of suspension is given by

$$x - \xi = \frac{p^2a}{n^2 - p^2} \cos pt = \frac{l}{l' - l} \xi. \quad \text{...(16)}$$

This has a bearing on the theory of seismographs, i.e. of instruments devised to record so far as possible the motion of the earth's surface during an earthquake. Every seismograph is virtually equivalent to a pendulum, and its indications depend on the *relative* motion, since the recording apparatus of necessity moves with the support. If p be large compared with n, or l' small compared with l, the relative displacement is $-\xi$, approximately, the 'bob' being nearly at rest. Hence vibrations which are comparatively rapid are all recorded on the same scale. For this reason the free period of the

* This illustration is due to Thomas Young [*S.* 137], who investigated in an elementary manner the theory of free and forced oscillations, and applied it to the theory of the tides, where the question of inversion of phase, in cases where the free period is longer than the period of the moon's disturbing force, is of great theoretical importance.

seismograph is made long, 15 or 20 seconds being a common amount. The record is however liable to be confused by the free vibrations which are set up; and for this reason damping appliances are introduced. See Chap. XII.

14. General Disturbing Force.

The solution of the more general equation

$$\frac{d^2x}{dt^2} + n^2x = f(t), \quad \text{.......................(1)}$$

where $f(t)$ represents the acceleration due to an arbitrary disturbing force, may be noticed, as illustrating the method of 'variation of parameters' in Differential Equations.

If at time t the disturbing force were to cease, the particle would proceed to execute a simple-harmonic vibration

$$x' = A \cos nt' + B \sin nt', \quad \text{.................(2)}$$

where t' denotes the time, reckoned from the same origin as t, in this hypothetical free vibration. The coefficients A and B are determined by the fact that when $t' = t$, the displacement and the velocity given by the formula (2) must coincide with those which obtain at that instant in the actual motion. In other words we must have

$$A \cos nt + B \sin nt = x, \quad \text{.....................(3)}$$

and

$$-nA \sin nt + nB \cos nt = \frac{dx}{dt}, \quad \text{..............(4)}$$

where x and dx/dt refer to the actual motion. The coefficients A and B are determined as functions of the time t, at which the disturbing force is imagined to cease, by the equations just written Hence differentiating (3), we have

$$\frac{dx}{dt} = \frac{dA}{dt} \cos nt + \frac{dB}{dt} \sin nt - nA \sin nt + nB \cos nt,$$

which reduces by (4) to

$$\frac{dA}{dt} \cos nt + \frac{dB}{dt} \sin nt = 0. \quad \text{..................(5)}$$

Again, differentiating (4),

$$\frac{d^2x}{dt^2} = -n^2A \cos nt - n^2B \sin nt - n\frac{dA}{dt} \sin nt + n\frac{dB}{dt} \cos nt$$

$$= -n^2x - n\frac{dA}{dt} \sin nt + n\frac{dB}{dt} \cos nt,$$

whence $$-n\frac{dA}{dt}\sin nt + n\frac{dB}{dt}\cos nt = f(t). \qquad \text{.............(6)}$$

Solving (5) and (6) for dA/dt and dB/dt, we find

$$\frac{dA}{dt} = -\frac{1}{n}f(t)\sin nt, \quad \frac{dB}{dt} = \frac{1}{n}f(t)\cos nt, \qquad \text{.........(7)}$$

whence $$A = -\frac{1}{n}\int f(t)\sin nt\,dt, \quad B = \frac{1}{n}\int f(t)\cos nt\,dt. \quad \text{...(8)}$$

Each of these integrals involves of course an arbitrary additive constant.

The formula $$x = A\cos nt + B\sin nt, \qquad \text{.................(9)}$$

with the values of A and B given by (8), constitutes the solution aimed at. It may easily be verified by differentiation.

If the particle be originally in equilibrium, the initial values of A and B are zero. If further, the disturbing force $f(t)$ be sensible only for a finite range of values of t, the lower limit of integration may be taken to be $-\infty$. For values of t, again, which are subsequent to the cessation of the force, the upper limit may be taken to be $+\infty$.

The effect of an impulse at time $t=0$, which generates instantaneously a velocity u_1, is of course given by

$$x = \frac{u_1}{n}\sin nt. \qquad \text{................................(10)}$$

If the same impulse be spread over a finite interval of time, the amplitude of the subsequent oscillation will be less, provided the force have always the same sign. To illustrate this, we may take the case of

$$f(t) = \frac{u_1\tau}{\pi(t^2+\tau^2)}. \qquad \text{..............................(11)}$$

It is true that the disturbing force has now no definite beginning or ending, but when t is moderately large compared with τ, whether it be positive or negative, the value of $f(t)$ is very small*. The factors in (11) have been chosen so as to make

$$\int_{-\infty}^{\infty} f(t)\,dt = u_1. \qquad \text{..............................(12)}$$

* The graph of the function in (11) is given, for another purpose, in Art. 95.

We have now*

$$\int_{-\infty}^{\infty} f(t) \sin nt\, dt = 0, \quad \int_{-\infty}^{\infty} f(t) \cos nt\, dt = u_1 e^{-n\tau}. \quad \text{.........(13)}$$

The residual vibration is therefore

$$x = \frac{u_1}{n} e^{-n\tau} \sin nt. \quad \text{..........................(14)}$$

The impulse is more diffused the larger the value of τ; the factor $e^{-n\tau}$ shews how this affects the amplitude. The effect is greater, for a given value of τ, the greater the natural frequency $(n/2\pi)$ of vibration.

15. Motion about Unstable Equilibrium.

The case of a particle subject to a *repulsive* force varying as the distance is typical of the motion of a body near a position of unstable equilibrium. The equation of motion is now of the form

$$\frac{d^2x}{dt^2} = n^2x. \quad \text{..........................(1)}$$

The solution is, as in Art. 5, Ex. 4,

$$x = Ae^{nt} + Be^{-nt}, \quad \text{..........................(2)}$$

where A, B are arbitrary.

It is possible so to adjust the initial conditions that $A = 0$, in which case the particle approaches asymptotically the position of equilibrium $(x = 0)$; but unless this be done the value of x will ultimately increase indefinitely.

It is to be remembered however that, in the application to practical cases, the equation (1) is merely an approximation, valid only so long as x does not exceed a certain limit. The solution (2) consequently ceases after a time to give a correct representation.

An example is supplied by an inverted pendulum (the string being replaced by a light rigid rod). If x be the horizontal displacement from the position of unstable equilibrium, the thrust in the rod is initially mg, where m is the mass of the bob, and its horizontal component is mgx/l outwards, if x be small. The equation (1) therefore applies, provided $n^2 = g/l$.

* The first integral vanishes by the cancelling of positive and negative elements of the same absolute value. The second depends on the formula

$$\int_0^{\infty} \frac{\cos ax}{x^2 + b^2} dx = \frac{\pi}{2b} \cdot e^{-ab},$$

which is proved in books on the Integral Calculus.

16. Motion under Variable Gravity.

If a particle be attracted towards the origin with a force varying inversely as the square of the distance, the equation of motion is

$$u\frac{du}{dx} = -\frac{\mu}{x^2}, \quad\text{.........................(1)}$$

where μ denotes the acceleration at unit distance. Integrating with respect to x, we have

$$\tfrac{1}{2}u^2 = \frac{\mu}{x} + C. \quad\text{.........................(2)}$$

If the particle start from rest at a distance c, we must have $u = 0$ for $x = c$, and therefore $C = -\mu/c$. Hence

$$u^2 = 2\mu\left(\frac{1}{x} - \frac{1}{c}\right). \quad\text{.........................(3)}$$

If c be very great this tends to the form

$$u^2 = \frac{2\mu}{x}, \quad\text{.........................(4)}$$

and the velocity thus determined is called the 'velocity from infinity' to the position x.

We may apply these results to the case of a particle falling directly to the earth, account being now taken of the variation of gravity with distance. Assuming, from the theory of Attractions, that the acceleration of gravity varies inversely as the square of the distance from the earth's centre, its value at a distance x will be ga^2/x^2, where a is the earth's radius. Hence, putting $\mu = ga^2$, we have, for the velocity from infinity,

$$u^2 = \frac{2ga^2}{x}. \quad\text{.........................(5)}$$

In particular, the velocity with which a particle starting from rest at a great distance would (if unresisted) strike the earth's surface is $\sqrt{(2ga)}$, i.e. it is equal to the velocity which a particle would acquire in falling from rest through a space equal to the earth's radius, if gravity were constant and equal to its surface value. If we put $a = 6{\cdot}38 \times 10^8$ cm., $g = 981$ cm./sec.^{2}, this velocity is found to be 11·2 kilometres, or about 7 miles, per second.

Returning to the more general case, it may be required to find the time of arriving at any given position. From (3) we have

$$\frac{dx}{dt} = u = -\sqrt{\left\{\frac{2\mu(c-x)}{cx}\right\}}, \qquad \text{.................(6)}$$

the minus sign being taken, since the motion is towards the origin. To integrate this we may put

$$x = c\cos^2\theta, \qquad \text{..........................(7)}$$

since x ranges from c to 0. This makes

$$2\cos^2\theta\frac{d\theta}{dt} = \sqrt{\left(\frac{2\mu}{c^3}\right)}, \qquad \text{.....................(8)}$$

or

$$\frac{dt}{d\theta} = \sqrt{\left(\frac{c^3}{2\mu}\right)}.(1+\cos 2\theta), \qquad \text{.................(9)}$$

whence

$$t = \sqrt{\left(\frac{c^3}{2\mu}\right)}.(\theta + \sin\theta\cos\theta), \qquad \text{..........(10)}$$

no additive constant being necessary if the origin of t be taken at the instant of starting, so that $t=0$ for $\theta=0$. This determines t as a function of θ, and therefore of x.

The substitution (7), and the result (10), may be interpreted geometrically as follows. If A be the starting point, we describe a circle on OA as diameter, and erect the ordinate PQ corresponding to any position P of the particle. If θ denote the angle AOQ, we have

$$OP = OQ\cos\theta = OA\cos^2\theta,$$

in agreement with (7). Also the area AOQ included between OA, OQ, and the arc AQ is easily seen to be

$$\tfrac{1}{4}c^2(\theta + \sin\theta\cos\theta).$$

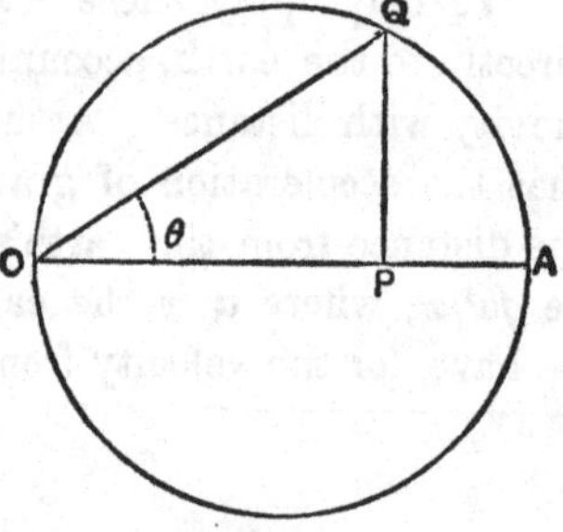

Fig. 11.

By comparison with (10) it appears that the area AOQ increases uniformly with the time, the rate per unit time being $\sqrt{(\frac{1}{8}\mu c)}$*.

It appears also from (7) and (10) that the space-time curve has

* Newton, *Principia*, lib. i., prop. xxxii.

the form of a cycloid, provided the scales of x and t be properly adjusted*.

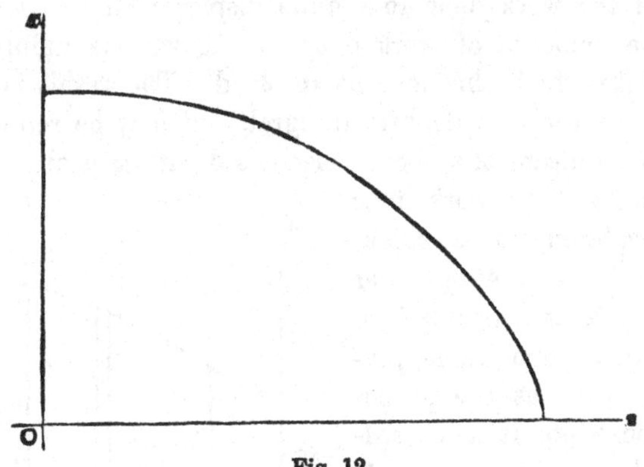

Fig. 12.

If in (10) we put $\theta = \frac{1}{2}\pi$, we obtain the time of arrival at the origin, viz.

$$t_0 = \pi \sqrt{\left(\frac{c^3}{8\mu}\right)}. \quad \text{.....................(11)}$$

This may be compared with the time of revolution of a particle in a circular orbit of radius c about the same centre of force, viz.

$$t_1 = 2\pi \sqrt{\left(\frac{c^3}{\mu}\right)}; \quad \text{.....................(12)}$$

(see Chap. x). We have

$$\frac{t_0}{t_1} = \frac{1}{8}\sqrt{2} = \cdot 177, \quad \text{.....................(13)}$$

nearly. For instance, if the orbital motion of the earth were arrested, it would fall into the sun in ·177 of a year, or about 65 days.

17. Work; Power.

The 'work' done by a force X acting on a particle, when the latter receives an infinitesimal displacement δx, is measured by the product $X\delta x$, and is therefore positive or negative according

* The coordinates of a point on a cycloid being expressible in the forms

$$x = a(2\psi + \sin 2\psi), \quad y = a(1 + \cos 2\psi),$$

if the axes be suitably chosen (*Inf. Calc.*, Art. 136).

as the directions of the force and the displacement are the same or opposite.

To find the work done in a *finite* displacement, we form the sum of the amounts of work done in the various infinitesimal elements into which this may be resolved. The result is therefore of the nature of a definite integral, and may be represented graphically by means of a curve constructed with x as abscissa and X as ordinate. The work done in any displacement is represented by the area swept over by the ordinate as x passes from its initial to its final value, provided this area has the proper sign attributed to it in accordance with the usual conventions*.

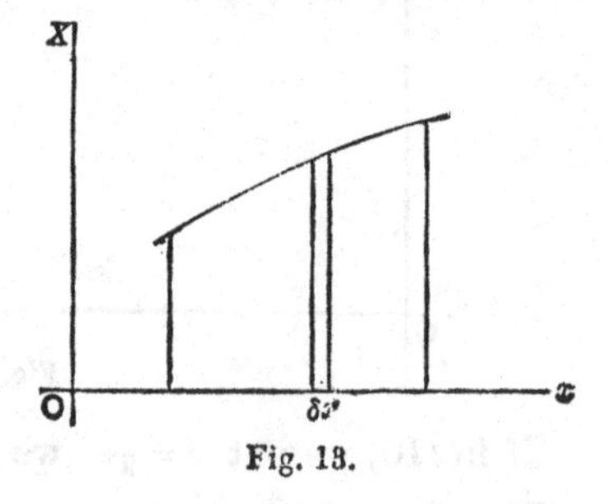

Fig. 13.

The absolute unit of work on the foot-pound-second system may be called the 'foot-poundal.' The corresponding unit on the C.G.S. system is known as the 'erg.' A certain multiple (10^7) of this is called the 'joule†.'

For engineering purposes, gravitational units such as the foot-pound, or the kilogramme-metre, are in common use.

The word 'power' is employed to denote the *rate* (per unit time) at which work is being done. A power of one joule per second is known as a 'watt‡.' For technical purposes a larger unit, viz. the 'kilowatt' (=1000 watts), is in use.

The conventional 'horse-power' of British engineers is a power of 33,000 foot-pounds per minute, or 550 foot-pounds per second. The corresponding French unit, the 'cheval-vapeur,' is 75 kilogramme-metres per second.

Assuming that the foot is equivalent to 30·48 cms., and the pound to 453·6 gms., we have

$$\frac{\text{ft.-lb.}}{\text{erg}} = 30{\cdot}48 \times 453{\cdot}6 \times 981 = 1{\cdot}356 \times 10^7.$$

* *Inf. Calc.*, Art. 99.

† After J. P. Joule (1818–89), who definitely established the relation between mechanical energy and heat.

‡ After James Watt (1736–1804) the inventor of the steam-engine.

The foot-pound is therefore equivalent to 1·356 joule; or the joule is ·7372 ft.-lbs. Hence, also,

$$\frac{\text{horse-power}}{\text{kilowatt}} = 550 \times 1{\cdot}356 \times 10^{-3} = {\cdot}746.$$

18. Equation of Energy.

We return now to the general equation of rectilinear motion, viz.

$$m\frac{du}{dt} = X. \quad \text{.........................(1)}$$

Multiplying by u we have

$$mu\frac{du}{dt} = Xu,$$

or

$$\frac{d}{dt}(\tfrac{1}{2}mu^2) = X\frac{dx}{dt}. \quad \text{.....................(2)}$$

The product $\frac{1}{2}mu^2$ is called the 'kinetic energy,' for a reason which will appear presently. Since $X\delta x$ denotes the work done by the force in an infinitesimal displacement δx, the product $X\,dx/dt$ measures the rate at which work is being done at the instant t. The equation (2) therefore expresses that the kinetic energy is at any instant increasing at a (positive or negative) rate equal to that at which work is being done on the particle. Hence, integrating with respect to t, we learn that the increment of the kinetic energy in any interval of time is equal to the total work done on the particle in that interval. In symbols,

$$\tfrac{1}{2}mu_2^2 - \tfrac{1}{2}mu_1^2 = \int_{t_1}^{t_2} X\frac{dx}{dt}\,dt, \quad \text{..................(3)}$$

where u_1, u_2 are the velocities at the instants t_1, t_2, respectively.

If the force is always the same in the same position, so that X is a definite function of x, the integral may be replaced by

$$\int_{x_1}^{x_2} X\,dx,$$

where x_1, x_2 are the initial and final positions. We have now a constant, and conservative, field of force [*S.* 49], and the conception of potential energy becomes applicable. The work required to be performed by an external agency in order to bring the particle from rest in some standard position (x_0) to rest in any assigned position x, against the action of the field, is the same in whatever way the transition be made, and is called the 'potential energy' corresponding to the position x. Denoting it by V we have

$$V = -\int_{x_0}^{x} X\,dx = \int_{x}^{x_0} X\,dx, \quad \text{..................(4)}$$

since $-X$ is the extraneous force required to balance the force of the field at each stage of the imagined process, which we may suppose to take place with infinite slowness*. If we consider a small displacement, we have

$$\delta V = -X\,\delta x, \quad \text{........................(5)}$$

or

$$X = -\frac{dV}{dx}. \quad \text{..........................(6)}$$

The work done by the force of the field as the particle passes from any position x_1 to any other position x_2 is

$$\int_{x_1}^{x_2} X\,dx = \int_{x_1}^{x_0} X\,dx + \int_{x_0}^{x_2} X\,dx = \int_{x_1}^{x_0} X\,dx - \int_{x_2}^{x_0} X\,dx = V_1 - V_2. \quad \text{......(7)}$$

Hence when there are no extraneous forces we have, from (3),

$$\tfrac{1}{2}mu_2^2 - \tfrac{1}{2}mu_1^2 = V_1 - V_2, \quad \text{..................(8)}$$

or

$$\tfrac{1}{2}mu_2^2 + V_2 = \tfrac{1}{2}mu_1^2 + V_1, \quad \text{..................(9)}$$

i.e. the sum of the kinetic and potential energies is constant.

If extraneous forces are operative we must add to the right-hand side of (8) the work done by them on the particle; in other words, the difference

$$(\tfrac{1}{2}mu_2^2 + V_2) - (\tfrac{1}{2}mu_1^2 + V_1)$$

is equal to the work so done. This work has therefore an equivalent in the (positive or negative) increment of the total energy of the particle. In some cases the same thing is more conveniently expressed by saying that the work done by the reaction of the particle on the external agency is equal to the *decrement* of the total energy.

* It will appear in a moment that this proviso is unnecessary.

The quantity $\frac{1}{2}mu^2 + V$, which we have termed the 'total energy,' therefore measures the amount of work which the particle can perform against external resistances in passing from its actual state as regards velocity and position to rest in the standard position. It is for this reason that the name 'energy' is applied to it, in the sense of 'capacity for doing work against resistance.' The two constituents of the energy have been variously called 'energy of motion,' or 'kinetic energy,' or 'actual energy,' and 'energy of position,' or 'statical energy,' or 'potential energy,' respectively*.

Ex. 1. The potential energy, as regards gravity, of a particle m at a height x above some standard level is

$$V = mgx. \qquad (10)$$

Hence in the unresisted vertical motion of a particle we have

$$\frac{1}{2}mu^2 + mgx = \text{const.} \qquad (11)$$

Ex. 2. The work required to stretch a helical spring from its natural length l to a length $l + x$ is, in the notation of Art. 10, Ex. 1,

$$\int_0^x Kx\,dx = \frac{1}{2}Kx^2. \qquad (12)$$

Hence in the case of a hanging weight we have

$$V = \frac{1}{2}Kx^2 - mgx, \qquad (13)$$

if x be measured downwards. The equation of energy is therefore

$$\frac{1}{2}mu^2 + \frac{1}{2}Kx^2 - mgx = \text{const.} \qquad (14)$$

Ex. 3. If the variation of gravity with distance from the earth's centre be taken into account, the work required to bring a particle from the surface to a distance x from the centre is

$$V = \int_a^x \frac{mga^2}{x^2}\,dx = mga^2\left(\frac{1}{a} - \frac{1}{x}\right), \qquad (15)$$

if a denote the radius. The equation of energy therefore gives, in the case of radial motion,

$$\frac{1}{2}mu^2 - \frac{mga^2}{x} = \text{const.}; \qquad (16)$$

cf. Art. 16.

* The product mu^2 had been known since the time of G. W. Leibnitz (1646–1716) as the 'vis viva'; the term 'energy' was substituted by Young. The name 'actual energy' for the expression $\frac{1}{2}mu^2$ was proposed by Rankine.

19. Dynamical Units and their Dimensions.

Any scalar quantity whatever can be specified by a number expressing the ratio which it bears to some standard or unit quantity of the same kind. This number will of course vary inversely as the magnitude of the unit chosen.

In any 'absolute' system of Dynamics the fundamental units are those of mass, length, and time. These may be fixed arbitrarily and independently, whilst all other units are derivative, and depend solely upon them. We denote their magnitudes by the symbols M, L, T, respectively.

We proceed to examine the dimensions of various derivative units, in addition to those of velocity and acceleration, which have been dealt with in Art. 3.

The unit *density* is that of unit mass diffused uniformly through unit volume; its symbol is therefore M/L^3, or ML^{-3}.

The unit *momentum* is that of unit mass moving with the unit velocity. Its symbol is ML/T, or MLT^{-1}.

The unit *force* is that which generates unit momentum per unit time, and is accordingly denoted by $(ML/T) \div T$, or ML/T^2, or MLT^{-2}.

The unit of *work*, or of *potential energy*, is the work done by the unit force acting through the unit length; its symbol is therefore $(ML/T^2) \times L$, or ML^2/T^2, or ML^2T^{-2}.

The unit of *kinetic energy* is the kinetic energy of unit mass moving with the unit velocity. Its symbol is $M \times (L^2/T^2)$, or ML^2/T^2, or ML^2T^{-2}.

The unit of *pressure-intensity*, or *stress* [*S.* 90, 136], in Hydrostatics or Elasticity, is an intensity of unit force per unit area. It is therefore denoted by $(ML/T^2) \div L^2$, or M/LT^2, or $ML^{-1}T^{-2}$. Since a *strain* is a mere ratio, these are also the dimensions (in absolute measure) of coefficients of elasticity, such as Young's modulus*.

In any *general* dynamical equation, i.e. one which is to hold whatever system of absolute units be adopted, the dimensions of

* The numerical values given in *Statics*, Art. 138, are in gravitational measure. They must be multiplied by g to reduce them to absolute measure.

each term must be the same. For otherwise a change in the fundamental units adopted would alter the numerical values of the various terms in different proportions. We have an illustration of this point in the equation of energy (Art. 18), where the dimensions of kinetic energy and potential energy are the same, viz. $\mathrm{ML}^2/\mathrm{T}^2$. The principle is exceedingly useful as a check on the accuracy of formulae.

The consideration of dimensions is also useful in another way, as helping us to forecast, to some extent, the manner in which the magnitudes involved in any particular problem will enter into the result.

For instance, if we assume that the period of a small oscillation of a given pendulum is a definite quantity, we see at once that it must vary as $\sqrt{(l/g)}$. For the only elements on which it can possibly depend are the mass (m) of the bob, the length (l) of the string, and the value of g at the place in question. And the above expression is the only combination of these symbols whose dimensions reduce to that of a time, simply.

Again, the time of falling from a distance c into a given centre of force varying inversely as the square of the distance will depend only on c and on the constant (μ) which denotes the acceleration at unit distance (Art. 16). The dimensions of μ, being such that μ/x^2 is an acceleration, must be $\mathrm{L}^3\mathrm{T}^{-2}$. Hence if we assume that the time in question varies as $c^p\mu^q$, where p, q are indices to be determined, the dimensions of the formula will be $\mathrm{L}^p(\mathrm{L}^3\mathrm{T}^{-2})^q$, or $\mathrm{L}^{p+3q}\mathrm{T}^{-2q}$. Hence we must have

$$p + 3q = 0, \quad -2q = 1,$$

or $p = \frac{3}{2}$, $q = -\frac{1}{2}$, and the required time will vary as $\sqrt{(c^3/\mu)}$.

The argument can be put in a more demonstrative form by the consideration of 'similar' systems, or (rather) of similar motions of similar systems* For example, we may consider the equations

$$\frac{d^2x}{dt^2} = -\frac{\mu}{x^2}, \quad \frac{d^2x'}{dt'^2} = -\frac{\mu'}{x'^2}, \quad \text{..................(1)}$$

* This line of argument originated, substantially, with J. B. Fourier (1768–1830) in his *Théorie analytique de la chaleur* (1822).

which are supposed to relate to two particles falling independently into two distinct centres of force varying inversely as the square

Mass ...	M	*gm.* (*gramme*) kg. (kilogramme)
Length ...	L	*cm.* (*centimetre*) m. (metre)
Time	T	*sec.* (*second*)
Velocity	L/T	*cm./sec.* m./sec.
Acceleration	L/T^2	*cm./sec.*^{2} m./sec.^{2}
Volume density ...	M/L^3	*gm./cm.*^{3}
Surface density ...	M/L^2	*gm./cm.*^{2}
Line density ...	M/L	*gm./cm.* kg./m.
Force ...	ML/T^2	*gm. cm./sec.*^{2} (*dyne*) kg. m./sec.^{2}
Work, Energy, Couple ...	ML^2/T^2	*gm. cm.*^{2}*/sec.*^{2} (*erg*) kg. m.^{2}/sec.^{2} (joule)
Power	ML^2/T^3	*gm. cm.*^{2}*/sec.*^{3} kg. m.^{2}/sec.^{3} (watt)
Stress, Elastic coefficients	M/LT^2	*gm./cm. sec.*^{2} kg./m. sec.^{2}
Moment of inertia	ML^2	*gm. cm.*^{2} kg. m.^{2}
Angular velocity ...	T^{-1}	*sec.*$^{-1}$
Angular momentum	ML^2/T	*gm. cm.*^{2}*/sec.* kg. m.^{2}/sec.

of the distance. It is obvious that we may have x' in a constant ratio to x, and t' in a constant ratio to t, provided that

$$\frac{x}{t^2} : \frac{x'}{t'^2} = \frac{\mu}{x^2} : \frac{\mu'}{x'^2}, \quad \text{.......................(2)}$$

and provided (of course) there is a suitable correspondence between the initial conditions. The relation (2) is equivalent to

$$t : t' = \frac{x^{\frac{3}{2}}}{\mu^{\frac{1}{2}}} : \frac{x'^{\frac{3}{2}}}{\mu'}, \quad \text{.......................(3)}$$

where t, t' are any two corresponding intervals of time, and x, x' any two corresponding distances from the centres of force. As a particular case, if t_1, t_1' be the times of falling into the centres from rest at distances c, c', respectively, we have

$$t_1 : t_1' = \frac{c^{\frac{3}{2}}}{\mu^{\frac{1}{2}}} : \frac{c'^{\frac{3}{2}}}{\mu'^{\frac{1}{2}}} \quad \text{.........................(4)}$$

The table on the opposite page gives a list of the more important kinds of magnitude which occur in Dynamics, with the dimensions which they have on any absolute system. An abbreviated mode of specification of various units on the C.G.S. and other metrical systems is indicated, together with such special names for particular units as are in current use. The C.G.S. units are distinguished by italics.

EXAMPLES. II.

1. A balloon whose virtual mass is M is falling with acceleration f; what amount of ballast must be thrown out of the car in order that it may have an upward acceleration f? (Neglect the frictional resistance of the air.)

[$2Mf/(f+g)$.]

2. A horizontal impulse applied to a mass m resting on a fixed horizontal board gives it a velocity u_0; find the time in which it will be brought to rest, and the space described, the coefficient of friction being μ.

If the board (of mass M) rests on a horizontal table and is free to move, prove that it will not be set in motion unless the coefficient of friction between the board and the table is less than

$$\frac{m}{M+m} \cdot \mu.$$

3. A rifle whose barrel is $2\frac{1}{2}$ ft. long discharges a bullet of 1 oz. weight with a velocity of 1000 ft. per sec., find (in ft.-lbs.) the energy of the bullet.

Also calculate (in horse-power) the rate at which the gases are doing work on the bullet just before the latter leaves the muzzle, on the assumption that the pressure on the bullet is constant during the discharge. [976·6 ; 710.]

4. If P be the tractive force in tons on a train weighing W tons, and R be the resistance, prove that the least time of travelling a distance s from rest to rest is

$$\sqrt{\left\{\frac{2s}{g}\cdot\frac{WP}{R(P-R)}\right\}},$$

and that the maximum velocity is

$$\sqrt{\left\{2gs\cdot\frac{R(P-R)}{WP}\right\}}.$$

Work out numerically for the case of $P=21$ tons, $W=800$ tons, $R=14$ tons, $s=1$ mile. [4 min.; 30 mile/hr.]

5. A railway truck will run at a constant speed down a track of inclination α. What will be its acceleration down a track of inclination $\beta\,(>\alpha)$, on the assumption that the frictional resistance bears the same ratio to the normal pressure in each case? $\left[\frac{\sin(\beta-\alpha)}{\cos\alpha}\cdot g.\right]$

6. At a distance l from a station a carriage is slipped from an express train going at full speed; prove that if the carriage come to rest at the station the rest of the train will then be at a distance $Ml/(M-m)$ beyond the station, M and m being the masses of the whole train and the carriage. (Assume that the pull of the engine is constant, and that the resistance on any portion is constant and proportional to its weight.)

7. Prove that the mean kinetic energy of a particle of mass m moving under a constant force, in any interval of time, is

$$\tfrac{1}{6}m\,(u_1^2+u_1u_2+u_2^2),$$

where u_1, u_2 are the initial and final velocities.

Prove that this is greater than the kinetic energy at the middle instant of the interval, but less than the kinetic energy of the particle when half-way between its initial and final positions.

8. A mass of 1 lb. is struck by a hammer which gives it a velocity of 1 ft./sec. Find the greatest pressure exerted during the impact, on the supposition that the pressure increases at a constant rate from zero to a maximum, and then falls at a constant rate from this maximum to zero, the whole duration of the impact being ·001 sec. [$62\tfrac{1}{2}$ lbs.]

9. If a curve be constructed with the kinetic energy of a particle as ordinate and the space described as abscissa, the force is represented by the gradient of the curve.

10. If work be done on a particle at a constant rate, prove that the velocity acquired in describing a space x from rest varies as $x^{\frac{1}{3}}$.

11. Two trains of equal mass are being drawn along smooth level lines by engines, one of which exerts a constant pull, while the other works at a constant rate. Prove that if they have equal velocities at two different instants, the second train will describe the greater space in the interval between these instants, and that they are working at equal rates at the middle instant of this interval.

EXAMPLES. III.

(Simple-Harmonic Motion; Pendulum, &c.)

1. A horizontal shelf is moved up and down with a simple-harmonic motion, of period $\frac{1}{2}$ sec. What is the greatest amplitude admissible in order that a weight placed on the shelf may not be jerked off? [2·4 in.]

2. A small constant horizontal force acts on the bob of a pendulum of length l, which is initially at rest, for a time t_1, and then ceases. Prove that the amplitude of the subsequent oscillation is

$$\frac{2f}{n^2} \cdot \sin \frac{1}{2} nt_1,$$

where f is the accelerating effect of the force, and $n = \sqrt{(g/l)}$.

3. Prove that if a straight tunnel were bored from London to Paris, a distance of (roughly) 200 miles, a train would traverse it under gravity alone in about 42 minutes, and that its maximum velocity would be about 446 miles per hour.

4. A mass of 5 lbs. hangs from a helical spring, and is observed to make 50 complete vibrations in 16·5 secs. Find in lbs. the force required to stretch the spring 1 inch. Also find the period of oscillation if an additional mass of 5 lbs. be attached. [4·67; ·47 sec.]

5. A mass M hangs by a helical spring from a point O; and when O is fixed the period of the vertical oscillations is one second. If O be made to execute a simple-harmonic oscillation in the vertical line, with an amplitude of 1 in., and a period of $\frac{1}{2}$ sec., find the amplitude of the forced oscillation of M. What is the relation between the phases of M and O?

6. A hydrometer floats upright in liquid; its displacement is 30 cm.^{3}, and the diameter of its stem is ·8 cm. Prove that the time of a small vertical oscillation is 1·55 sec.

7. Prove that in simple-harmonic motion if the initial displacement be x_0, and the initial velocity u_0, the amplitude will be

$$\sqrt{\left(x_0^2 + \frac{u_0^2}{n^2}\right)},$$

and the initial phase

$$-\tan^{-1} \frac{u_0}{nx_0}.$$

8. Prove that, in simple-harmonic motion, if the velocity be instantaneously altered from v to $v+\delta v$ the changes of amplitude (a) and phase (ϕ) are given by

$$\delta a = -\frac{\delta v}{n}\sin\phi, \quad \delta\phi = -\frac{\delta v}{na}\cos\phi.$$

9. Prove that in the small oscillations of a pendulum the mean kinetic and potential energies are equal.

10. Prove that in a half-swing of a pendulum from rest to rest the mean velocity of the bob is ·637 of the maximum velocity.

11. A pendulum is taken to an altitude of one mile above the earth's surface; by what fraction must its length be diminished in order that it may oscillate in the same period as before? [$\frac{1}{2000}$.]

12. A weight hangs from a fixed point by a string 100 ft. long. If it be started from its lowest position with a velocity of $2\frac{1}{2}$ ft./sec., how far will it swing before coming to rest? Also, how many seconds will it take to describe the first $2\frac{1}{2}$ ft.? [4·42 ft.; 1·06 sec.]

13. A simple pendulum hangs from the roof of a railway carriage and remains vertical while the train is running smoothly at 30 miles an hour. When the brakes are put on, the pendulum swings through an angle of 3°. Prove that the train will come to rest in about 385 yards, the resistance being assumed constant.

14. A mass M, hanging from the end of a horizontal cantilever of length l, and sectional area ω, makes vertical oscillations whose period is T. Prove that the value of Young's modulus for the cantilever is

$$\frac{4\pi^2 Ml^3}{3\omega\kappa^2 T^2},$$

in absolute measure, where κ is the radius of gyration of the cross-section about a horizontal line through its centre.

15. A mass M is suspended from the middle point of a horizontal bar of length l, supported at the ends. If the period of the vertical oscillations of M be T secs., find the flexural rigidity of the bar, neglecting its inertia. [S. 146.] [$\frac{1}{12}\pi^2 Ml^3/T^2$.]

16. A string whose ends are fixed is stretched with a tension P. Prove that the work required to produce a small lateral deflection x at a given point by a force applied there is

$$\frac{a+b}{2ab}Px^2,$$

where a, b are the distances of the point from the ends.

Verify that this is equal to the work required to produce the actual increase of length of the string against the tension P.

17. A particle m is attached to a light wire which is stretched tightly between two fixed points with a tension P. If a, b be the distances of the particle from the two ends, prove that the period of a small transverse oscillation of m is

$$2\pi\sqrt{\left(\frac{m}{P}\cdot\frac{ab}{a+b}\right)}.$$

Prove that for a wire of given length the period is longest when the particle is attached at the middle point.

18. A particle is disturbed from a position of unstable equilibrium; sketch the various forms of space-time curve for the initial stage of the motion.

19. If in Art. 14 the disturbing force $f(t)$ be proportional to e^{-t^2/τ^2}, prove that the resulting simple-harmonic vibration is given by

$$x = \frac{u_1}{n} e^{-\frac{1}{4}n^2\tau^2} \sin nt,$$

where u_1 is the velocity generated by an instantaneous impulse of the same amount.

(This depends on the formula

$$\int_0^\infty e^{-x^2/a^2} \cos 2bx\,dx = \tfrac{1}{2}\sqrt{\pi}\,a e^{-a^2b^2}.)$$

EXAMPLES. IV.

(Variable Gravity, &c.)

1. A particle is projected upwards from the earth's surface with a velocity which would, if gravity were uniform, carry it to a height h. Prove that if variation of gravity be allowed for, but the resistance of the air neglected, the height reached will be greater by $h^2/(a-h)$, where a is the earth's radius.

2. A particle is projected vertically upwards with a velocity just sufficient to carry it to infinity. Prove that the time of reaching a height h is

$$\frac{1}{3}\sqrt{\left(\frac{2a}{g}\right)}\cdot\left\{\left(1+\frac{h}{a}\right)^{\frac{3}{2}}-1\right\},$$

where a is the earth's radius.

3. If a particle be shot upwards from the earth's surface with a velocity of 1 mile per sec., find roughly the difference between the heights it will attain, (1) on the hypothesis of constant gravity, (2) on the hypothesis that gravity varies inversely as the square of the distance from the earth's centre.

[1·72 miles.]

4. If a particle fall to the earth from rest at a great distance the times of traversing the first and second halves of this distance are as 9 to 2 very nearly.

5. If a particle moves in a straight line under a central attraction varying inversely as the cube of the distance, prove that the space-time curve is a conic.

Under what condition is it an ellipse, parabola, or hyperbola, respectively?

Examine the case where the force is repulsive.

6. A particle moves from rest at a distance a towards a centre of force whose accelerative effect is $\mu/(\text{dist.})^3$; prove that the time of falling in is $a^2/\sqrt{\mu}$.

7. Prove that the time of falling from rest at a distance a into a centre of attractive force whose accelerative effect is $\mu \times (\text{dist.})^n$ varies as

$$\frac{a^{\frac{1}{2}(1-n)}}{\sqrt{\mu}}.$$

8. A particle is subject to two equal centres of force whose accelerative effect is $\mu/(\text{dist.})^2$. If it starts from rest at equal distances a from the two centres, describe the subsequent motion; and prove that the maximum velocity is

$$2\sqrt{\left(\mu \cdot \frac{a-b}{ab}\right)},$$

if $2b$ be the distance between the centres.

CHAPTER III

TWO-DIMENSIONAL KINEMATICS

20. Velocity.

The total displacement which a moving point undergoes in any given interval of time is of the nature of a *vector*, and is represented by the straight line drawn from the initial to the final position of the point, or by any equal and parallel straight line drawn in the same sense. Successive displacements are obviously compounded by the law of addition of vectors [*S.* 2].

If the displacements in equal intervals of time are equal in every respect, the velocity is said to be constant, and may be specified by the vector which gives the displacement per unit time.

To obtain a definition of 'velocity' in the general case, let P, P' be the positions of the moving point at the instants t, $t + \delta t$, respectively, where δt is not as yet assumed to be small. Produce PP' to a point U such that

$$\mathrm{PU} = \frac{\mathrm{PP'}}{\delta t} \quad \text{.........(1)}$$

The vector PU* then represents what we may call the 'mean velocity' of the point in the interval δt. That is to say, a point moving with a constant velocity equal to this would undergo the same displacement PP′ in the same interval δt. If this interval

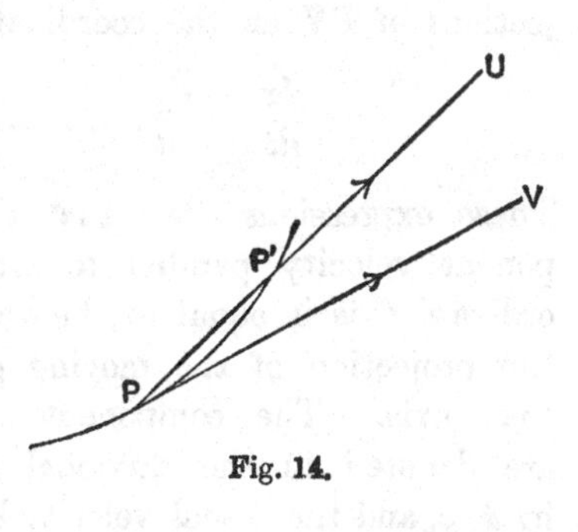

Fig. 14.

* Roman type is used whenever a vector is indicated by the terminal letters of a line; the order of the letters is of course important [*S.* 1].

be varied, the vector will assume different values, and if we imagine δt to be diminished without limit, it will in all cases which it is necessary to consider tend to a definite limiting value PV. This limiting vector is adopted as the definition of the velocity of the moving point at the instant t. Briefly, the 'velocity at the instant t' is the mean velocity in an infinitely short interval beginning at the instant t. Its direction is that of the tangent to the path at P; moreover, if s denote the arc of the curve, measured from some fixed point on it, the chord PP' will ultimately be in a ratio of equality to the arc δs, and the velocity at P is therefore given as to magnitude and sign by the formula

$$v = \frac{ds}{dt} \text{..............................(2)}$$

For purposes of calculation, velocities, like other vectors, may be specified by means of their components referred to some system of coordinates. Thus if, in two dimensions, x, y denote the Cartesian coordinates of P, and $x + \delta x$, $y + \delta y$ those of P', the projections of the vector PP′ on the coordinate axes are δx, δy, and those of the mean velocity PU are accordingly

$$\frac{\delta x}{\delta t}, \quad \frac{\delta y}{\delta t}.$$

Hence the 'component velocities' at the instant t, i.e. the projections of PV on the coordinate axes, will be

$$\frac{dx}{dt}, \quad \frac{dy}{dt}. \text{..............(3)}$$

These expressions shew that the component velocity parallel to either coordinate axis is equal to the velocity of the projection of the moving point on that axis. The component velocities are denoted in the fluxional notation by $\dot{x}$, $\dot{y}$, and the actual velocity by $\dot{s}$. It should be noticed that at present there is no restriction to rectangular axes.

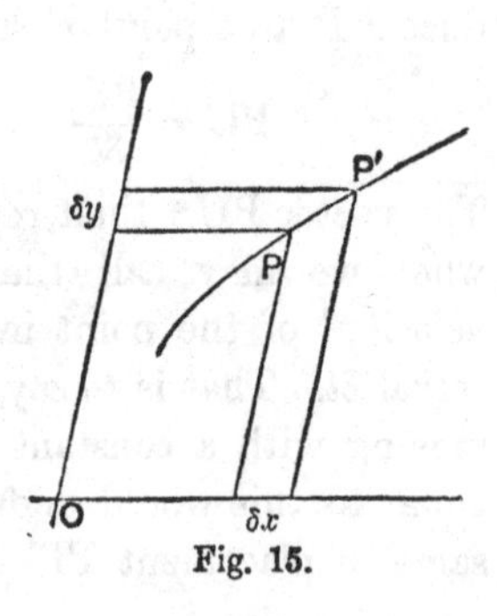

Fig. 15.

If, however, the axes are rectangular, then denoting by ψ the angle which the direction of motion makes with the axis of x, the component velocities will be

$$\dot{x} = \frac{dx}{ds}\frac{ds}{dt} = \dot{s}\cos\psi, \quad \dot{y} = \frac{dy}{ds}\frac{ds}{dt} = \dot{s}\sin\psi, \quad \text{.........(4)}$$

whence
$$\dot{s}^2 = \dot{x}^2 + \dot{y}^2, \quad \tan\psi = \dot{y}/\dot{x}. \quad \text{..................(5)}$$

These relations are otherwise obvious by orthogonal projection.

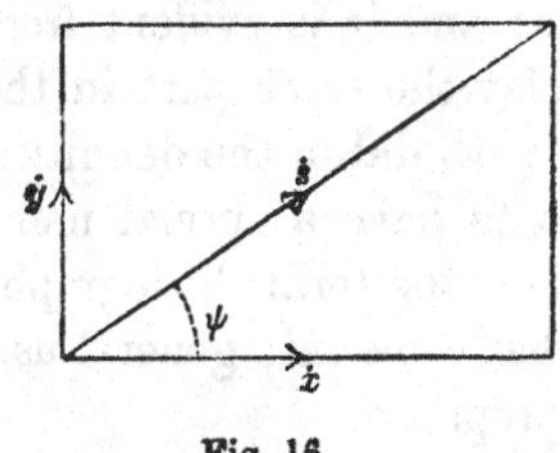

Fig. 16.

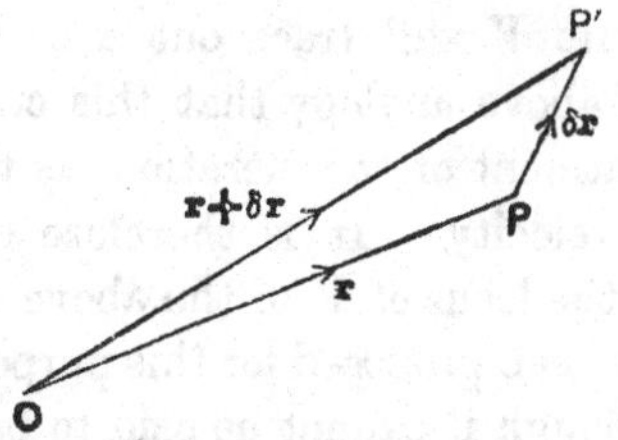

Fig. 17.

The definition of velocity has its most concise expression in the language of vectors. If $\mathbf{r}$ denote the position-vector [*S.* 5] of the moving point relative to any fixed origin O, we may write

$$\text{OP} = \mathbf{r}, \quad \text{OP}' = \mathbf{r} + \delta\mathbf{r}, \text{....................(6)}$$

and therefore

$$\text{PP}' = \text{PO} + \text{OP}' = \text{OP}' - \text{OP} = \delta\mathbf{r}.$$

The mean velocity in the interval δt is therefore denoted by

$$\frac{\delta\mathbf{r}}{\delta t},$$

and the velocity at the instant t is expressed by

$$\mathbf{v} = \frac{d\mathbf{r}}{dt} = \dot{\mathbf{r}}. \quad \text{..........................(7)}$$

Hence velocity may be defined as rate of change of position, provided we understand the word 'change' to refer to *vectorial* increment.

If x, y denote the projections of OP on coordinate axes through O, we have

$$\mathbf{r} = x\mathbf{i} + y\mathbf{j}, \quad \text{..........................(8)}$$

where $\mathbf{i}$, $\mathbf{j}$ are unit vectors parallel to the respective axes. Hence

$$\mathbf{v} = \dot{\mathbf{r}} = \dot{x}\mathbf{i} + \dot{y}\mathbf{j}, \quad \text{........................(9)}$$

which shews (again) that the projections of $\mathbf{v}$ are $\dot{x}$, $\dot{y}$.

21. Hodograph. Acceleration.

Just as velocity may be defined briefly as rate of change of position, so 'acceleration' may be described as rate of change of velocity. Like velocity it is a vector, having both direction and magnitude.

If from a fixed point O vectors OV be drawn, as in Fig. 18, to represent the velocities of a moving point at different instants, the points V will trace out a certain curve; and it is evident from the above analogy that this curve will play the same part in the treatment of 'acceleration' as the actual path did in the definition of velocity. It is therefore convenient to have a special name for the locus of V in the above construction; the term 'hodograph' has been proposed for this purpose*, and has come into general use, although it cannot be said to be very appropriate.

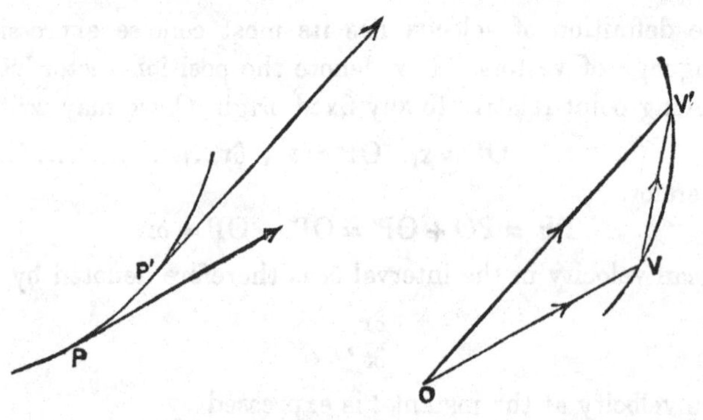

Fig. 18.

If OV, OV′ represent the velocities at the beginning and end of any interval, the vector VV′, which is their difference, indicates completely the change which the velocity has undergone in the interval. If this change is always the same in equal intervals, the moving point is said to have a constant acceleration, and this acceleration is specified by the vector which gives the change of velocity per unit time. See Fig. 26, p. 75, where, on the right hand, OA, OB, OC, OD represent velocities at equidistant times.

* By Sir W. R. Hamilton (1805–65), Professor of Astronomy at Dublin 1827–65, inventor of the Calculus of Quaternions.

In the general case, if OV, OV′ represent the velocities at the instants t, $t+\delta t$, the vector

$$\frac{\mathrm{VV'}}{\delta t}$$

gives what may be called the 'mean acceleration' in the interval δt; in the sense that a constant acceleration equal to this would produce the same change of velocity VV′ in the same interval.

If δt be diminished without limit, this mean acceleration tends in general to a definite limiting value, which is adopted as the definition of the 'acceleration at the instant t.' Its direction is that of the tangent to the hodograph at V, and its magnitude is given by the velocity of V along the hodograph. For instance, in the case of a constant acceleration, the hodograph is a straight line described with constant velocity.

If we introduce Cartesian coordinates, rectangular or oblique, the projections of OV on the axes will be denoted by $\dot{x}$, $\dot{y}$, and those of VV′ accordingly by $\delta\dot{x}$, $\delta\dot{y}$. The projections of the mean acceleration in the interval δt will therefore be

$$\frac{\delta\dot{x}}{\delta t},\ \frac{\delta\dot{y}}{\delta t}.$$

Hence the 'component accelerations,' i.e. the projections of the acceleration at the instant t, will be $\ddot{x}$, $\ddot{y}$, or

$$\frac{d^2x}{dt^2},\ \frac{d^2y}{dt^2}, \qquad \text{.........................(1)}$$

in the more usual notation. They are therefore identical with the accelerations of the projections of the moving point on the coordinate axes.

In the notation of vectors we write

$$\mathrm{OV} = \mathbf{v}, \quad \mathrm{OV'} = \mathbf{v} + \delta\mathbf{v},$$

and therefore

$$\mathrm{VV'} = \delta\mathbf{v}.$$

The mean acceleration in the interval δt is accordingly denoted by $\delta\mathbf{v}/\delta t$, and the acceleration at the instant t by

$$\mathbf{a} = \frac{d\mathbf{v}}{dt} = \dot{\mathbf{v}}. \qquad \text{.........................(2)}$$

Again, since

$$\dot{\mathbf{v}} = \dot{x}\mathbf{i} + \dot{y}\mathbf{j}, \qquad (3)$$

by Art. 20 (9), we have

$$\mathbf{a} = \ddot{x}\mathbf{i} + \ddot{y}\mathbf{j}. \qquad (4)$$

If the acceleration $\mathbf{a}$ be constant, we have by integration of (2)

$$\dot{\mathbf{r}} = \mathbf{v} = \mathbf{a}t + \mathbf{b}, \qquad (5)$$

$$\mathbf{r} = \tfrac{1}{2}\mathbf{a}t^2 + \mathbf{b}t + \mathbf{c}, \qquad (6)$$

where the vectors $\mathbf{b}$, $\mathbf{c}$ are arbitrary. Properly interpreted, this shews that the path is a parabola.

It is to be remarked that the preceding definitions are in no way restricted to the case of motion in two dimensions. The formulæ involving Cartesian coordinates are easily generalized by the introduction of a third coordinate z.

The formulæ for velocities and accelerations in terms of *polar* coordinates are investigated in Art. 86.

Ex. 1. Let a point P describe a circle of radius a about a fixed point O with the constant angular velocity n.

Relatively to rectangular axes through O we have

$$x = a\cos\theta, \quad y = a\sin\theta, \qquad (7)$$

where

$$\theta = nt + \epsilon. \qquad (8)$$

Hence

$$\dot{x} = -na\sin\theta, \quad \dot{y} = na\cos\theta, \qquad (9)$$

shewing that the velocity is at right angles to OP, and equal to na.

Also

$$\ddot{x} = -n^2 a\cos\theta = -n^2 x, \quad \ddot{y} = -n^2 a\sin\theta = -n^2 y. \qquad (10)$$

Since $-x$, $-y$ are the projections of PO, the acceleration is represented by the vector n^2. PO.

Ex. 2. Let P describe an ellipse in such a way that the eccentric angle increases uniformly with the time.

Referring to the principal axes, we have

$$x = a\cos\phi, \quad y = b\sin\phi, \qquad (11)$$

where

$$\phi = nt + \epsilon. \qquad (12)$$

Hence

$$\dot{x} = -na\sin\phi, \quad \dot{y} = nb\cos\phi. \qquad (13)$$

The resultant velocity is therefore

$$\dot{s} = n\sqrt{(a^2\sin^2\phi + b^2\cos^2\phi)} = n \,.\, OD, \qquad (14)$$

if OD be the semi-diameter conjugate to OP. The velocity is therefore represented by the vector n . OD; and the hodograph is (on a certain scale) coincident with the locus of D, i.e. with the ellipse itself.

Again $\ddot{x} = -n^2 a \cos\phi = -n^2 x, \quad \ddot{y} = -n^2 b \sin\phi = -n^2 y.$ ....(15)

The acceleration is therefore represented by the vector n^2. PO.

This type of motion is called 'elliptic harmonic.' It presents itself from the converse point of view in Art. 28.

22. Relative Motion.

The vector PQ which indicates the position of a point Q relative to a point P is of course the difference of the position-vectors of Q and P relative to any origin O. Hence if P, Q be regarded as points in motion, the displacement of Q relative to P in any interval of time will be the geometric difference of the displacements of Q and P. This follows at once from the theory of addition of vectors, but may be illustrated by a figure. If from a fixed point C we draw lines CR equal and parallel to PQ in its

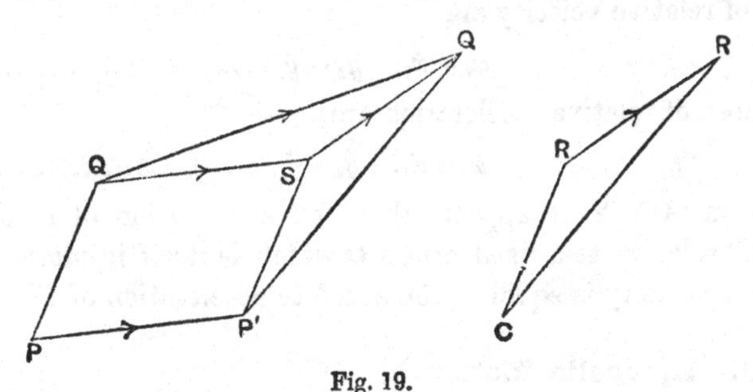

Fig. 19.

various positions, the locus of R will give the path of Q relative to P. Hence if P, Q and P', Q' denote two pairs of simultaneous positions of the moving points, and if we draw CR, CR' parallel and equal to PQ, $P'Q'$ respectively, RR′ will represent the relative displacement in the interval. But if we complete the parallelogram $PP'SQ$ determined by PP' and PQ, the triangles CRR' and $P'SQ'$ will be congruent. Hence

$$RR' = SQ' = QQ' - QS = QQ' - PP',$$

which was to be proved.

Further, considering displacements per unit time, it appears that the velocity of Q relative to P is the geometric difference of the velocities of Q and P. And considering changes of velocity per unit time, we get a similar rule for the relative acceleration.

In vector notation, if $\mathbf{r}_1$, $\mathbf{r}_2$ be the position vectors of P, Q relatively to a fixed point O, a relative displacement is expressed by

$$\delta(\mathbf{r}_2 - \mathbf{r}_1) = \delta\mathbf{r}_2 - \delta\mathbf{r}_1\,; \quad \text{.....................(1)}$$

the relative velocity is

$$\frac{d}{dt}(\mathbf{r}_2 - \mathbf{r}_1) = \dot{\mathbf{r}}_2 - \dot{\mathbf{r}}_1\,; \quad \text{.....................(2)}$$

and the relative acceleration is

$$\frac{d}{dt}(\dot{\mathbf{r}}_2 - \dot{\mathbf{r}}_1) = \ddot{\mathbf{r}}_2 - \ddot{\mathbf{r}}_1. \quad \text{.....................(3)}$$

In Cartesian coordinates, the components of a relative displacement are

$$\delta x_2 - \delta x_1, \quad \delta y_2 - \delta y_1\,; \quad \text{.....................(4)}$$

those of relative velocity are

$$\dot{x}_2 - \dot{x}_1, \quad \dot{y}_2 - \dot{y}_1\,; \quad \text{.....................(5)}$$

and those of relative acceleration are

$$\ddot{x}_2 - \ddot{x}_1, \quad \ddot{y}_2 - \ddot{y}_1. \quad \text{.....................(6)}$$

As in Art. 2, it appears that the acceleration of a moving point P relative to a fixed origin O which is itself in motion with constant velocity is equal to the absolute acceleration of P.

23. Epicyclic Motion.

An important illustration of relative motion is furnished by the theory of epicyclic motion. If a point Q describes a circle about a fixed centre O with constant angular velocity, whilst P describes a circle relative to Q with constant angular velocity, the path of P is called an 'epicyclic.'

The coordinates of Q relatively to rectangular axes through O will have the forms

$$x = a\cos(nt + \epsilon), \quad y = a\sin(nt + \epsilon), \quad \text{............(1)}$$

where n is the angular velocity of Q in its circle, and a the radius of this circle. Again, if we complete the parallelogram $OQPQ'$, the coordinates of Q' will be

$$x = a'\cos(n't + \epsilon'), \quad y = a'\sin(n't + \epsilon'), \quad \text{........(2)}$$

where $a' = OQ' = QP$, and n' is the angular velocity of QP or OQ'.

Since $$\mathrm{OP} = \mathrm{OQ} + \mathrm{QP} = \mathrm{OQ} + \mathrm{OQ}', \qquad\qquad (3)$$
the coordinates of P will be

$$\left.\begin{aligned} x &= a\cos(nt+\epsilon) + a'\cos(n't+\epsilon'), \\ y &= a\sin(nt+\epsilon) + a'\sin(n't+\epsilon'). \end{aligned}\right\} \qquad\qquad (4)$$

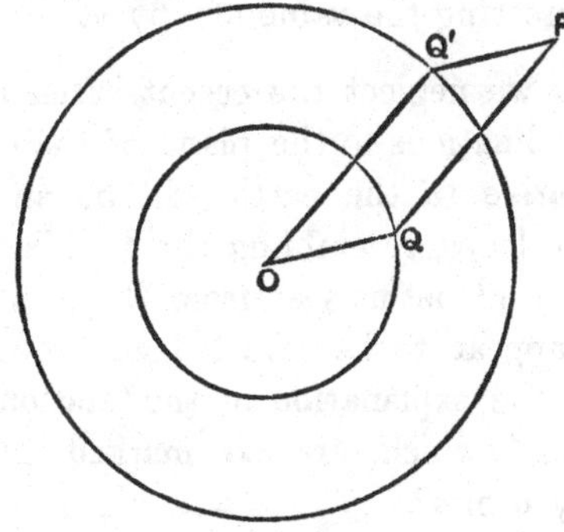

Fig. 20.

It is evident that the path of P may also be regarded as that of a point which describes a circle relatively to Q', whilst Q' describes a circle about O, the angular velocities being constant and equal to n, n', respectively.

If the angular velocities n, n' have the same sign, i.e. if the revolutions are in the same sense, the epicyclic is said to be

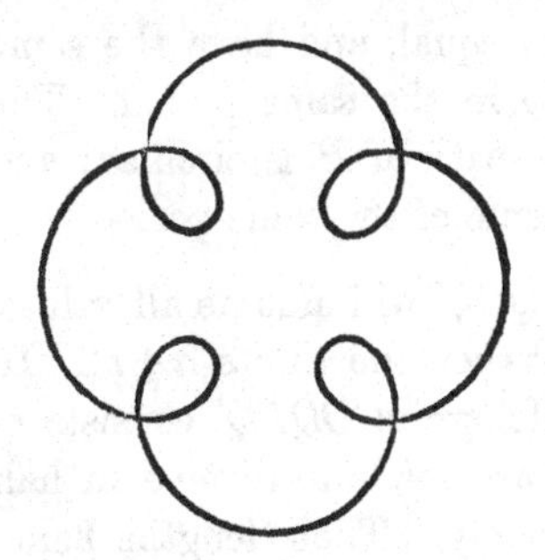

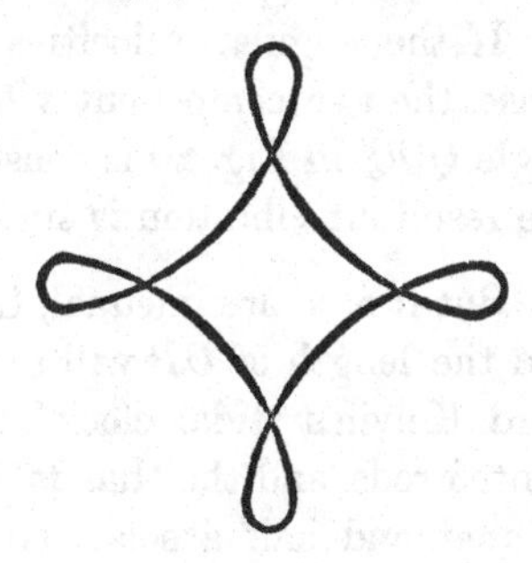

Fig. 21.

'direct'; in the opposite case it is said to be 'retrograde.' It will be a re-entrant, or closed, curve if the angular velocities n, n' are in the ratio of two integers, but not otherwise. The annexed Fig. 21 shews two cases, corresponding to $n = 5n'$ and $n = -3n'$ respectively.

The relative orbit of two points which are describing concentric circles with constant angular velocities is also an epicyclic. Thus, in Fig. 20, the vector QQ′ is the geometric difference of OQ′ and OQ, and its projections on the coordinate axes will therefore be obtained by reversing the sign of a in (4), or (what comes to the same thing) by increasing the value of ϵ by π.

For example, if we neglect the eccentricities of the planetary orbits, and their inclinations to the plane of the ecliptic, the path of any planet relative to the earth will be an epicyclic. The relative orbits have loops, resembling the first case of Fig. 21, so that the motion of a planet as seen from the earth, and projected on the sky, will appear to be arrested and reversed at regular intervals. This is the explanation of the 'stationary points' and 'retrograde motions' which are so marked a feature of the apparent planetary orbits*.

24. Superposition of Simple-Harmonic Vibrations.

The result of superposing two simple-harmonic vibrations in the same straight line is to produce a motion which is the orthogonal projection of motion in an epicyclic. This is evident from the geometrical representation of Art. 10, or from the formulæ (4) of Art. 23.

If the angular velocities n, n' are equal, and have the same sense, the two component vibrations have the same period. The angle QOQ' in Fig. 20 is constant; the path of P is circular; and the resultant vibration is simple-harmonic of the same period.

But if n, n' are unequal, the angle QOQ' will assume all values, and the length of OP will oscillate between the limits $a \pm a'$. In Lord Kelvin's 'tidal clock,' the parallelogram $OQPQ'$ consists of jointed rods, and the 'hands' OQ, OQ' are made to revolve in half a lunar and half a solar day, respectively. Their lengths being made proportional to the amplitudes of the lunar and solar semi-diurnal tides, the projection of P on a straight line through O will give the tide-height due to the combination of these.

* They were explained by the hypothesis of epicyclics, by Ptolemy of Alexandria (d. A.D. 168) in his *Almagest*.

If the angular velocities n, n' are nearly, but not exactly equal, the angle QOQ' will vary very little in the course of a single revolution of OQ or OQ', and the resultant vibration may be described, somewhat loosely, as a simple-harmonic vibration whose amplitude fluctuates between the limits $a \pm a'$. The period of this fluctuation is the interval in which one arm gains or loses four right angles relatively to the other, and is therefore $2\pi/(n-n')$. In other words, the *frequency* $(n-n')/2\pi$ of the fluctuations is the difference of the frequencies of the two primary vibrations.

As an illustration we may point to the alternation of 'spring' and 'neap' tides, which occur when the phases of the lunar and solar semi-diurnal components are respectively coincident or opposed. In Acoustics we have the phenomenon of 'beats' between two simple tones of nearly the same pitch.

The annexed figure shews the space-time curve for the case of $a = 2a'$, $9n = 10n'$.

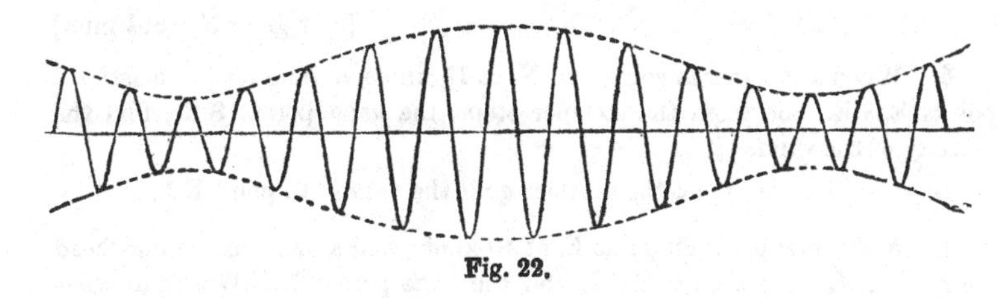

Fig. 22.

The fluctuations are of course most marked when the amplitudes a, a' of the primary vibrations are equal. The amplitude of the resultant then varies between 0 and $2a$.

Analytically, we have

$$x = a\cos\theta + a'\cos\theta', \qquad\qquad (1)$$

where

$$\theta = nt + \epsilon, \quad \theta' = n't + \epsilon'. \qquad\qquad (2)$$

Hence

$$x = (a+a')\cos\tfrac{1}{2}(\theta+\theta')\cos\tfrac{1}{2}(\theta-\theta') - (a-a')\sin\tfrac{1}{2}(\theta+\theta')\sin\tfrac{1}{2}(\theta-\theta'). \qquad (3)$$

If we put

$$\left.\begin{aligned} r\cos\phi &= (a+a')\cos\tfrac{1}{2}(\theta-\theta'),\\ r\sin\phi &= (a-a')\sin\tfrac{1}{2}(\theta-\theta'),\end{aligned}\right\} \qquad\qquad (4)$$

this may be written

$$x = r\cos\{\tfrac{1}{2}(\theta+\theta') + \phi\} \qquad\qquad (5)$$

the values of r and ϕ being given by

$$r^2 = a^2 + 2aa' \cos(\theta - \theta') + a'^2, \quad \text{......................(6)}$$

$$\tan\phi = \frac{a - a'}{a + a'} \tan\tfrac{1}{2}(\theta - \theta'). \quad \text{......................(7)}$$

If n and n' are nearly equal, the formula (5) may be regarded as representing a simple-harmonic vibration whose amplitude r varies slowly between $a+a'$ and $a-a'$, whilst there is also a slow variation of phase. The dotted curve in Fig. 22 shews the variation of amplitude.

EXAMPLES. V.

1. One ship is approaching a port at speed u, and another is leaving it at a speed v, the courses being straight and making an angle α with one another. Prove that when the distance between them is least their distances from the port are as

$$v + u\cos\alpha : u + v\cos\alpha.$$

2. In what direction must a boat whose speed is 8 knots be steered in order to reach a point 10 miles distant in a N.N.E. direction, there being a S.E. current of 2 knots? Also what will be the time required?

[9° 8′ E. of N.; $85\frac{1}{2}$ min.]

3. When a steamer is going due N. at 15 knots, a vane on the masthead points E.N.E., and when the steamer stops, the vane points S.E.; find the velocity of the wind.

In what direction must the steamer go if the vane is to point E.?

4. A steamer is running due E. at 10 knots, and a vane on the masthead points N.N.E.; she then turns N. and the vane points N.N.W.; from what quarter does the wind blow, and with what velocity? [N.W.; 10 knots.]

5. If P, Q be two points on the spokes of a carriage wheel, find the direction and magnitude of the velocity of Q relative to P, having given the radius a of the wheel and the speed u of the carriage.

6. Having given the hodograph of a moving point, and the law of its description, shew how to find the path. What is the effect of a change in the position of the pole?

If the hodograph be a circle described with constant velocity, the path is a trochoidal curve.

7. A boat is impelled across a stream with a constant velocity relatively to the water, this velocity being equal to that of the stream. Prove that if it be steered towards a fixed point on the bank, its path will be a parabola having this point as focus, and that the actual landing point will be the vertex of the parabola.

8. If in the preceding Question the ratio of the velocity of the stream to the (relative) velocity of the boat be $1/n$, prove that the polar equation of the path is

$$r = \frac{a(\sin\frac{1}{2}\theta)^{n-1}}{(\cos\frac{1}{2}\theta)^{n+1}}.$$

Examine the cases of $n<1$ and $n>1$, respectively.

9. Prove that, in the notation of Art. 23, the velocity at any point of an epicylic is

$$\sqrt{\{n^2a^2+2nn'aa'\cos(\theta-\theta')+n'^2a'^2\}}.$$

10. Prove that in the notation of Art. 23, the direction of motion in an epicyclic passes through the origin when

$$\cos\{(n-n')t+\epsilon-\epsilon'\} = -\frac{na^2+n'a'^2}{(n+n')aa'}.$$

Prove that this cannot occur unless na is numerically less than $n'a'$, where a is the greater of the two radii a, a'.

11. Prove that if an epicyclic pass through the centre its polar equation is of the form

$$r = a\cos m\theta,$$

where $m \lesseqgtr 1$ according as the epicyclic is direct or retrograde.

12. A point subject to an acceleration which is constant in magnitude and direction passes through three points P_1, P_2, P_3 at the instants t_1, t_2, t_3, respectively. Prove that the acceleration is represented as to direction and magnitude by the vector

$$\frac{2(\mathrm{P_2P_3}\,.\,t_1+\mathrm{P_3P_1}\,.\,t_2+\mathrm{P_1P_2}\,.\,t_3)}{(t_2-t_3)(t_3-t_1)(t_1-t_2)}.$$

13. A point has an acceleration which is constant in magnitude and direction. Prove that its path relative to a point moving with constant velocity in a straight line is a parabola.

14. Prove by differentiation of the equations

$$\dot{x} = v\cos\psi, \quad \dot{y} = v\sin\psi$$

that the acceleration of a point describing a plane curve is made up of a component dv/dt along the tangent and a component $vd\psi/dt$ towards the centre of curvature.

15. Prove by differentiation of the equations

$$x = r\cos\theta, \quad y = r\sin\theta$$

that in polar coordinates the component velocities along and at right angles to the radius vector are

$$\dot{r}, \quad r\dot{\theta},$$

and that the component accelerations in the same directions are

$$\ddot{r}-r\dot{\theta}^2, \quad r\ddot{\theta}+2\dot{r}\dot{\theta},$$

respectively.

16. Prove that if two simple-harmonic vibrations of the same period, in the same line, whose amplitudes are a, b, and whose phase differs by ϵ, be compounded, the amplitude of the resultant vibration is

$$\sqrt{(a^2+2ab\cos\epsilon+b^2)}.$$

17. Prove that an epicyclic in which the angular velocities are equal and opposite is an ellipse.

18. The ends P, Q of a rod are constrained to move on two straight lines OA, OB at right angles to one another. If the velocity of P be constant, prove that the acceleration of any point on the rod is at right angles to OA and varies inversely as the cube of the distance from OA.

19. A string is unwound with constant angular velocity ω from a fixed reel, the free portion being kept taut in a plane perpendicular to the axis. Prove that the acceleration of the end P of the string is along PR and equal to $\omega^2 . PR$, where R is on the radius through the point of contact, at double the distance from the centre.

20. Prove that epicyclic motion is equivalent to motion in an ellipse which revolves uniformly about its centre, the relative motion in the ellipse following the elliptic-harmonic law (Art. 28).

CHAPTER IV

DYNAMICS OF A PARTICLE IN TWO DIMENSIONS. CARTESIAN COORDINATES

25. Dynamical Principle.

The 'momentum' of a particle is the product of the mass, which is a scalar quantity, into the velocity, and is therefore to be regarded as a *vector*, having at each instant a definite magnitude and direction. The hodograph of the particle may in fact be used to represent, on the appropriate scale, the variations in the momentum.

The 'change of momentum' in any interval of time is that momentum which must be compounded by geometrical addition with the initial momentum in order to produce the final momentum. In other words it is the vector difference of the final and initial momenta.

The 'impulse' of a force in any infinitely small interval δt is the product of the force into δt; this again is to be regarded as a vector. The 'total,' or 'integral,' impulse in any finite interval is the geometric sum of the impulses in the infinitesimal elements δt of which the interval in question may be regarded as made up.

The fundamental assumption which we now make, is the same as in Art. 7, but in an extended sense. It asserts that change of momentum is proportional to the impulse, and therefore *equal* to the impulse if the absolute system of force-measurement be adopted. This is, as before, a physical postulate which can only be justified by a comparison of theoretical results with experience. It is a statement as to equality of vectors, and accordingly implies identity of direction as well as of magnitude. It is immaterial whether the interval of time considered be finite or infinitesimal; the statement in either form involves the other as a consequence.

It is, moreover, assumed that when two or more forces act simultaneously, the changes of momentum in an infinitesimal interval δt due to the several forces may be calculated separately, and the results combined by geometrical addition to obtain the actual change of momentum. Since the changes of momentum due to the several forces are in the directions of these forces, and proportional to them, it follows that the actual change of momentum is the same as would be produced by a single force which is the geometric sum of the given forces. It appears, then, that our assumptions include the law of composition of forces on a particle which is known in Statics as the 'polygon of forces.'

If m be the mass of the particle, $\mathbf{v}$ its velocity, $\mathbf{P}$ the geometric sum of the forces acting on it, we have, in vector notation,

$$\delta(m\mathbf{v}) = \mathbf{P}\delta t,$$

or

$$m\frac{d\mathbf{v}}{dt} = \mathbf{P}. \quad \text{.........................(1)}$$

Integrating this over a finite time we have

$$m\mathbf{v}_2 - m\mathbf{v}_1 = \int_{t_1}^{t_2} \mathbf{P}\,dt, \quad \text{.....................(2)}$$

where the definite integral is to be interpreted as the limit of the sum of an infinite series of infinitesimal vectors. The form (2) corresponds to Newton's formulation of his Second Law of Motion*.

If $\mathbf{r}$ denote the position-vector of the particle, we have

$$\mathbf{v} = \dot{\mathbf{r}}, \quad \text{..............................(3)}$$

and therefore

$$m\ddot{\mathbf{r}} = \mathbf{P}. \quad \text{..............................(4)}$$

The solution of this equation will involve two arbitrary vectors, which may be determined in terms of given initial conditions as to position and velocity.

* "Mutationem motus proportionalem esse vi motrici impressae, et fieri secundum lineam rectam qua vis illa imprimitur." Change of motion [momentum] is proportional to the impressed force [impulse], and takes place in the direction of the straight line in which that force is impressed.

26. Cartesian Equations.

For purposes of calculation some system of coordinates is necessary. Considering (for simplicity) the case of motion in two dimensions, and using Cartesian coordinates (rectangular or oblique), let us denote by u, v the components of the velocity at the instant t, and by X, Y those of the force. The components of momentum will therefore be mu, mv, and those of the change of momentum in the interval δt will be $\delta(mu)$, $\delta(mv)$. The components of impulse will be $X\delta t$, $Y\delta t$. Since the parallel projections of equal vectors are equal, we must have

$$\delta(mu) = X\delta t, \quad \delta(mv) = Y\delta t, \quad \text{...............(1)}$$

or

$$m\frac{du}{dt} = X, \quad m\frac{dv}{dt} = Y. \quad \text{..................(2)}$$

Since

$$u = \frac{dx}{dt}, \quad v = \frac{dy}{dt}, \quad \text{...............(3)}$$

the equations may be written

$$m\frac{d^2x}{dt^2} = X, \quad m\frac{d^2y}{dt^2} = Y. \quad \text{..................(4)}$$

The solution of these equations in any particular case will involve four arbitrary constants, which may be adjusted to satisfy given initial conditions as to the values of x, y, dx/dt, dy/dt.

The extension to three dimensions, where the solution involves six arbitrary constants, is obvious.

Ex. A particle slides up a line of greatest slope on a plane of inclination α, under gravity and the reaction of the plane.

If the axis of x be drawn upwards along a line of greatest slope, and the axis of y normal to the plane, we have

$$X = -mg\sin\alpha - F, \quad Y = -mg\cos\alpha + R, \quad \text{..............(5)}$$

where F and R are the tangential and normal components of the reaction. Since, by hypothesis, $y=0$, we have

$$R = mg\cos\alpha, \quad \text{..................................(6)}$$

and

$$m\frac{d^2x}{dt^2} = -mg\sin\alpha - F. \quad \text{...........................(7)}$$

If the plane be smooth, we have $F=0$, and the retardation is $g\sin\alpha$. If the plane be rough, we have $F=\mu R$, on the usual law of friction, and the retardation is

$$g(\sin\alpha+\mu\cos\alpha)=g\frac{\sin(\alpha+\lambda)}{\cos\lambda}, \quad\text{..........(8)}$$

if λ be the angle of friction.

27. Motion of a Projectile.

A simple application of the preceding equations is to the unresisted motion of a particle under gravity.

If the axis of y be drawn vertically upwards, then, whether the axis of x be horizontal or not, we have

$$X=0, \quad Y=-mg, \quad\text{..........(1)}$$

and therefore

$$\ddot{x}=0, \quad \ddot{y}=-g. \quad\text{..........(2)}$$

Integrating, we find

$$\dot{x}=A, \quad \dot{y}=-gt+B, \quad\text{..........(3)}$$

$$x=At+C, \quad y=-\tfrac{1}{2}gt^2+Bt+D, \quad\text{..........(4)}$$

where A, B, C, D are arbitrary.

For instance, suppose that the particle is projected from the origin of coordinates with the velocity (u_0, v_0) at the instant $t=0$. We have then

$$A=u_0, \quad B=v_0, \quad C=0, \quad D=0, \quad\text{..........(5)}$$

whence

$$\dot{x}=u_0, \quad \dot{y}=v_0-gt, \quad\text{..........(6)}$$

$$x=u_0t, \quad y=v_0t-\tfrac{1}{2}gt^2. \quad\text{..........(7)}$$

Eliminating t we have the equation of the path, viz.

$$y=\frac{v_0}{u_0}x-\frac{g}{2u_0^2}x^2. \quad\text{..........(8)}$$

This represents a parabola whose axis is vertical.

If for a moment we take the origin at the vertex of the path, and the axis of x horizontal, we have $v_0=0$, and therefore

$$x=u_0t, \quad y=-\tfrac{1}{2}gt^2, \quad\text{..........(9)}$$

whence

$$y=-\frac{g}{2u_0^2}x^2. \quad\text{..........(10)}$$

The latus-rectum is therefore $2u_0^2/g$, where u_0 is the (constant) horizontal velocity. Also if q be the velocity at time t,

$$q^2 = \dot{x}^2 + \dot{y}^2 = u_0^2 + g^2t^2 = u_0^2 - 2gy. \quad \text{..........(11)}$$

If we put

$$y' = \frac{u_0^2}{2g} - y, \quad \text{......................(12)}$$

so that y' denotes depth below the directrix, we have

$$q^2 = 2gy'. \quad \text{.........................(13)}$$

Hence the velocity at any point is that which would be acquired by a particle falling vertically from rest at the level of the directrix.

We return to the formulæ (7), where the origin is any point on the path, and the axis of x may have any direction, whilst that

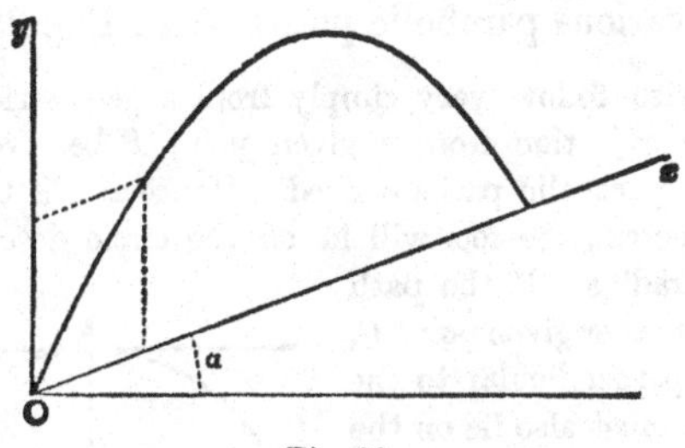

Fig. 23.

of y is vertical, as in Fig. 23. To find where the path meets the axis of x again we put $y = 0$, and obtain

$$t = \frac{2v_0}{g}, \quad x = \frac{2u_0v_0}{g}. \quad \text{..................(14)}$$

This gives the range on an inclined plane, and the corresponding time of flight, if the axis of x be taken along the plane. If α be the inclination of the plane to the horizontal the initial velocity (q_0, say) is given by

$$q_0^2 = u_0^2 + 2u_0v_0 \sin \alpha + v_0^2. \quad \text{..............(15)}$$

Writing this in the form

$$q_0^2 = (u_0 - v_0)^2 + 2u_0v_0 (1 + \sin \alpha) \quad \text{...........(16)}$$

we see that if q_0 be fixed, the product u_0v_0 is greatest when $u_0 = v_0$, i.e. when the direction of projection bisects the angle between the line of slope of the plane and the vertical. The line of slope then contains the focus of the parabola. The maximum range is, moreover, from (14) and (16),

$$\frac{q_0^2}{g(1+\sin\alpha)}. \quad \text{.......................(17)}$$

If we denote this by r, and write

$$\theta = \tfrac{1}{2}\pi - \alpha, \quad l = q_0^2/g, \quad \text{................(18)}$$

we have

$$\frac{l}{r} = 1 + \cos\theta. \quad \text{.......................(19)}$$

This is the polar equation of a parabola, whose focus is at the point of projection, referred to the vertical as initial line. It marks out the limits which can be reached in different directions from the origin, with the given velocity of projection. Since the semi-latus-rectum is q_0^2/g it touches at the vertex the common directrix of the various parabolic paths. See Fig. 25.

These results also follow very simply from a geometrical construction. If the velocity of projection from a given point P be given, the common directrix of all the parabolic paths is fixed. Hence if PA be drawn perpendicular to this directrix, the foci will lie on the circle described with P as centre and PA as radius. If the path is to pass through another given point Q, and QB be drawn perpendicular to the directrix, the focus must also lie on the circle having Q as centre and radius QB.

If the two circles intersect, the intersections S, S' are the possible positions of the focus. There are in this case two possible paths from P to Q, and the corresponding directions of projection from P bisect the angles APS, APS', respectively.

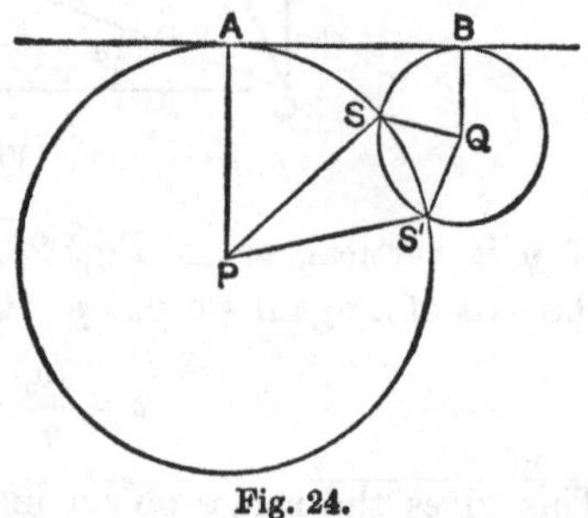

Fig. 24.

If the circles do not intersect, the point Q is out of range from P.

If the circles touch, as in Fig. 25, the two paths coincide, and the focus is at the point of contact S. The point Q is then just within range from P; it is in fact a point on the envelope of the various parabolas through P having the given directrix. If we produce PA to X, making $AX = PA$, and also produce QB to meet the horizontal line through X in M, we have

$$PQ = PS + SQ = PA + QB = QM, \quad \text{.....................(20)}$$

and the envelope is therefore a parabola with P as focus and A as vertex, as already found analytically.

Since the change of velocity per unit time is in the direction of the downward vertical, and of constant amount, the hodograph

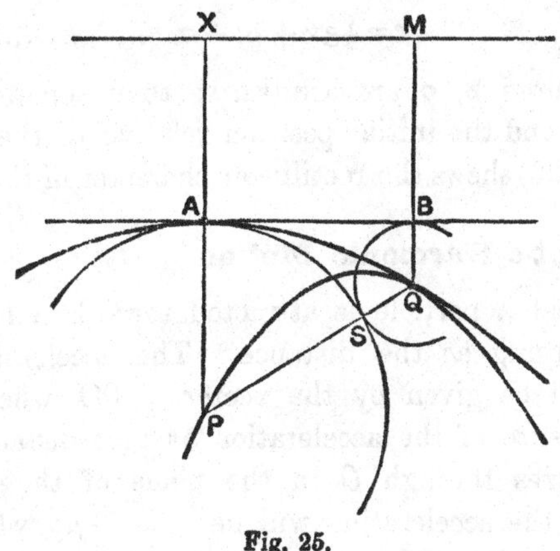

Fig. 25.

(Art. 21) is a vertical straight line described with constant velocity (g). This is illustrated by Fig. 26, where the velocities are shewn (on the right) for a series of equidistant instants.

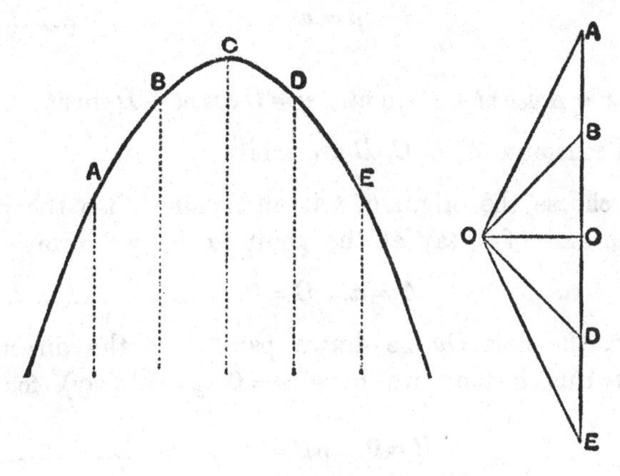

Fig. 26.

In the vector treatment of the question we have

$$\ddot{\mathbf{r}} = \mathbf{g}, \qquad \text{.........................(21)}$$

where $\mathbf{g}$ is the vector representing the change of velocity in unit time. Hence

$$\mathbf{v} = \dot{\mathbf{r}} = \mathbf{g}t + \mathbf{b}, \quad \text{.......................(22)}$$

$$\mathbf{r} = \tfrac{1}{2}\mathbf{g}t^2 + \mathbf{b}t + \mathbf{c}, \quad \text{.....................(23)}$$

where the vectors $\mathbf{b}$, $\mathbf{c}$ are arbitrary; they denote in fact the initial velocity and the initial position relative to the origin of $\mathbf{r}$. The equation (22) shews the rectilinear character of the hodograph.

28. Elliptic-Harmonic Motion.

Suppose that a particle is attracted towards a fixed point O by a force varying as the distance. The acceleration in any position P will be given by the vector μ . PO, where μ is the numerical measure of the acceleration at unit distance. Hence, relatively to axes through O in the plane of the motion, the components of the acceleration will be $-\mu x$, $-\mu y$, where x, y are the coordinates of P. Hence

$$\frac{d^2x}{dt^2} = -\mu x, \qquad \frac{d^2y}{dt^2} = -\mu y. \quad \text{..............(1)}$$

These equations can be solved independently. Putting

$$\mu = n^2, \quad \text{..............................(2)}$$

we have

$$x = A\cos nt + B\sin nt, \quad y = C\cos nt + D\sin nt, \quad \text{...(3)}$$

where the constants A, B, C, D are arbitrary.

If we choose the origin of t at an instant when the particle crosses the axis of x, say at the point $(a, 0)$, we have

$$A = a, \quad C = 0. \quad \text{.......................(4)}$$

If, further, the axis Oy be drawn parallel to the direction of motion at this instant, we have $\dot{x} = 0$, $\dot{y} = v_0$ (say), for $t = 0$ whence

$$B = 0, \quad nD = v_0. \quad \text{......................(5)}$$

Hence, relatively to these special axes we have

$$x = a\cos nt, \quad y = b\sin nt, \quad \text{.................(6)}$$

where

$$b = v_0/n. \quad \text{.........................(7)}$$

Eliminating t we obtain

$$\frac{x^2}{a^2}+\frac{y^2}{b^2}=1, \qquad\qquad (8)$$

which is the equation of an ellipse referred to a pair of conjugate diameters. Moreover, since any point on the path may be regarded as the starting point, the formula (7) shews that the velocity at any point P varies as the length of the semidiameter (OD, say) conjugate to OP (cf. Art. 21, Ex. 2). In other words, the hodograph is similar to the locus of D, i.e. to the elliptic orbit itself.

If we refer the orbit to its principal axes, so that the coordinates x, y are rectangular, the angle nt in (6) becomes identical with the

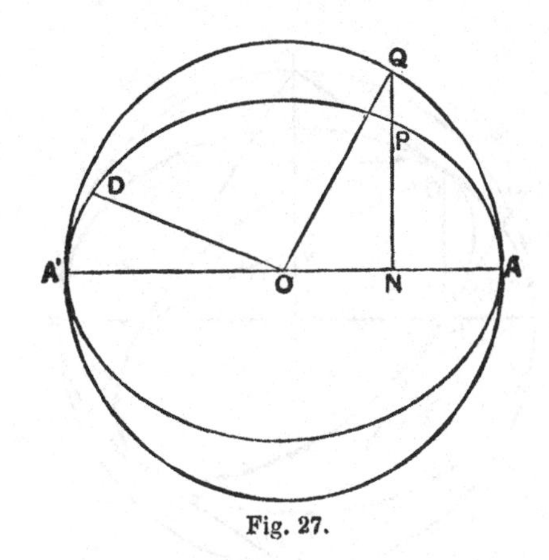

Fig. 27.

'eccentric angle' of P; and the law of description is that this angle increases at a constant rate. Moreover, since the areas swept over by corresponding radii of an ellipse and its auxiliary circle are in a constant ratio, the above statement is equivalent to this, that the radius vector OP sweeps over equal areas in equal times.

This type of motion is called 'elliptic harmonic.' The period of a complete revolution of P is

$$T=\frac{2\pi}{n}=\frac{2\pi}{\sqrt{\mu}}, \qquad\qquad (9)$$

and is therefore independent of the initial circumstances.

The solution in vectors is very compact. We have

$$\ddot{\mathbf{r}} = -n^2\mathbf{r}, \quad \text{..........................(10)}$$

and therefore

$$\mathbf{r} = \mathbf{a} \cos nt + \mathbf{b} \sin nt, \quad \text{..................(11)}$$

where the vectors **a**, **b** are arbitrary. This is the equation of an ellipse; moreover, at the instant $t = 0$, we have

$$\mathbf{r} = \mathbf{a}, \quad \mathbf{v} = \dot{\mathbf{r}} = n\mathbf{b}. \quad \text{..................(12)}$$

These results are equivalent to (3) and (7).

Ex. To find the envelope of the paths described by different particles projected from a given point P in different directions with the same velocity.

Since the velocity of projection is given, the semidiameter (OD) conjugate to OP is determinate in length; and the sum of the squares of the principal

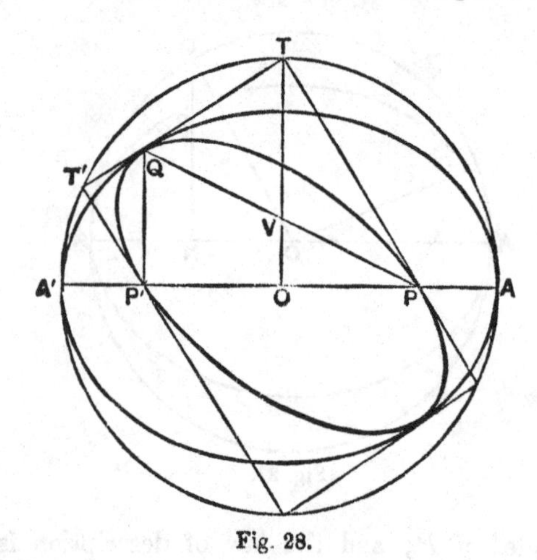

Fig. 28.

semiaxes, being equal to $OP^2 + OD^2$, is therefore the same for all the orbits. Hence the various orbits have the same director-circle (locus of intersection of perpendicular tangents). If the tangent at P to any one of the orbits meets this circle at T, the perpendicular TQT' to PT at T will also touch this orbit. If Q be the point of contact, OT will bisect PQ (in V, say), and will therefore be parallel to $P'Q$, where P' is the point on the orbit opposite to P. Hence

$$PQ + P'Q = 2TV + 2OV = 2OT = AA'.$$

Moreover, since PQ and $P'Q$ are parallel to OT' and OT, respectively, they are equally inclined to TT'. The orbit therefore touches at Q the ellipse described on AA' as major axis, with P, P' as foci. This ellipse is therefore the required envelope.

29. Spherical Pendulum. Blackburn's Pendulum.

The motion of the bob of a 'spherical pendulum,' i.e. of a simple pendulum whose oscillations are not confined to a vertical plane, comes under the preceding investigation, provided the extreme inclination of the string to the vertical be small. As in the case of Art. 11, the vertical motion may be ignored, and the tension of the string equated to mg, where m is the mass of the suspended particle. The acceleration of the bob is therefore directed towards the vertical through the point of suspension, and is equal to gr/l, where r is the distance from this vertical, and l is the length of the string. The preceding investigation therefore applies, with $n^2 = g/l$. The path is approximately a horizontal ellipse, described in the period

$$T = 2\pi \sqrt{\frac{l}{g}}. \qquad \text{(1)}$$

The above problem is obviously identical with that of the oscillations of a particle in a smooth spherical bowl, in the neighbourhood of the lowest point, the normal reaction of the bowl playing the same part as the tension of the string.

In Blackburn's* pendulum a weight hangs by a string CP from a point C of another string ACB whose ends A, B are fixed. If we neglect the inertia of the strings the point P will always be in the same plane with A, B, C.

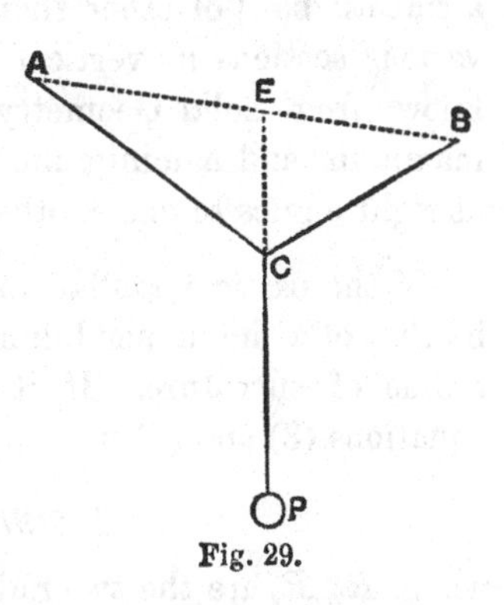

Fig. 29.

It is evident that if the particle make small oscillations in the vertical plane through AB, the motion will be that of a simple pendulum of length CP, whilst if it oscillates at right angles to that plane the motion will be that of a pendulum of length EP, where E is the point of AB which is vertically above the equilibrium position of P. Hence if x, y denote small displacements in the aforesaid planes, and if we write

$$p^2 = g/CP, \quad q^2 = g/EP, \qquad \text{(2)}$$

* H. Blackburn, Professor of Mathematics at Glasgow 1849–79.

the equations appropriate to the two types of oscillation will be

$$\frac{d^2x}{dt^2} = -p^2x, \quad \frac{d^2y}{dt^2} = -q^2y. \quad \text{..................(3)}$$

Again, it is clear that the restoring force on the bob parallel to x will not be altered, to the first order of small quantities, by a small displacement parallel to y, so that the equations (3) may be taken to hold when displacements of both types are superposed. Hence we have, for the most general small motion of the bob,

$$x = A_1 \cos pt + B_1 \sin pt, \quad y = A_2 \cos qt + B_2 \sin qt. \quad \text{...(4)}$$

The curves obtained by compounding two simple-harmonic vibrations of different periods in perpendicular directions are of importance in experimental Acoustics, and are usually associated with the name of Lissajous*, who studied them in great detail from this point of view. If the two periods $2\pi/p$, $2\pi/q$ are commensurable, the values of x and y in (4) will both recur whenever t increases by the least common multiple of these periods, and the curves are accordingly re-entrant. Many mechanical and optical appliances have been devised for producing the curves.

The equations (3) also apply to the oscillations of a particle in a smooth bowl of other than spherical shape. If we consider the various sections by vertical planes through the lowest point, it is known from Solid Geometry that the curvature at this point is a maximum and a minimum, respectively, for two definite sections at right angles to one another.

If the particle oscillates in one of these planes, the period will be that of a simple pendulum of length equal to the corresponding radius of curvature. If it oscillate in any other manner, the equations (3) and (4) will apply, provided we put

$$p^2 = g/R_1, \quad q^2 = g/R_2, \quad \text{..................(5)}$$

where R_1, R_2 are the two radii of curvature in question†.

* J. A. Lissajous, *Étude optique des mouvements vibratoires*, 1873.

† The locus of the point P in Blackburn's pendulum is an 'anchor ring' generated by the revolution of a circle with centre C and radius CP about the line AB.

30. Equation of Energy.

The general equations of Art. 26, viz.

$$m\frac{du}{dt}=X,\quad m\frac{dv}{dt}=Y,\qquad\text{(1)}$$

hold, as has been stated, whether the coordinate axes be rectangular or oblique, but for the present purpose we assume the axes to be rectangular.

If we multiply the above equations by u, v, respectively, and add, we obtain

$$m\left(u\frac{du}{dt}+v\frac{dv}{dt}\right)=Xu+Yv,\qquad\text{(2)}$$

or

$$\frac{d}{dt}\cdot\tfrac{1}{2}m(u^2+v^2)=X\frac{dx}{dt}+Y\frac{dy}{dt}.\qquad\text{(3)}$$

If q denote the resultant velocity we have

$$q^2=u^2+v^2,\qquad\text{(4)}$$

and the product $\frac{1}{2}m(u^2+v^2)$ is accordingly the kinetic energy. Again, the work done by the force (X, Y) during a small displacement $(\delta x, \delta y)$ is $X\delta x+Y\delta y$, and the right-hand member of (3) therefore represents the *rate* at which work is being done on the particle at the instant t. The equation therefore asserts that the kinetic energy is increasing at a rate equal to that at which work is being done. Integrating, we infer as in Art. 18 that the increment of the kinetic energy in any interval of time is equal to the total work done on the particle. In symbols,

$$\tfrac{1}{2}mq_2^2-\tfrac{1}{2}mq_1^2=\int_{t_1}^{t_2}\left(X\frac{dx}{dt}+Y\frac{dy}{dt}\right)dt,\qquad\text{(5)}$$

where, of course, X, Y are supposed expressed as functions of t.

If, however, we have a constant field of force [*S.* 49], i.e. the force acting on the particle is always the same in the same position, so that X, Y are given as functions of x, y, independent of t, the definite integral may be replaced by

$$\int(Xdx+Ydy),\qquad\text{(6)}$$

it being understood that the expression $X\delta x+Y\delta y$ is to be calculated for all the infinitesimal elements of the path, and the results added. An equivalent form is

$$\int_{s_1}^{s_2}\left(X\frac{dx}{ds}+Y\frac{dy}{ds}\right)ds, \qquad (7)$$

where s denotes the arc of the curve, measured from some fixed point on it.

Since $(-X, -Y)$ is the force which would balance the force of the field, the expression

$$-\int(Xdx+Ydy), \qquad (8)$$

taken between the proper limits, gives the amount of work which would have to be performed by an external agency in order to bring the particle with infinite slowness from the first position to the second. If the constitution of the field be quite arbitrary, this amount will in general depend on the nature of the path [S. 49], and not merely on the initial and final positions. If, however, it depends on the terminal points alone, the field is said to be 'conservative.' It is with conservative fields that we are chiefly concerned in the case of natural phenomena.

In a conservative field, the work required to be performed by extraneous forces in order to bring the particle (with infinite slowness) from some standard position A to any other position P is a definite function of the coordinates of P. It is called the 'potential energy,' and is denoted usually by the letter V. The work done by the forces of the field alone in the passage from A to P is accordingly $-V$. The work which these same forces do in the passage from any position P_1 to any other position P_2 is therefore $V_1 - V_2$, since the passage may be supposed made first from P_1 to A and then from A to P_2.

Hence, in the case of a particle moving in a conservative field, with no extraneous forces, we have

$$\tfrac{1}{2}mq_2^2 - \tfrac{1}{2}mq_1^2 = V_1 - V_2, \qquad (9)$$

or

$$\tfrac{1}{2}mq_2^2 + V_2 = \tfrac{1}{2}mq_1^2 + V_1, \qquad (10)$$

i.e., in words, the sum of the kinetic and potential energies is constant.

If there are extraneous forces, in addition to those of the field, the work which they do in the passage from P_1 to P_2 must be added to the right-hand member of (9), and we learn that the sum of the kinetic and potential energies is increased by an amount equal to the work of the extraneous forces. Cf. Art. 18.

It need hardly be said that these conclusions are not restricted to the case of motion in two dimensions. The vector equation of motion, which is quite independent of such limitations, is, as in Art. 25 (4),

$$m\ddot{\mathbf{r}} = \mathbf{P}, \qquad (11)$$

whence

$$m\dot{\mathbf{r}}\ddot{\mathbf{r}} = \mathbf{P}\dot{\mathbf{r}}, \qquad (12)$$

the products of vectors being of the type called 'scalar' [*S.* 63]. Hence

$$\tfrac{1}{2}m\dot{\mathbf{r}}^2 = \int \mathbf{P}\dot{\mathbf{r}}\,dt, \qquad (13)$$

or, if $\mathbf{P}$ is a function of $\mathbf{r}$ only,

$$\tfrac{1}{2}m\dot{\mathbf{r}}^2 = \int \mathbf{P}d\mathbf{r}. \qquad (14)$$

Since the scalar square of a vector is the square of the absolute value of the vector, the left-hand member is the kinetic energy. Also the scalar product $\mathbf{P}\delta\mathbf{r}$ is [*S.* 63] the work done by the force $\mathbf{P}$ in a small displacement $\delta\mathbf{r}$.

Ex. 1. In the case of ordinary gravity, if y denote altitude above some fixed level we may write

$$V = mgy. \qquad (15)$$

Hence in the free motion of a projectile we have

$$\tfrac{1}{2}mq^2 + mgy = \text{const.}, \qquad (16)$$

in agreement with Art. 27. The same formula applies to motion on a smooth curve under gravity, since the normal reaction of the curve does no work.

Ex. 2. If a particle is attracted towards a fixed point O by a force $\phi(r)$ which is a function of the distance r only, we have

$$V = \int_a^r \phi(r)\,dr, \qquad (17)$$

where the lower limit a refers to the standard position.

Thus if

$$\phi(r) = Kr, \qquad (18)$$

we may put

$$V = \tfrac{1}{2}Kr^2, \qquad (19)$$

omitting an arbitrary constant. Hence in a free orbit under this force we have

$$\tfrac{1}{2}mq^2 + \tfrac{1}{2}Kr^2 = \text{const.} \quad \text{.........................(20)}$$

We have seen in Art. 28 that the orbit is an ellipse, and that the velocity is

$$q = nr', \quad \text{.................................(21)}$$

where r' is the semidiameter conjugate to the radius vector r, and $n = \sqrt{(K/m)}$. The formula (10) is thus verified, since the sum $r^2 + r'^2$ is constant in the ellipse.

Ex. 3. In the case of the simple pendulum (Art. 11) the potential energy may be calculated from the work done by the horizontal force mgx/l necessary to produce the deflection. This makes

$$V = \tfrac{1}{2}mgx^2/l. \quad \text{.................................(22)}$$

On the other hand, considering the work done against gravity we have

$$V = mgy, \quad \text{.....................................(23)}$$

where y denotes altitude above the level of the equilibrium position. Since

$$x^2 = y(2l - y) \quad \text{.................................(24)}$$

these formulæ agree, to the order of approximation required.

This verification applies of course also to the spherical pendulum.

31. Properties of a Conservative Field of Force.

The force on a particle in a conservative field can be naturally expressed in terms of the potential energy V. If we displace the particle through a small space PP' ($= \delta s$) in any given direction, the work done by the extraneous force required to balance the force of the field, in the imagined process of the preceding Art., is $-F\delta s$, where F denotes the component force of the field in the direction PP'. Hence

$$\delta V = -F\delta s, \quad \text{..........................(1)}$$

or

$$F = -\frac{\partial V}{\partial s}. \quad \text{..........................(2)}$$

The symbol of *partial* differentiation is employed, because the space-gradient of V in one out of an infinite number of possible directions is taken.

In particular, if we take PP' parallel to the two (rectangular) coordinate axes in succession, we have

$$X = -\frac{\partial V}{\partial x}, \quad Y = -\frac{\partial V}{\partial y}, \quad \text{..................(3)}$$

for the components of force at the point (x, y).

A line along which V is constant is called an 'equipotential' line*. If in (1) we take the element δs along such a line, we have $\partial V/\partial s = 0$; the resultant force at any point is therefore normal to the equipotential line through that point. Also if we draw the equipotential lines

$$V = C \quad \text{.............................(4)}$$

for a series of equal infinitesimal increments δC of the constant C, and if δn denote the perpendicular distance between two consecutive curves, the resultant force (R) is given by

$$R \,.\, \delta n = -\,\delta V = -\,\delta C. \quad \text{.......................(5)}$$

The intensity of the force therefore varies inversely as δn. The system of equipotential lines, drawn as above, therefore indicate by their degree of closeness the greater or lesser intensity of the force.

A line drawn from point to point so that its direction is everywhere that of the resultant force is called a 'line of force.' The lines of force are orthogonal to the equipotential lines wherever the force is neither zero nor infinite. In the case of ordinary gravity the equipotential lines and lines of force are horizontal and vertical respectively. In the case of a central attraction they are concentric circles and radial straight lines.

32. Oscillations about Equilibrium. Stability.

The coordinates x, y of the possible positions of equilibrium in a conservative field (which is here for simplicity taken to be two-dimensional) are determined by the conditions $X = 0$, $Y = 0$, or

$$\frac{\partial V}{\partial x} = 0, \quad \frac{\partial V}{\partial y} = 0. \quad \text{.......................(1)}$$

The equilibrium positions are therefore characterised by the property that the potential energy is stationary for all infinitesimal displacements. This follows also immediately from Art. 31 (1).

To investigate the nature of the equilibrium in any case, let us suppose the origin to be transferred to the position in question.

* In three dimensions we have of course equipotential *surfaces*. When the field is due to a distribution of gravitating matter they are also called 'level surfaces.'

The value of V at points in the immediate neighbourhood may be supposed expanded in powers of x, y, thus

$$V = V_0 + \alpha x + \beta y + \tfrac{1}{2}(ax^2 + 2hxy + by^2) + \ldots \quad \ldots\ldots(2)$$

Since the equations (1) must be satisfied for $x = 0$, $y = 0$, the coefficients α, β must vanish, so that

$$V - V_0 = \tfrac{1}{2}(ax^2 + 2hxy + by^2) + \ldots \quad \ldots\ldots\ldots\ldots(3)$$

Hence, neglecting the terms of the third and higher degrees, we may say that the equipotential lines in the immediate neighbourhood of the origin are the system of concentric and similar conics

$$ax^2 + 2hxy + by^2 = \text{const.} \quad \ldots\ldots\ldots\ldots\ldots\ldots(4).$$

If the coordinate axes be chosen to coincide with the principal axes of these conics the formula (3) takes the simpler shape

$$V - V_0 = \tfrac{1}{2}(ax^2 + by^2) + \ldots, \quad \ldots\ldots\ldots\ldots\ldots\ldots(5)$$

where the values of a, b are of course altered. The equations of motion of a particle m therefore reduce, for small values of x, y, to the forms

$$\left.\begin{aligned} m\frac{d^2x}{dt^2} &= -\frac{\partial V}{\partial x} = -ax, \\ m\frac{d^2y}{dt^2} &= -\frac{\partial V}{\partial y} = -by. \end{aligned}\right\} \quad \ldots\ldots\ldots\ldots\ldots\ldots\ldots(6)$$

Hence if the coefficients a, b in (5) are both positive, the motion will consist of two superposed simple-harmonic vibrations in perpendicular directions, of periods

$$2\pi\sqrt{(m/a)}, \quad 2\pi\sqrt{(m/b)}, \quad \ldots\ldots\ldots\ldots\ldots\ldots(7)$$

respectively, as in the theory of Blackburn's pendulum, which is indeed merely a particular case of the present investigation. It follows that if the initial displacements and velocities be sufficiently small, the particle will oscillate about the equilibrium position, which is therefore reckoned as 'stable.'

If on the other hand either a or b is negative, the solution of the corresponding differential equation will involve real exponentials, as in Art. 15, and a disturbance, however small, will in general tend to increase until the approximation is no longer valid. The equilibrium position is then reckoned as 'unstable.'

It appears from (5) that a and b will both be positive if, and only if, the value of V in any position sufficiently near to the equilibrium position is greater than at this position itself. In other words, the potential energy is an absolute *minimum* at a position of stable equilibrium. This is the necessary and sufficient condition for stability from the present point of view*.

It is otherwise obvious without analysis that if V increases in all directions from the origin O we can describe a closed contour about O such that at every point on it $V - V_0$ will have a certain positive value, E. If the particle be started anywhere within the region thus bounded, with a *total* energy less than $V_0 + E$, its subsequent path will be confined within this region. For if it were to reach the boundary its potential energy would be $V_0 + E$, and its total energy would therefore exceed $V_0 + E$, contrary to the hypothesis†.

No such simple reasoning is available to prove that the minimum condition is a *necessary* one in order that the particle may remain in the neighbourhood of the origin. But if there are extraneous forces of resistance, however slight, which are called into play by any *motion* of the particle‡, the total energy will continually decrease so long as the motion continues. Hence if the particle start from rest in any position where the potential energy is less than V_0, the total energy, and therefore *à fortiori* the potential energy, will continually decrease. This means that the particle, unless it come to rest in some new equilibrium position, must deviate more and more from the position O.

Ex. 1. A particle is attracted to several centres of force $O_1, O_2, \ldots$ by forces $K_1 r_1, K_2 r_2, \ldots$ proportional to its distances $r_1, r_2, \ldots$ from these points, respectively.

We may put $$V = \tfrac{1}{2}(K_1 r_1^2 + K_2 r_2^2 + \ldots). \qquad (8)$$

It is known [*S.* 74] that there is only one position of the particle for which this expression is stationary in value, viz. the mass-centre (G) of a system of

* Cases where the coefficients a, h, b in the development (3) all vanish are not here considered.

† This argument, which is seen to apply to any conservative mechanical system, is due to P. Lejeune Dirichlet (1846).

‡ Statical friction is left out of account. Its effect is to render positions of equilibrium more or less indeterminate.

particles of masses proportional to $K_1, K_2, \ldots$, and situate at $O_1, O_2, \ldots$, respectively, and that V is then a minimum.

It is otherwise evident that the resultant of all the given forces is a force $\Sigma(K) . \bar{r}$ towards G, where $\bar{r}$ denotes the distance of the particle from G. Hence the particle, however started, will describe an ellipse about G, in the period

$$2\pi \sqrt{\left(\frac{m}{\Sigma(K)}\right)}.$$

Ex. 2. A particle subject to gravity, and constrained to lie on a smooth surface of any form, will be in equilibrium at a point where the tangent plane is horizontal. If the surface, in the immediate neighbourhood, lies altogether above this plane, the equilibrium is stable; if altogether below, the equilibrium is unstable. If the surface crosses the tangent plane, as in the case of a saddle-shaped surface, the equilibrium is stable for some displacements and unstable for others, and therefore on the whole unstable.

If z denote altitude above the tangent plane we may put

$$V = mgz. \qquad \text{.................................(9)}$$

If the origin be at the point in question, and x, y be horizontal coordinates in the two principal planes of curvature, we have

$$2z = \frac{x^2}{\rho_1} + \frac{y^2}{\rho_2} + \ldots, \qquad \text{..........................(10)}$$

where ρ_1, ρ_2 are positive or negative according as the principal sections to which they relate are concave or convex upwards. Cf. Art. 29 (5).

33. Rotating Axes.

It has been remarked (Art. 22) that the equations of motion obtained on the supposition that the axes of reference are fixed retain the same form if the axes are supposed to have any constant velocity of translation. The case is altered if the axes have a motion of rotation.

To illustrate this we may form the equations relative to (rectangular) axes which are rotating about the origin with angular velocity ω. Let Ox, Oy be the positions of the axes at the instant t, and Ox', Oy' their positions at the instant $t + \delta t$, the angle xOx' being therefore equal to $\omega\,\delta t$. The position P of a moving point at time t is specified by its coordinates (x, y) relative to Ox, Oy, and the position P' of the same point at time $t+\delta t$ by its coordinates $(x+\delta x, y+\delta y)$ relative to Ox', Oy'. Hence, relative to Ox, Oy the coordinates of P' will be

$$(x+\delta x)\cos\omega\,\delta t-(y+\delta y)\sin\omega\,\delta t,$$
$$(x+\delta x)\sin\omega\,\delta t+(y+\delta y)\cos\omega\,\delta t,$$

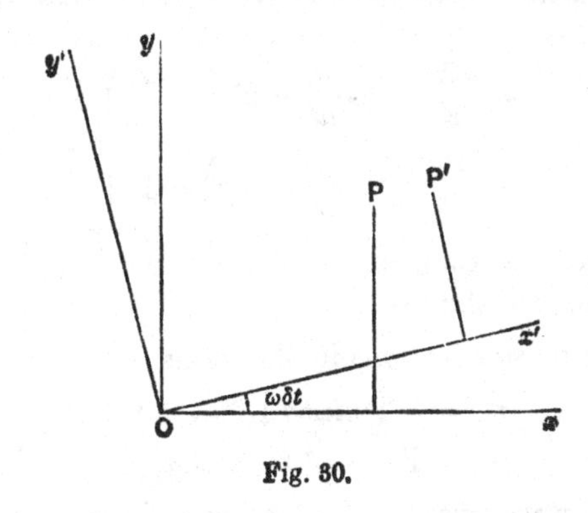

Fig. 30.

by ordinary formulæ for transformation of coordinates. Neglecting small quantities of the second order, these may be written

$$x+\delta x-\omega y\,\delta t,\quad y+\delta y+\omega x\,\delta t,$$

so that the projections of PP′ on Ox, Oy are

$$\delta x-\omega y\,\delta t,\quad \delta y+\omega x\,\delta t,\quad \text{..................(1)}$$

ultimately. The component velocities of P parallel to Ox, Oy are therefore

$$u=\frac{dx}{dt}-\omega y,\quad v=\frac{dy}{dt}+\omega x,\quad \text{...............(2)}$$

respectively.

If OV, OV′ be vectors representing the velocity at the instants t and $t+\delta t$, the same method can be applied to find the projections of VV′ on Ox, Oy. Thus, denoting by $(u+\delta u, v+\delta v)$ the projections of OV′ on Ox', Oy', the projections of VV′ on Ox, Oy will be

$$\delta u-\omega v\,\delta t,\quad \delta v+\omega u\,\delta t,\quad \text{..................(3)}$$

in analogy with (1). The component accelerations are therefore

$$\alpha=\frac{du}{dt}-\omega v,\quad \beta=\frac{dv}{dt}+\omega u.\quad \text{...............(4)}$$

If the angular velocity ω be constant, we have on substitution of the values of u, v from (2)

$$\alpha = \frac{d^2x}{dt^2} - 2\omega\frac{dy}{dt} - \omega^2 x, \quad \beta = \frac{d^2y}{dt^2} + 2\omega\frac{dx}{dt} - \omega^2 y. \quad \text{...(5)}$$

The dynamical equations relative to the rotating axes are therefore

$$\left.\begin{aligned} m\left(\frac{d^2x}{dt} - 2\omega\frac{dy}{dt} - \omega^2 x\right) &= X, \\ m\left(\frac{d^2y}{dt^2} + 2\omega\frac{dx}{dt} - \omega^2 y\right) &= Y, \end{aligned}\right\} \quad \text{..............(6)}$$

where X, Y are the components of force parallel to the instantaneous directions of the axes.

If we write these equations in the forms

$$\left.\begin{aligned} m\ddot{x} &= X + m\omega^2 x + 2m\omega\dot{y}, \\ m\ddot{y} &= Y + m\omega^2 y - 2m\omega\dot{x}, \end{aligned}\right\} \quad \text{..................(7)}$$

an interpretation presents itself. The particle is apparently acted on by certain forces in addition to the true force (X, Y). We have in the first place the components $m\omega^2 x$, $m\omega^2 y$, which are those of an apparent 'centrifugal force' $m\omega^2 r$, where r denotes distance from the origin. In addition we have an apparent force whose components are

$$2m\omega\dot{y} = 2m\omega\dot{s}\sin\psi, \quad -2m\omega\dot{x} = -2m\omega\dot{s}\cos\psi,$$

where ψ is the inclination of the apparent path (i.e. the path relative to the rotating axes) to the axis of x. The magnitude of this force is $2m\omega\dot{s}$, and its direction is obtained from that of the apparent velocity $\dot{s}$ by a rotation through a right angle in the sense opposite to that of the angular velocity ω*.

If we multiply the equations (7) by $\dot{x}$, $\dot{y}$, respectively and add, we have

$$m(\dot{x}\ddot{x} + \dot{y}\ddot{y}) = X\dot{x} + Y\dot{y} + m\omega^2(x\dot{x} + y\dot{y}), \quad \text{........(8)}$$

whence, integrating with respect to t,

$$\tfrac{1}{2}m(\dot{x}^2 + \dot{y}^2) = \int (X\dot{x} + Y\dot{y})\,dt + \tfrac{1}{2}m\omega^2(x^2 + y^2) + C. \quad \text{...(9)}$$

* This was called by G. Coriolis (1831) the 'force centrifuge composée,' to distinguish it from the 'force centrifuge ordinaire' $m\omega^2 r$.

If the force (X, Y) be due to a field which rotates unchanged with the coordinate axes, we may replace the definite integral by

$$\int (X\,dx + Y\,dy).$$

If V denote the potential energy due to this field we have

$$\tfrac{1}{2}m(\dot{x}^2+\dot{y}^2) + V - \tfrac{1}{2}m\omega^2(x^2+y^2) = \text{const.} \quad \text{......(10)}$$

which is the form now taken by the equation of energy.

The expression

$$V - \tfrac{1}{2}m\omega^2(x^2+y^2), \quad \text{.....................(11)}$$

the latter part of which may be called the potential energy in relation to centrifugal force, accordingly plays the same part as the true potential energy in a stationary field.

Ex. 1. If a point be at rest, its motion relative to the rotating axes will be given by

$$\dot{x} = \omega y, \quad \dot{y} = -\omega x. \quad \text{.............................(12)}$$

Hence

$$\ddot{x} = \omega\dot{y} = -\omega^2 x, \quad \text{.............................(13)}$$

the solution of which is

$$x = c\cos(\omega t + \epsilon). \quad \text{.............................(14)}$$

The former of equations (12) then gives

$$y = \frac{1}{\omega}\dot{x} = -c\sin(\omega t+\epsilon). \quad \text{.........................(15)}$$

Hence

$$x^2+y^2=c^2, \quad \frac{y}{x} = -\tan(\omega t+\epsilon). \quad \text{.........................(16)}$$

The relative path is therefore a circle described with constant velocity in the sense opposite to that of ω, as is otherwise obvious.

Ex. 2. If Blackburn's pendulum be made to rotate with angular velocity ω about the vertical through E (Fig. 29, p. 79), the equations of motion, referred to horizontal axes in and perpendicular to the vertical plane through AB, will be of the forms

$$\left.\begin{aligned} \ddot{x} - 2\omega\dot{y} - \omega^2 x &= -p^2 x, \\ \ddot{y} + 2\omega\dot{x} - \omega^2 y &= -q^2 y, \end{aligned}\right\} \quad \text{.........................(17)}$$

where

$$p^2 = g/CP, \quad q^2 = g/EP. \quad \text{.........................(18)}$$

These are satisfied by

$$x = A\cos(nt+\epsilon), \quad y = B\sin(nt+\epsilon), \quad \text{..................(19)}$$

provided

$$\left.\begin{aligned} (n^2+\omega^2-p^2)\,A + 2n\omega B &= 0, \\ 2n\omega A + (n^2+\omega^2-q^2)\,B &= 0. \end{aligned}\right\} \quad \text{.........................(20)}$$

Eliminating the ratio A/B we have

$$(n^2+\omega^2-p^2)(n^2+\omega^2-q^2)-4n^2\omega^2=0, \quad\text{..................(21)}$$

or

$$n^4-(p^2+q^2+2\omega^2)n^2+(p^2-\omega^2)(q^2-\omega^2)=0, \quad\text{...........(22)}$$

which is a quadratic in n^2. The square of the difference of the roots is

$$(p^2+q^2+2\omega^2)^2-4(p^2-\omega^2)(q^2-\omega^2)=(p^2-q^2)^2+8\omega^2(p^2+q^2), \text{...(23)}$$

and since this is positive both roots are real. Again the product of the roots is positive unless ω^2 lie between p^2 and q^2. Since the sum of the roots is positive we infer that unless ω^2 lie between p^2 and q^2 both values of n^2 will be positive, and we shall have two independent solutions of the types

$$x=A_1\cos(n_1t+\epsilon_1), \quad y=B_1\sin(n_1t+\epsilon_1), \quad\text{..............(24)}$$

and

$$x=A_2\cos(n_2t+\epsilon_2), \quad y=B_2\sin(n_2t+\epsilon_2), \quad\text{..............(25)}$$

where the ratios B_1/A_1 and B_2/A_2 are determined by either of the equations (20), with the appropriate value of n^2 inserted. Since the equations are linear, these solutions may be superposed, and we thus obtain a solution which is complete, since it involves four arbitrary constants A_1, A_2, ϵ_1, ϵ_2.

In the excepted case one value of n^2 (say n_1^2) is still positive, and the solution (24) is still valid. The remaining solution will be of the form

$$x=Ce^{\lambda t}+C'e^{-\lambda t}, \quad y=De^{\lambda t}+D'e^{-\lambda t}, \quad\text{.................(26)}$$

but the working out may be left to the student.

We infer that the vertical position of the pendulum is stable unless the period $(2\pi/\omega)$ of the rotation be intermediate to one of the free periods $(2\pi/p, 2\pi/q)$ of the pendulum when there is no rotation. This conclusion is however liable to be modified by the operation of dissipative forces. (See Art. 96.)

EXAMPLES. VI.

(Projectiles.)

1. Prove by the method of 'dimensions' that the range of a projectile having a given initial elevation varies as v^2/g, where v is the velocity of projection.

2. The resistance of the air being neglected, a shot would have a maximum range of 2000 yds. What would be the range with an elevation of 30°?

Also, what would be the elevations with which an object at a horizontal distance of 1500 yds. could be hit? [1732 yds.; 24° 18′, 65° 42′.]

3. A particle is projected at an elevation θ, measured from a plane of inclination α through the point of projection. Prove that, if it strike the plane at right angles,

$$\tan\theta=\tfrac{1}{2}\cot\alpha.$$

4. Particles are projected from a point O, in a vertical plane, with velocity $\sqrt{(2gk)}$; prove that the locus of the vertices of their paths is the ellipse

$$x^2 + 4y(y-k) = 0.$$

5. If at any point P on the path of a projectile the direction of motion be slightly changed, without change of velocity, the new path will intersect the old one at the other extremity of the focal chord through P.

6. Particles are projected simultaneously from a point, in different directions, with equal velocities V; prove that after t seconds they will lie on the surface of a sphere of radius Vt, and that the centre of this sphere has a downward acceleration g.

7. If OR be the horizontal range of a projectile, and the line joining O to any position P of the particle meets the vertical through R in Q, the point Q descends with a constant velocity numerically equal to the initial vertical component of the velocity of the particle.

8. Prove that in the parabolic path of a projectile the direction of motion is at any instant changing at the rate $\frac{1}{2} u_0/y$, where u_0 is the horizontal velocity, and y denotes depth below the directrix.

9. Prove the following construction for finding the horizontal range and greatest altitude of a particle projected with given velocity v, in any direction, from a point O:

Draw OA upwards, and equal to $2v^2/g$, and describe a sphere on OA as diameter. Through O draw a chord OP in the direction of projection, and draw PN perpendicular to the horizontal plane through O. Then ON is the range, and the greatest altitude is $\frac{1}{4}PN$.

10. Adapt the above construction to find the range on an *inclined* plane through O.

11. Two particles are projected with the same velocity in different directions, but so as to have the same horizontal range. Prove that the geometric mean of their greatest altitudes is one-quarter the range, and that the arithmetic mean of the same quantities is one-quarter the maximum horizontal range corresponding to the given velocity of projection.

12. Obtain by *dynamical* reasoning the following properties of the path of a projectile:

(1) The vertical through any point P on the path bisects all chords parallel to the tangent at P;

(2) If the vertical through the intersection T of the tangents at any two points Q, Q' meets the curve in P, and the chord QQ' in V, then $QV = VQ'$, and $TP = PV$;

(3) The subnormal is constant and equal to u_0^2/g, where u_0 is the horizontal velocity;

(4) The velocity varies as the normal.

13. If AB be a focal chord of the parabolic path of a projectile, the time from A to B is equal to the time a particle would take in falling vertically from rest through a space equal to AB.

14. Prove that if TP, TQ be two tangents to the path of a projectile, the velocities at P and Q are in the ratio of TP to TQ.

15. If OA, OB be vectors representing the velocities at any two points P, Q of the path of a projectile, and C be the middle point of AB, prove that OC represents the *mean* velocity between P and Q.

Prove that if PQ be a focal chord the mean square of the kinetic energy between P and Q is one-third the sum of the kinetic energies at P and Q.

16. Two particles are projected from the same point at the same instant with equal velocities v, at elevations α, α'. Prove that the time that elapses between their transits through the point where the paths intersect is

$$\frac{2v}{g} \cdot \frac{\sin \frac{1}{2}(\alpha - \alpha')}{\cos \frac{1}{2}(\alpha + \alpha')}.$$

17. A particle is projected so as to have a range R on a horizontal plane through the point of projection, and the greatest height attained by it is h. Prove that the maximum horizontal range with the same velocity of projection is

$$2h + \tfrac{1}{8}R^2/h.$$

18. A fort is on the edge of a cliff of height h. Prove that the greatest horizontal distance at which a gun in the fort can hit a ship is $2\surd\{k(k+h)\}$, and that the greatest horizontal distance at which a gun in a ship can hit the fort is $2\surd\{k(k-h)\}$, if $\surd(2gk)$ be the muzzle-velocity of the shot in each case.

19. A particle is projected with the velocity $\surd(2gk)$ from a point at height h above a plane of inclination α. Prove that the maximum ranges up and down the plane are increased by

$$2 \sec \alpha \surd\{k(h + k \sec^2 \alpha)\} - 2k \sec^2 \alpha.$$

Find the limiting forms of the result when the ratio h/k is small, and shew how it might have been foreseen.

20. Prove that the area which is within range from a given point O on a plane of inclination α, when the velocity of projection has a given value, is bounded by an ellipse of eccentricity $\sin \alpha$ with O as focus.

21. If at any stage in the flight of a projectile the velocity is reduced by piercing a thin board, shew by a sketch how the path is altered.

Explain generally the effect of the continual resistance of the air on the shape of the path.

EXAMPLES. VII.

(Elliptic Harmonic Motion, &c.)

1. If the coordinates of a moving point are

$$x = a\cos(nt+\alpha), \quad y = b\cos(nt+\beta),$$

the equation of the path is

$$\frac{x^2}{a^2} - \frac{2xy}{ab}\cos(\alpha-\beta) + \frac{y^2}{b^2} = \sin^2(\alpha-\beta).$$

2. A particle is projected from a given point, in a given direction, and with a given velocity, under a central attractive force varying as the distance; give a geometrical construction for finding the principal axes of the orbit.

3. Prove that in elliptic harmonic motion the mean kinetic and potential energies are equal.

4. A point is describing an ellipse under an acceleration $\mu \cdot CP$ to the centre C. Prove that the rate at which the direction of motion is changing is

$$\frac{\sqrt{\mu} \cdot ab}{CD^2},$$

where a, b are the semi-axes, and CD is the semi-diameter conjugate to CP.

5. The ends of a rod which rotates with constant angular velocity move on two intersecting straight lines at right angles; prove that any other point on the rod executes an elliptic harmonic motion.

6. A lamina rotates with constant angular velocity in its own plane, and two given points on it are constrained to move on two fixed straight lines. Prove that any other point on the lamina executes an elliptic (or rectilinear) harmonic motion.

7. Prove that in elliptic harmonic motion the time-average of the kinetic energy is equal to the arithmetic mean of the greatest and least values of the kinetic energy.

8. Two points are executing elliptic harmonic motions (not necessarily in the same plane) about the same centre, with the same period. Prove that their relative motion is elliptic harmonic.

9. A particle moves under a repulsive force varying as the distance from a fixed point; prove that the orbit is one branch of a hyperbola, and that the velocity at any point varies as the conjugate semi-diameter.

10. Also prove that if the hyperbola is rectangular, the angle θ which the radius vector makes with the transverse axis is connected with the time t from the vertex by a relation of the form

$$\sin 2\theta = \tanh 2nt.$$

11. A particle is acted on by several centres of force, attractive or repulsive, each varying as the distance; find the nature of the orbit, and the period (when the orbit is closed).

In what case is the orbit a parabola?

12. If a particle describing an ellipse under an acceleration to the centre receive at any instant a blow in the direction towards or from the centre, it will proceed to describe a new ellipse of equal area with the former orbit.

13. Prove that in the spherical pendulum, if the extreme inclinations α, β of the string to the vertical be small, the total energy is

$$\tfrac{1}{2} mgl (\alpha^2 + \beta^2),$$

where l is the length of the string, and m the mass of the bob.

14. If in Blackburn's pendulum one period be double the other, prove that a possible form of the path is an arc of a parabola described backwards and forwards.

Also find the equation of that form of path which has two axes of symmetry.

$$\left[\frac{y^2}{b^2} = \frac{4x^2}{a^2}\left(1 - \frac{x^2}{a^2}\right).\right]$$

15. Apply the equations of relative motion in Art. 33 to shew that if a particle be subject to a central acceleration $\omega^2 r$, its path relative to axes rotating with angular velocity ω will be a circle described with the constant angular velocity 2ω.

EXAMPLES. VIII.

1. A particle moves in a plane under an attractive force which is always perpendicular to a fixed straight line, and varies as the distance from that line; prove that the path is a curve of sines.

2. A particle is subject to a force constant in direction; prove that if the field be conservative the force must be uniform over any plane perpendicular to this direction.

If the axis of y be parallel to the direction of the force, prove that the differential equation of the path is of the form

$$c^2 \frac{d^2 y}{dx^2} = \phi(y),$$

where $\phi(y)$ is the acceleration.

3. If $\phi(b) = 0$, prove that the rectilinear path $y = b$ is stable or unstable according as $\phi'(b)$ is negative or positive. Express this criterion in terms of energy.

4. If, in Question 2, $\phi(y) = \mu/y^3$, prove that the path is a conic.

5. Find the law of force parallel to an asymptote under which a rectangular hyperbola can be described.

6. A number n (>2) of centres of force attracting according to the law of the inverse square are arranged symmetrically round the circumference of a circle of radius a, the force at unit distance being μ for each. Prove that the potential energy of a particle in the plane of the circle at a small distance r from the centre is given by

$$V = V_0 - \frac{n\mu}{4a^3} r^2,$$

approximately.

If the particle be on the axis of the circle at a small distance z from the centre the potential energy is

$$V = V_0 + \frac{n\mu}{2a^3} z^2.$$

7. Prove that if in a constant plane field of force the components X, Y of force at any point be given as functions of the coordinates x, y, the work done on a particle which describes the contour of a rectangular element $\delta x\, \delta y$ in the positive sense is

$$\left(\frac{\partial Y}{\partial x} - \frac{\partial X}{\partial y}\right) \delta x\, \delta y.$$

8. A particle is subject to a force perpendicular to a straight line AB and varying inversely as the square of the distance from AB. Prove that if it be projected with the velocity due to a fall from rest at infinity, its path will be a cycloid.

9. A particle moves under the influence of a number of centres of attractive force varying in each case as the distance. Prove that its motion will be elliptic harmonic.

10. The axes of x, y are rotating with constant angular velocity ω, and the velocities of a particle parallel to Ox and Oy are

$$\frac{a^2 - b^2}{a^2 + b^2} \omega y \quad \text{and} \quad \frac{a^2 - b^2}{a^2 + b^2} \omega x,$$

respectively. Prove that its motion relative to the axes is elliptic harmonic, and find the period.

11. Prove that in the case of a field of force which rotates uniformly with the axes the equations (6) of Art. 33 have the integral

$$\tfrac{1}{2}(\dot{x}^2 + \dot{y}^2) - \tfrac{1}{2}\omega^2 (x^2 + y^2) + V = \text{const.},$$

where V is the potential energy (per unit mass) due to the field.

CHAPTER V

TANGENTIAL AND NORMAL ACCELERATIONS. CONSTRAINED MOTION

34. Tangential and Normal Accelerations.

It is often convenient to resolve the acceleration of a moving point in the directions of the tangent and normal, which are intrinsic to the path and do not involve any arbitrary system of coordinates. This is specially the case when the path is prescribed, as in the case of a particle constrained to move on a given curve.

Let P, P' be the positions of the point at the instants t, $t + \delta t$, respectively, and let the normals at P, P' meet in C, making an angle $\delta\psi$. Let the velocities at P and P' be denoted by v and $v + \delta v$. If s be the arc of the curve, measured as usual from some fixed point on it, we have by Art. 20 (2)

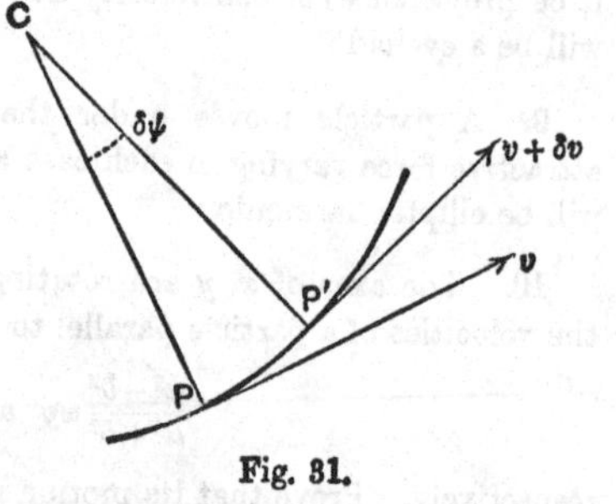

Fig. 31.

$$v = \frac{ds}{dt}. \quad \text{.........(1)}$$

In the interval δt the velocity parallel to the tangent at P changes from v to $(v + \delta v) \cos \delta\psi$, and the increment is therefore

$$(v + \delta v) \cos \delta\psi - v, \quad \text{or} \quad \delta v,$$

to the first order of small quantities, since $\cos \delta\psi$ differs from unity by a small quantity of the *second* order. The mean acceleration parallel to the tangent at P is therefore $\delta v / \delta t$, or ultimately

$$\frac{dv}{dt}. \quad \text{..................................(2)}$$

Again, the velocity parallel to the normal at P changes from 0 to $(v+\delta v)\sin\delta\psi$, or $v\delta\psi$, to the first order. The mean acceleration in the direction of the normal at P is therefore $v\delta\psi/\delta t$, or ultimately

$$v\frac{d\psi}{dt}. \qquad \text{.............................(3)}$$

Some other expressions for the components are important. Thus for the tangential acceleration we have

$$\frac{dv}{dt}=\frac{d^2s}{dt^2}, \qquad \text{.........................(4)}$$

from (1). Again

$$\frac{dv}{dt}=\frac{dv}{ds}\frac{ds}{dt}=v\frac{dv}{ds}, \qquad \text{....................(5)}$$

if v be now regarded as a function of s.

For the normal acceleration we have

$$v\frac{d\psi}{dt}=v\frac{d\psi}{ds}\frac{ds}{dt}=\frac{v^2}{\rho}, \qquad \text{... ....(6)}$$

if $\rho\ (=ds/d\psi)$ be the radius of curvature.

The above results follow also from a consideration of the hodograph. Thus in Fig. 18, p. 58, the vectors OV and OV′ represent the velocities v and $v+\delta v$, respectively, and the angle VOV' is equal to $\delta\psi$. The velocity VV′ generated in the time δt may be resolved into two components along and perpendicular to OV, i.e. along the tangent and normal to the path at P. The former component is ultimately equal to δv, and the latter to $v\delta\psi$. Dividing by δt we obtain, in the limit, the formulæ (2), (3).

This proof has the advantage that it is easily extended to the case of motion in three dimensions, where the path and the hodograph are both tortuous curves. The tangents to the path at P and P' do not in general meet, but the plane VOV' which is parallel to them has a definite limiting position, viz. it is parallel to what is called the 'osculating plane' of the path at P. The resultant acceleration is therefore in the osculating plane, and its components along the tangent and the 'principal normal,' i.e. the normal to the curve in this plane, are still given by the formulæ (2) and (3), provided $\delta\psi$ be used to denote the inclination of consecutive tangents to the path. The expression $d\psi/ds$ then coincides with what is called in Solid Geometry the 'principal curvature' at P, and denoted (usually) by $1/\rho$. The form (6) for the normal acceleration is therefore also applicable.

In vector notation, if $\mathbf{t}$, $\mathbf{n}$ denote two *unit* vectors in the directions of the tangent and normal respectively, the velocity $\mathbf{v}$ is given by

$$\mathbf{v} = v\mathbf{t}. \qquad (7)$$

We have then for the acceleration

$$\dot{\mathbf{v}} = \dot{v}\mathbf{t} + v\dot{\mathbf{t}}. \qquad (8)$$

Now the angle between $\mathbf{t}$ and $\mathbf{t} + \delta\mathbf{t}$ is $\delta\psi$, and since these vectors are of unit length $\delta\mathbf{t}$ is ultimately at right angles to both, and therefore parallel to the (principal) normal; its length is moreover $\delta\psi$. Hence $\delta\mathbf{t} = \mathbf{n}\,\delta\psi$, and

$$\dot{\mathbf{v}} = \dot{v}\mathbf{t} + v\dot{\psi}\mathbf{n}. \qquad (9)$$

The acceleration is thus expressed as the geometric sum of a tangential component $\dot{v}$ and a normal component $v\dot{\psi}$.

Ex. 1. In the case of a circular orbit of radius a we may write $s = a\psi$, whence

$$v = a\frac{d\psi}{dt} = a\omega, \qquad (10)$$

if $\omega\ (= d\psi/dt)$ be the angular velocity of the radius through the moving point. The tangential acceleration is therefore

$$\frac{dv}{dt} = a\frac{d^2\psi}{dt^2} = a\frac{d\omega}{dt}; \qquad (11)$$

and the normal acceleration is

$$v\frac{d\psi}{dt} = a\omega^2. \qquad (12)$$

Ex. 2. The formula (6) may be used conversely to find the curvature of a path when the velocity and acceleration are known.

Thus in the case of epicyclic motion (Art. 23), when the tracing point P in Fig. 20 is at its greatest distance from the centre, the inward acceleration, being made up of the acceleration of Q and the acceleration of P relative to Q, is $n^2a + n'^2a'$, by (12). The velocity is in a similar manner seen to be $na + n'a'$. The curvature is therefore

$$\frac{1}{\rho} = \frac{n^2a + n'^2a'}{(na + n'a')^2}. \qquad (13)$$

The curvature when the tracing point is nearest to the centre is found, in the case of $a' > a$, by changing the sign of a, viz. it is

$$\frac{1}{\rho} = \frac{n'^2a' - n^2a}{(n'a' - na)^2}. \qquad (14)$$

This will be negative if $n'^2a' < n^2a$.

To apply this to the orbit of the moon (P) relative to the sun (O) we may put $n = 13n'$, $a' = 400a$, roughly. It appears that ρ is positive; the orbit is in fact everywhere concave to the sun.

But if we suppose the symbols n', a' to refer to the motion of the sun (Q) relative to the earth (O), whilst n, a refer to the motion of an inferior planet (P) relative to the sun, we have

$$n'^2a' - n^2a = n^2a^3\left(\frac{1}{a'^2} - \frac{1}{a^2}\right), \qquad (15)$$

since $n^2a^3 = n'^2a'^3$, by Kepler's Third Law (Art. 80). Since $a' > a$, this expression is negative; the orbit of the planet relative to the earth has in fact loops, and resembles the first of the two types of epicyclic shewn in Fig. 21, p. 63. The case of a superior planet is included if we imagine P to refer to the planet and Q to the sun.

35. Dynamical Equations.

If the force acting on a particle m be resolved at each instant in the directions of the tangent and normal respectively, then denoting the two components by $\mathfrak{T}$ and $\mathfrak{N}$, we have

$$mv\frac{dv}{ds} = \mathfrak{T}, \quad \frac{mv^2}{\rho} = \mathfrak{N}. \qquad (1)$$

The former of these equations leads at once to the equation of energy; thus, integrating with respect to s, we have

$$\tfrac{1}{2}mv_1^2 - \tfrac{1}{2}mv_0^2 = \int_{s_0}^{s_1} \mathfrak{T}\,ds. \qquad (2)$$

The integral on the right hand denotes the total work done on the particle; for the work of the tangential component in an infinitesimal displacement is $\mathfrak{T}\,\delta s$, and that of the normal component vanishes.

In three dimensions a third equation of motion is required, expressing that the resultant force normal to the osculating plane vanishes. The equation (2) will of course still hold.

If $\mathfrak{T} = 0$, i.e. if the resultant force be always in the direction of the normal, we have $dv/dt = 0$, so that the velocity is constant.

This is the case of a particle moving on a smooth curve under no force except the reaction of the curve. The second of the equations (1) then shews that the reaction varies as the curvature $1/\rho$.

Ex. 1. To find the condition that a particle m attached to a fixed point by a string of length l may describe a horizontal circle.

If θ be the constant inclination of the string to the vertical, ω the requisite angular velocity in the circle, and T the tension, then since there is by hypothesis no vertical motion, we have

$$T\cos\theta = mg. \qquad\qquad (3)$$

Also, resolving along the radius of the circle,

$$m\omega^2 l \sin\theta = T\sin\theta, \quad \text{or} \quad m\omega^2 l = T. \qquad\qquad (4)$$

Hence

$$\omega^2 = \frac{g}{l\cos\theta}. \qquad\qquad (5)$$

The period $2\pi/\omega$ of a revolution is therefore the same as the period of small oscillation of a simple pendulum of length $l\cos\theta$. If θ be small this is equal to l, practically. The projection of the particle on a vertical plane will then move like the bob of a simple pendulum of length l.

Ex. 2. To find the deviation of the plumb-line at any place, due to the earth's rotation.

When the plumb-line is in apparent (i.e. relative) equilibrium, the resultant of the tension T of the string and of the true gravity of the suspended mass m must be a force $m\omega^2 r$ towards the earth's axis, where ω is the earth's angular velocity, and r is the radius of the diurnal circle described by m. We denote the true acceleration of gravity by g', and the apparent acceleration by g, so that $T = mg$.

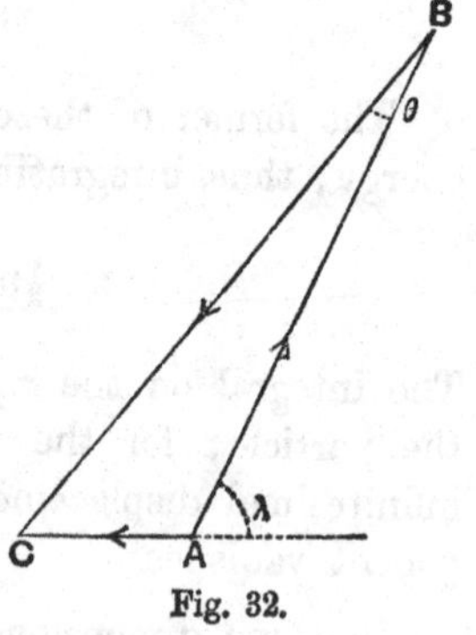

Fig. 32.

In the annexed figure, AB represents the tension mg, BC the true gravity mg', and AC the resultant $m\omega^2 r$. Hence if λ be the latitude of the place as determined astronomically, i.e. the angle which the plumb-line makes with the plane of the equator, we have

$$\frac{\sin\theta}{\sin(\lambda-\theta)} = \frac{AC}{AB} = \frac{\omega^2 r}{g}. \qquad\qquad (6)$$

The numerical data shew that the fraction on the right hand is always very small. Hence θ is always a small angle, and

$$\sin\theta = \frac{\omega^2 r \sin\lambda}{g},$$

approximately. We may also write $r=a\cos\lambda$, where a is the earth's radius, without serious error. Thus

$$\sin\theta=\frac{\omega^2 a}{g}\sin\lambda\cos\lambda. \qquad \text{.........................(7)}$$

Again, we have

$$\frac{g}{g'}=\frac{AB}{BC}=\frac{\sin(\lambda-\theta)}{\sin\lambda}=\cos\theta-\frac{\omega^2 a}{g}\cos^2\lambda,$$

or

$$g=g'\left(1-\frac{\omega^2 a}{g}\cos^2\lambda\right), \qquad \text{.........................(8)}$$

with sufficient approximation.

In terms of the centimetre and second, we have

$$\frac{2\pi}{\omega}=86164,\quad a=6\cdot38\times10^8,\quad g=981,$$

whence

$$\frac{\omega^2 a}{g}=\cdot00346=\frac{1}{289}.$$

The maximum value of θ is when $\lambda=45°$, and is about $6'$.

If the value g' of true gravity were the same over the earth's surface, the formula (8) would give the variation of apparent gravity with latitude. Actually g' is itself variable, the earth's surface, and the strata of equal density inside it, not being exactly spherical*. The attraction therefore deviates both in magnitude and direction from that of a symmetrical spherical body.

36. Motion on a Smooth Curve.

In the case of motion, under gravity, on a smooth curve in a vertical plane, the forces are

$$\mathfrak{T}=-mg\sin\psi,\quad \mathfrak{N}=-mg\cos\psi+R, \qquad \text{.........(1)}$$

where ψ denotes the inclination of the tangent, drawn in the direction of s increasing, to the horizontal, and R is the pressure exerted by the curve, reckoned positive when it acts towards the centre of curvature. The equations of motion are therefore

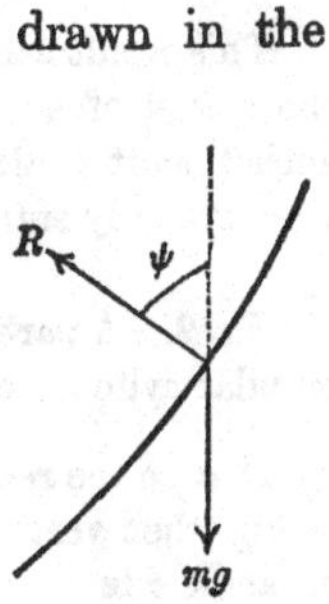

$$mv\frac{dv}{ds}=-mg\sin\psi,\quad \frac{mv^2}{\rho}=-mg\cos\psi+R. \qquad (2)$$

Fig. 33.

If we take axes of x and y which are horizontal and vertical, respectively, the positive direction of y being upwards, we have

$$\cos\psi=\frac{dx}{ds},\quad \sin\psi=\frac{dy}{ds}, \qquad \text{.................(3)}$$

* The deviation from spherical symmetry being due originally to the rotation.

and therefore
$$mv\frac{dv}{ds} = -mg\frac{dy}{ds}. \qquad \text{.........................(4)}$$

Hence
$$\tfrac{1}{2}mv^2 + mgy = \text{const.}, \qquad \text{.....................(5)}$$

as might have been written down at once from the principle of energy, since the reaction of the smooth curve does no work. This formula is usually applied in the form

$$v^2 = C - 2gy. \qquad \text{.........................(6)}$$

When the arbitrary constant has been determined, the second of the equations (2) gives the pressure R.

Ex. 1. Let the curve have the form of a parabola with its axis vertical, and concavity downwards. This could, as we know, be described freely, if the particle were properly started. Now if v' refer to the free motion we have from (2)

$$\frac{mv'^2}{\rho} = -mg\cos\psi. \qquad \text{..............................(7)}$$

Hence if v be the velocity in the constrained motion, we have

$$R = \frac{m(v^2 - v'^2)}{\rho}. \qquad \text{..............................(8)}$$

But, from (6),
$$v'^2 = C' - 2gy, \qquad \text{..............................(9)}$$

and therefore
$$R = \frac{m(C - C')}{\rho}; \qquad \text{..............................(10)}$$

i.e. the normal pressure varies as the curvature.

This result can easily be extended to the case of a particle moving, under the action of any given forces whatever, on a smooth curve whose shape is such that it could be described freely under the same forces, if the particle were properly started.

Ex. 2. A particle hangs tangentially from the circumference of a horizontal circular cylinder by a string wrapped round it.

If a be the radius of the cylinder, l the length of the free portion of the string when vertical, the potential energy when the string is deflected through an angle θ is

$$V = mg\{a\sin\theta - (l + a\theta)\cos\theta\} + mgl, \qquad \text{.................(11)}$$

if the zero value of V be supposed to correspond to $\theta = 0$. The equation of energy is therefore

$$\tfrac{1}{2}mv^2 + mg\{a\sin\theta - (l + a\theta)\cos\theta\} + mgl = \tfrac{1}{2}mv_0^2, \qquad \text{.........(12)}$$

if v_0 denote the velocity of the particle when passing through its lowest position. For small values of θ this reduces to

$$v^2 = v_0^2 - g\,(l\theta^2 + \tfrac{2}{3}a\theta^3), \quad \text{.........................(13)}$$

approximately, terms of the order θ^4 being neglected.

To find the positions of rest we put $v=0$. Assuming the extreme inclinations of the string to be small, we write for the sake of comparison with the circular pendulum

$$n^2 = g/l, \quad v_0 = nl\alpha, \quad \text{.............................(14)}$$

and the equation becomes

$$\theta^2 + \frac{2}{3}\frac{a}{l}\theta^3 = \alpha^2, \quad \text{.............................(15)}$$

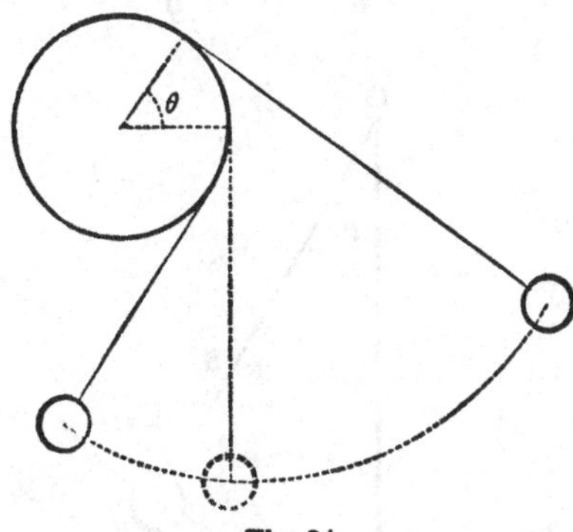

Fig. 34.

which may be solved by approximation. Thus for a first approximation we have $\theta = \pm\alpha$, and for a second

$$\theta^2 = \alpha^2 \mp \frac{2}{3}\frac{a}{l}\alpha^3, \quad \text{.............................(16)}$$

whence

$$\theta = \pm\alpha\left(1 \mp \frac{1}{3}\frac{a\alpha}{l}\right). \quad \text{.........................(17)}$$

The oscillations are accordingly not symmetrical, but extend to equal distances on each side of the position

$$\theta = -\frac{1}{3}\frac{a\alpha^2}{l}. \quad \text{.................................(18)}$$

37. The Circular Pendulum.

Let l be the length of the pendulum, θ the inclination at any instant to the vertical. The tangential acceleration being $l\,d^2\theta/dt^2$, by Art. 34, we have

$$l\frac{d^2\theta}{dt^2} = -g\sin\theta, \quad \text{.......................(1)}$$

or
$$\frac{d^2\theta}{dt^2} + n^2 \sin\theta = 0, \quad\text{.......................(2)}$$
if
$$n^2 = g/l. \quad\text{.........................(3)}$$

If the angle θ is always small we may replace $\sin\theta$ by θ, approximately, and the solution of the equation (2) as thus modified is
$$\theta = \alpha\cos(nt + \epsilon), \quad\text{.......................(4)}$$
where α and ϵ are arbitrary. The period of a complete oscillation is therefore
$$T = \frac{2\pi}{n} = 2\pi\sqrt{\frac{l}{g}}, \quad\text{....................(5)}$$
as in Art. 11.

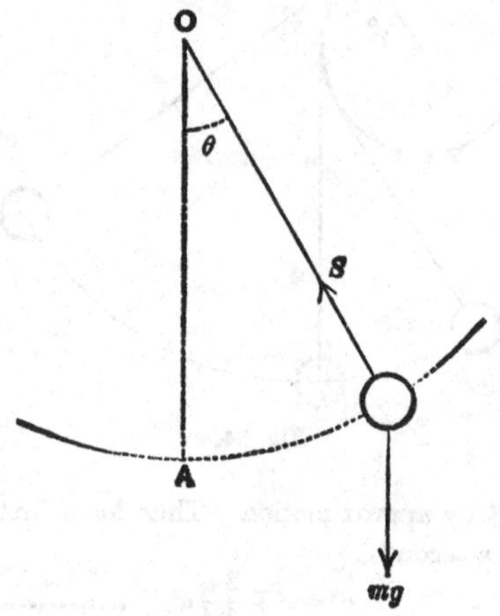

Fig. 35.

We have, however, now to consider the case of oscillation through a finite angle α on each side of the vertical. The equation of energy (Art. 36 (5)) gives
$$l\left(\frac{d\theta}{dt}\right)^2 = 2g\cos\theta + C, \quad\text{....................(6)}$$
as appears also from (1) if we multiply by $d\theta/dt$, and integrate with respect to t.

If we assume that $d\theta/dt = 0$ for $\theta = \alpha$, we have $C = -2g\cos\alpha$, and
$$\left(\frac{d\theta}{dt}\right)^2 = 2n^2(\cos\theta - \cos\alpha) = 4n^2(\sin^2\tfrac{1}{2}\alpha - \sin^2\tfrac{1}{2}\theta). \quad\text{...(7)}$$

Hence, considering the outward swing from the vertical, in the positive direction of θ, we have

$$\frac{n\,dt}{d\theta}=\frac{1}{2\sqrt{(\sin^2\frac{1}{2}\alpha-\sin^2\frac{1}{2}\theta)}}\text{..................(8)}$$

The integration in finite form involves the use of elliptic functions. To effect it in the standard form of such functions we introduce a new variable ϕ such that

$$\sin\tfrac{1}{2}\theta=\sin\tfrac{1}{2}\alpha\sin\phi.\text{ ....(9)}$$

In the outward swing ϕ therefore increases from 0 to $\frac{1}{2}\pi$. We have, then,

$$\frac{n\,dt}{d\theta}=\frac{1}{2\sin\frac{1}{2}\alpha\cos\phi},\text{....................(10)}$$

and, from differentiation of (9),

$$\frac{d\theta}{d\phi}=\frac{2\sin\frac{1}{2}\alpha\cos\phi}{\cos\frac{1}{2}\theta}.\text{ ....(11)}$$

Hence
$$\frac{n\,dt}{d\phi}=\frac{n\,dt}{d\theta}\frac{d\theta}{d\phi}=\frac{1}{\cos\frac{1}{2}\theta}=\frac{1}{\sqrt{(1-\sin^2\frac{1}{2}\alpha\sin^2\phi)}}\text{......(12)}$$

The time t of swinging through an arc θ is therefore given by

$$nt=\int_0^\phi\frac{d\phi}{\sqrt{(1-\sin^2\frac{1}{2}\alpha\sin^2\phi)}},\text{ ....(13)}$$

the upper limit being related to θ by the formula (9).

In the notation of elliptic integrals [*S.* 127]

$$nt=F(\sin\tfrac{1}{2}\alpha,\ \phi).\text{ ....(14)}$$

To find the time of a complete oscillation we must put $\phi=\frac{1}{2}\pi$, and multiply the resulting value of t by 4. The period is therefore

$$T=\frac{4}{n}\int_0^{\frac{1}{2}\pi}\frac{d\phi}{\sqrt{(1-\sin^2\frac{1}{2}\alpha\sin^2\phi)}}=\frac{4}{n}F_1(\sin\tfrac{1}{2}\alpha),\text{ ......(15)}$$

where F_1 denotes the 'complete' elliptic integral of the first kind to the modulus $\sin\frac{1}{2}\alpha$. The ratio which this bears to the period in an infinitely small arc is

$$\frac{2}{\pi}F_1(\sin\tfrac{1}{2}\alpha).\text{ ....(16)}$$

This is tabulated below for a series of values of α, by means of the tables of elliptic integrals. The result is exhibited graphically in Fig. 36.

α	$\frac{2}{\pi}F_1(\sin\frac{1}{2}\alpha)$	α	$\frac{2}{\pi}F_1(\sin\frac{1}{2}\alpha)$
0°	1·000 000	20°	1·007 669
1°	1·000 019	30°	1·017 409
2°	1·000 076	45°	1·039 973
3°	1·000 170	60°	1·073 182
4°	1·000 305	90°	1·180 340
5°	1·000 476	120°	1·372 880
10°	1·001 907	150°	1·762 204
		180°	∞

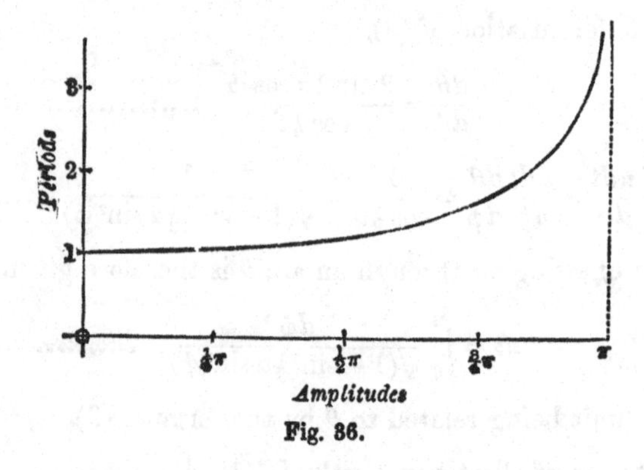

Fig. 36.

An expression for the period can also be obtained in the form of an infinite series, as follows. We have

$$T=\frac{4}{n}\int_0^{\frac{1}{2}\pi}(1-\sin^2\tfrac{1}{2}\alpha\sin^2\phi)^{-\frac{1}{2}}\,d\phi$$

$$=\frac{4}{n}\int_0^{\frac{1}{2}\pi}\left(1+\tfrac{1}{2}\sin^2\tfrac{1}{2}\alpha\sin^2\phi+\frac{1\,.\,3}{2\,.\,4}\sin^4\tfrac{1}{2}\alpha\sin^4\phi+\ldots\right)d\phi$$

$$=\frac{2\pi}{n}\left(1+\frac{1^2}{2^2}\sin^2\tfrac{1}{2}\alpha+\frac{1^2\,.\,3^2}{2^2\,.\,4^2}\sin^4\tfrac{1}{2}\alpha+\ldots\right).\quad\ldots\ldots\ldots\ldots(17)^*$$

* In virtue of the formula

$$\int_0^{\frac{1}{2}\pi}\sin^{2s}\phi\,d\phi=\frac{1\,.\,3\,.\,5\ldots(2s-1)}{2\,.\,4\,.\,6\ldots2s}\cdot\frac{\pi}{2}.$$

If α be small the terms rapidly diminish in value. A first approximation is $T = 2\pi/n$, or $2\pi\sqrt{(l/g)}$, and a second is

$$T = 2\pi\sqrt{\frac{l}{g}}\,.\,(1 + \tfrac{1}{4}\sin^2\tfrac{1}{2}\alpha). \quad \text{..............(18)}$$

The correction amounts to one part in a thousand when

$$\sin^2\tfrac{1}{2}\alpha = \cdot 004, \text{ or } \sin\tfrac{1}{2}\alpha = \cdot 0632, \text{ or } \alpha = 7^\circ\, 12', \text{ about.}$$

There is one case in which the equation (8) can be integrated in a finite form in terms of ordinary functions, viz. when $\alpha = \pi$, i.e. when the pendulum just swings into the position of unstable equilibrium. We have then

$$n\frac{dt}{d\theta} = \frac{1}{2\cos\frac{1}{2}\theta}, \quad \text{.....................(19)}$$

whence $$nt = \log\tan(\tfrac{1}{4}\pi + \tfrac{1}{4}\theta), \quad \text{................(20)}$$

no additive constant being needed if $t = 0$ for $\theta = 0$.

The tension (S, say) of the string (or light rod) supporting the bob is found, for any position, by considering the normal acceleration. We have

$$ml\left(\frac{d\theta}{dt}\right)^2 = S - mg\cos\theta. \quad \text{.................(21)}$$

On substituting from (6) we find

$$S = m(3g\cos\theta + C). \quad \text{.................(22)}$$

If the pendulum comes to rest when $\theta = \alpha$, we have $C = -2g\cos\alpha$, and

$$S = mg(3\cos\theta - 2\cos\alpha). \quad \text{..............(23)}$$

If ω denote the angular velocity of the pendulum when $\theta = 0$, we have, by (6),

$$\omega^2 l = 2g + C, \quad \text{..................................(24)}$$

so that $$\left(\frac{d\theta}{dt}\right)^2 = \omega^2 - 2n^2(1 - \cos\theta)$$

$$= \omega^2 - 4n^2\sin^2\tfrac{1}{2}\theta. \quad \text{..........................(25)}$$

If the pendulum make complete revolutions, this must be positive when $\theta = \pi$, which requires that $\omega^2 > 4n^2$. Hence, putting

$$k = 2n/\omega, \quad \text{......................................(26)}$$

we have $$\frac{\omega dt}{d\theta} = \frac{1}{\sqrt{(1 - k^2\sin^2\frac{1}{2}\theta)}}, \quad \text{.......................(27)}$$

or $$\omega t = 2\int_0^{\frac{1}{2}\theta}\frac{d\phi}{\sqrt{(1 - k^2\sin^2\phi)}} = 2F(k, \tfrac{1}{2}\theta). \quad \text{...................(28)}$$

This gives the time from the lowest point to any position up to $\theta=\pi$. The time of arriving at the highest position is therefore

$$\frac{2}{\omega}F_1\left(\frac{2n}{\omega}\right). \qquad \text{.....(29)}$$

The tension is given by

$$S=3mg\cos\theta-2mg+m\omega^2 l. \qquad \text{.....(30)}$$

In order that this may be positive when $\theta=\pi$ we must have

$$\omega^2>5g/l. \qquad \text{.....(31)}$$

An interesting variation of the problem of the present Art. is presented by the case of a particle moving on a smooth circle in a plane making any given angle β with the horizontal. The gravity of the particle may then be resolved into a component $mg\sin\beta$ along a line of greatest slope on the plane, and a component $mg\cos\beta$ normal to the plane. The latter component affects only the pressure on the plane. The motion in the circle is therefore covered by the preceding analysis, provided we replace g by $g\sin\beta$. In particular, if a be the radius of the circle, the period of a small oscillation will be

$$2\pi\sqrt{\frac{a}{g\sin\beta}}, \qquad \text{.....(33)}$$

the same as for a pendulum of length

$$l=\frac{a}{\sin\beta}. \qquad \text{.....(34)}$$

By making β very small, the length of the equivalent pendulum can be made very great. This is the theory of the so-called 'horizontal pendulum' as used in instruments for recording earthquakes, or for the measurement of the lunar disturbance of gravity. In practice the pendulum consists of a relatively heavy mass carried by a rod or 'boom,' which is free to rotate about an axis making a small angle (β) with the vertical. Cf. Art. 67, Fig. 62.

38. The Cycloidal Pendulum.

In any case of motion on a smooth curve in a vertical plane, under gravity, we have, resolving along the tangent,

$$\frac{d^2s}{dt^2}=-g\sin\psi, \qquad \text{.....(1)}$$

where ψ denotes the inclination of the curve to the horizontal, and s is the arc, which we will assume to be measured from a position of equilibrium, where $\psi = 0$.

For small inclinations we may write

$$\sin\psi = \psi = s/\rho_0, \quad \text{.................(2)}$$

where ρ_0 is the radius of curvature at the equilibrium position, reckoned positive when the concavity is upwards. Hence in a small oscillation we have

$$\frac{d^2s}{dt^2} = -\frac{g}{\rho_0}s, \quad \text{............................(3)}$$

the same equation as for a pendulum of length ρ_0.

If in (1) $\sin\psi$ were *accurately* proportional to s, so that

$$s = k\sin\psi, \quad \text{..........................(4)}$$

say, we should have

$$\frac{d^2s}{dt^2} + \frac{g}{k}s = 0. \quad \text{......................(5)}$$

The variations of s would therefore follow exactly the simple-harmonic law, and the period, viz.

$$2\pi\sqrt{\frac{k}{g}} \quad \text{..........................(6)}$$

would be the same for all amplitudes, large or small. The oscillations in such a case are said to be 'isochronous.'

To ascertain the nature of the curve whose intrinsic equation is of the form (4), we take axes of x and y which are horizontal and vertical, respectively, the positive direction of y being upwards. Then

$$\left.\begin{aligned} \frac{dx}{d\psi} &= \frac{dx}{ds}\frac{ds}{d\psi} = k\cos^2\psi = \tfrac{1}{2}k(1+\cos 2\psi), \\ \frac{dy}{d\psi} &= \frac{dy}{ds}\frac{ds}{d\psi} = k\sin\psi\cos\psi = \tfrac{1}{2}k\sin 2\psi. \end{aligned}\right\} \quad \text{......(7)}$$

Hence, integrating with respect to ψ,

$$x = \tfrac{1}{4}k(2\psi + \sin 2\psi), \quad y = \tfrac{1}{4}k(1-\cos 2\psi), \quad \text{.....(8)}$$

provided the additive constants of integration be adjusted so as to make $x=0$, $y=0$ for $\psi=0$. These equations are seen to coincide with the usual formulæ for the cycloid, viz.*

$$x=a(\theta+\sin\theta),\quad y=a(1-\cos\theta), \qquad \text{...........(9)}$$

where a is the radius of the rolling circle, and θ is the angle through which it has turned from the central position.

It is known† that the evolute of a cycloid is an equal cycloid whose cusps correspond to the vertices of the original curve, and *vice versâ*. Hence the bob of a pendulum will move accurately in

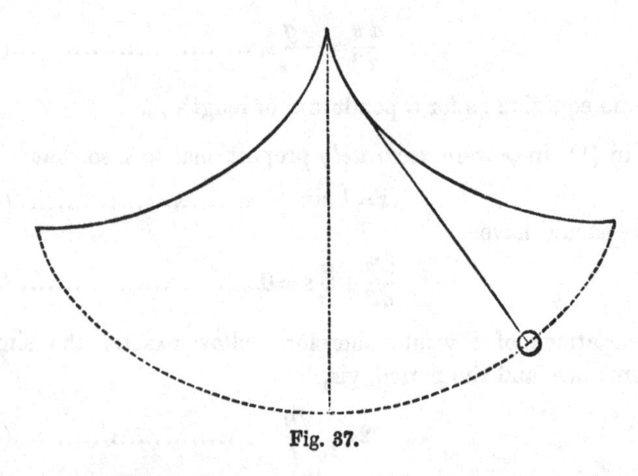

Fig. 37.

a cycloid if it be suspended by a string of suitable length which wraps itself alternately on two cycloidal arcs as shewn in the figure. For oscillations of moderate amplitude these arcs need not extend to more than a short distance from the cusp on each side. This device was proposed by Huygens, as a means of securing constancy of rate in a clock, in spite of variations in the amplitude of oscillation. Subsequent designers have gone on a different plan; and have directed their efforts to secure constancy of amplitude by careful regulation of the driving force, whose office it is to make good the losses of energy by frictional and other resistances.

* *Inf. Calc.*, Art. 186. The isochronous property of the cycloid was discovered by Christian Huygens (1629–95), the inventor of the pendulum clock. His chief dynamical work is the *Horologium Oscillatorium*, Paris, 1673.

† *Inf. Calc.*, Art. 159.

39. Oscillations on a Smooth Curve; Finite Amplitude.

We return to the general equation (1) of Art. 38, relating to the oscillations of a particle on a smooth curve of any form. For sufficiently small amplitudes the approximation in (2) of that Art. can be continued. As the question furnishes a good illustration of methods which are often useful in Dynamics, a little space may be given to it.

The height of the particle above the level of the equilibrium position may be supposed expressed, for small values of s, in the form

$$y = \tfrac{1}{2}cs^2 + \tfrac{1}{6}c's^3 + \tfrac{1}{24}c''s^4 + \ldots, \qquad \text{(1)}$$

the first two terms of the expansion being absent since $y=0$, $dy/ds=0$ for $s=0$. The meanings of the coefficients can be found by differentiation. Thus

$$\sin\psi = cs + \tfrac{1}{2}c's^2 + \tfrac{1}{6}c''s^3 + \ldots, \qquad (2)$$

$$\cos\psi\frac{d\psi}{ds} = c + c's + \tfrac{1}{2}c''s^2 + \ldots, \qquad (3)$$

$$\cos\psi\frac{d^2\psi}{ds^2} - \sin\psi\left(\frac{d\psi}{ds}\right)^2 = c' + c''s + \ldots, \qquad (4)$$

$$\cos\psi\frac{d^3\psi}{ds^3} - 3\sin\psi\frac{d\psi}{ds}\frac{d^2\psi}{ds^2} - \cos\psi\left(\frac{d\psi}{ds}\right)^3 = c'' + \ldots. \qquad (5)$$

Putting $\psi=0$, $s=0$ in these formulæ we find that c denotes the curvature (ρ^{-1}) at the origin, whilst

$$c' = \frac{d.\rho^{-1}}{ds}, \quad c'' = \frac{d^2.\rho^{-1}}{ds^2} - c^3, \qquad \text{(6)}$$

where, again, the values of the differential coefficients at the origin are to be understood.

The equation of motion (Art. 38 (1)) now takes the form

$$\frac{d^2s}{dt^2} = -g\,(cs + \tfrac{1}{2}c's^2 + \tfrac{1}{6}c''s^3 + \ldots), \qquad \text{(7)}$$

which is to be solved by successive approximation.

Neglecting s^2 we have the solution

$$s = \beta\cos(nt+\epsilon), \qquad \text{(8)}$$

provided $$n^2 = gc\,; \qquad \text{..........................(9)}$$
this is of course equivalent to our previous result.

For a second approximation we write the equation in the form

$$\frac{d^2s}{dt^2} + n^2 s = -\tfrac{1}{2}gc's^2 = -\tfrac{1}{2}gc'\beta^2\cos^2(nt+\epsilon)$$
$$= -\tfrac{1}{4}gc'\beta^2\{1+\cos 2(nt+\epsilon)\}, \qquad \text{......................(10)}$$

the error involved in substituting the approximate value of s in the small term of the second order being of the order s^3. The solution of (10) which is consistent with the first approximation is

$$s = \beta\cos(nt+\epsilon) - \frac{gc'\beta^2}{4n^2} + \frac{gc'\beta^2}{12n^2}\cos 2(nt+\epsilon), \qquad \text{......(11)}$$

as is verified immediately on differentiation. Although the motion is no longer accurately simple-harmonic, the interval $2\pi/n$ between two successive passages through the same position in the same direction is, to this order of approximation, unaltered.

The extreme positions are found by putting $nt+\epsilon = 0$ and π, respectively, and are therefore

$$s_1 = \beta - \frac{gc'\beta^2}{6n^2}, \quad s_2 = -\beta - \frac{gc'\beta^2}{6n^2}, \qquad \text{...........(12)}$$

the arithmetic mean of which is

$$s = -\frac{gc'\beta^2}{6n^2} = -\frac{c'\beta^2}{6c}. \qquad \text{.....................(13)}$$

To apply this investigation to the case of Art. 36, Ex. 2, we must put

$$c = 1/l, \quad c' = -a/l^3,$$

as is easily proved. The formulæ (12) can then be shewn to agree with (17) of the Art. cited.

When we proceed to a third approximation a new point arises. This will be sufficiently illustrated, and the calculations somewhat shortened, if we assume the curve to be symmetrical with respect to the position of equilibrium, so that $c' = 0$. The equation to be solved is then

$$\frac{d^2s}{dt^2} + n^2 s = -\tfrac{1}{6}gc''s^3. \qquad \text{......................(14)}$$

Substituting in the second member the value of s from (8) we have

$$\frac{d^2s}{dt^2} + n^2s = -\tfrac{1}{24}gc''\beta^3 \{3\cos(nt+\epsilon) + \cos 3(nt+\epsilon)\}. \quad \text{...(15)}$$

If we proceed to integrate this by the ordinary method, a term

$$-\frac{gc''\beta^3}{16n} t \sin(nt+\epsilon)$$

makes its appearance in the result, as in Art. 13 (11). It appears then that owing to the continual increase of the factor t, the solution thus obtained will sooner or later become inconsistent with the fundamental assumption as to the smallness of s. This was indeed to be expected from the particular case of the circular pendulum. We have seen that such a pendulum would get more and more out of step with a pendulum of the same length vibrating in an infinitely small arc, owing to the increase of period with amplitude.

This remark indicates that we should take as our first approximation

$$s = \beta \cos(\nu t + \epsilon), \quad \text{......................(16)}$$

where ν differs slightly from n, to an extent to be determined. The symbol n, where it occurs on the *right*-hand side of (15), is then to be replaced by ν. If we now assume

$$s = \beta\cos(\nu t+\epsilon) + C\cos 3(\nu t+\epsilon), \quad \text{...........(17)}$$

we find, on substitution, that the equation, as thus modified, will be satisfied provided

$$(n^2 - \nu^2)\beta = -\tfrac{1}{8}gc''\beta^3, \quad (n^2 - 9\nu^2) C = -\tfrac{1}{24}gc''\beta^3, \quad \text{...(18)}$$

i.e. provided

$$\nu^2 = n^2 + \tfrac{1}{8}gc''\beta^2, \quad \text{......................(19)}$$

and

$$C = \frac{gc''\beta^3}{192n^2}, \quad \text{............................(20)}$$

approximately. The former of these may be written

$$\nu^2 = n^2\left(1 + \frac{c''\beta^2}{8c}\right), \quad \text{......................(21)}$$

and the altered period is therefore

$$T = \frac{2\pi}{\nu} = \frac{2\pi}{n}\left(1 - \frac{c''\beta^2}{16c}\right), \text{.................}(22)$$

approximately*.

In the case of the circular pendulum we find

$$c = 1/l, \quad c'' = -1/l^3, \quad \text{.................}(23)$$

and the formula makes

$$T = 2\pi\sqrt{\frac{l}{g}}\cdot\left(1 + \frac{\beta^2}{16l^2}\right), \quad \text{..............}(24)$$

which agrees with Art. 37 (18), to our order of approximation. In the cycloidal pendulum $c'' = 0$, and the correction vanishes.

EXAMPLES. IX.

1. A ship is steaming along the equator at 30 km. per hour. Prove that the value of gravity on board will be apparently increased or diminished by ·12 cm./sec.^{2}, according as she is going W. or E. respectively.

What effect will this have on the height (760 mm.) of the barometer, assuming $g = 978$? [$\pm$·094 mm.]

2. A weight is suspended by two equal strings from two points at the same level, the inclination of each string to the vertical being α. If one string be cut, prove that the tension of the other is instantaneously reduced in the ratio $1 : 2\cos^2\alpha$.

3. A weight hangs in equilibrium by two strings whose inclinations to the vertical are α, β. Prove that if the second string be cut the tension of the first is instantaneously diminished in the ratio

$$\frac{\sin\beta}{\sin(\alpha+\beta)\cos\alpha}.$$

4. A wheel of radius a is rolling along the ground; prove that the horizontal and vertical accelerations of a point on the rim at an angular distance θ from the lowest point are

$$a(1-\cos\theta)\frac{d\omega}{dt} + a\omega^2\sin\theta, \quad a\sin\theta\frac{d\omega}{dt} + a\omega^2\cos\theta,$$

respectively, where ω is the angular velocity.

5. Apply the formula v^2/ρ for normal acceleration to find the radius of curvature at the vertex of a parabola.

Also at either vertex of an ellipse.

* These investigations are taken with slight alteration from Lord Rayleigh's *Theory of Sound*, Cambridge, 1899, Art. 67.

6. A point P describes relatively to a plane lamina a curve with constant velocity v. If the lamina has a velocity u which is constant in magnitude and direction, the radius of curvature of the absolute path of P is

$$\frac{(u^2+2uv\cos\theta+v^2)^{\frac{3}{2}}}{(u\cos\theta+v)\,v^2}\,\rho,$$

where ρ is the radius of curvature on the lamina, and θ is the angle between the directions of u and v.

7. A simple pendulum is started so as to make complete revolutions in a vertical plane. If ω_1, ω_2 be the greatest and least angular velocities, prove that the angular velocity when the pendulum makes an angle θ with the vertical is

$$\sqrt{(\omega_1^2\cos^2\tfrac{1}{2}\theta+\omega_2^2\sin^2\tfrac{1}{2}\theta)},$$

and that the tension of the string is

$$T_1\cos^2\tfrac{1}{2}\theta+T_2\sin^2\tfrac{1}{2}\theta,$$

where T_1, T_2 are the greatest and least tensions.

8. A particle oscillates on a smooth parabola whose axis is vertical, coming to rest at the extremities of the latus rectum. Find the pressure of the particle on the curve when passing through its lowest position. [$2mg$.]

9. A mass of 10 lbs. is attached to one end of a light rod which can turn freely about the other end as a fixed point. The mass is just started from its position of unstable equilibrium; find at what inclination to the vertical the thrust in the rod vanishes.

Also find (in gravitation measure) the thrust or tension in the rod, (i) when it is horizontal, (ii) when the mass is passing through its lowest position.

[48°; 20 lbs.; 50 lbs.]

10. A particle is constrained to move, under gravity, on a smooth parabola whose axis is vertical and vertex upwards. Prove that the pressure on the curve is

$$m\,(u_0^2-gl)/\rho,$$

where u_0 is the velocity at the vertex, l the semi-latus-rectum, and ρ the radius of curvature.

11. Prove that if a particle describe a given curve freely under the action of given forces, a chain having the form of this curve can be in equilibrium under the same forces (per unit mass) *reversed*, provided the line-density of the chain vary inversely as the velocity of the particle.

Illustrate this by the case of a projectile.

12. Prove that if the driving force and the resistance of a rocket were so related that the velocity was constant, the path would be the 'catenary of uniform strength' (inverted).

13. A disk is maintained in steady rotation ω about a vertical axis. From a point on the lower surface of the disk at a distance a ($< l$) from the axis is suspended a simple pendulum of length l. If $\sin^3\beta = a/l$, prove that the inclination θ of the pendulum to the vertical, in relative equilibrium, is given by

$$\cos\theta + \sin^3\beta\cot\theta = g/\omega^2 l.$$

Prove that if $\omega^2 > g/l \,.\, \sec^3\beta$, this equation has *three* solutions between $\pm\frac{1}{2}\pi$, of which two are negative, one being numerically greater, and the other numerically less than β.

Illustrate by a figure.

14. A mass m is attached to a fixed point O by a string of length l, and a mass m' is attached to m by a string of length l'. Prove that if the strings lie always in a vertical plane which revolves with constant angular velocity ω, and make constant angles θ, θ' respectively with the vertical, the following conditions must be satisfied:

$$\frac{\omega^2 l}{g} = (1+\mu)\sec\theta - \mu\,\mathrm{cosec}\,\theta\tan\theta',$$

$$\frac{\omega^2 l}{g}\sin\theta + \frac{\omega^2 l'}{g}\sin\theta' = \tan\theta',$$

where $\mu = m'/m$.

15. Taking the case where θ, θ' are both small, prove that there are two distinct solutions of the problem, and that the corresponding values of ω^2 are given by the equation

$$\omega^4 - (1+\mu)g\left(\frac{1}{l} + \frac{1}{l'}\right)\omega^2 + (1+\mu)\frac{g^2}{ll'} = 0.$$

EXAMPLES. X.

(Pendulum; Cycloid.)

1. If the Earth's rotation were arrested, would a seconds pendulum at the equator go faster or slower than before? What would be the gain or loss in a day?

2. A pendulum beats seconds when vibrating through an infinitely small arc. How many beats will it lose per day if it oscillates through an angle of 5° on each side the vertical? [41·3.]

3. A pendulum of length l is making complete revolutions in a vertical plane, and its least velocity is large compared with $\sqrt{(2gl)}$. Prove that the time of describing an angle θ from the lowest position is given by

$$\omega t = \theta - \frac{g}{\omega^2 l}\sin\theta,$$

approximately, where ω is the angular velocity when the string is horizontal.

4. If a simple pendulum just reaches the unstable position, prove that the angle θ, and the time t, from the lowest point, are connected by the relations

$$\tan\tfrac{1}{2}\theta = \sinh nt, \quad \cos\tfrac{1}{2}\theta = \operatorname{sech} nt, \quad \sin\tfrac{1}{2}\theta = \tanh nt, \quad \tan\tfrac{1}{4}\theta = \tanh\tfrac{1}{2}nt,$$

where $n = \surd(g/l)$.

5. If a pendulum whose period of small oscillation is 2 sec. just makes a complete revolution, find the time of moving through 90° from the lowest position. [·28 sec.]

6. A simple pendulum oscillates through an angle α on each side of the vertical; find the polar equation of the hodograph, and sketch the forms of the curve in the cases $\alpha < \frac{1}{2}\pi$, $\alpha = \frac{1}{2}\pi$, $\pi > \alpha > \frac{1}{2}\pi$.

7. A particle oscillates on a smooth cycloid, from rest at a cusp, the axis being vertical, and the vertex downwards. Prove the following properties:

(1) The angular velocity of the generating circle is constant;

(2) The hodograph consists of a pair of equal circles, touching one another;

(3) The resultant acceleration is constant and equal to g, and is in the direction of the radius of the generating circle;

(4) The pressure on the curve is $2mg\cos\psi$, where ψ is the inclination of the tangent to the horizontal.

8. Prove that in the case of the cycloidal pendulum the hodograph is the inverse of a conic with respect to its centre, the conic being an ellipse or a hyperbola according as the maximum velocity is greater or less than $\surd(gl)$, where l is the length of the pendulum.

9. A particle rests in unstable equilibrium on the vertex of a smooth inverted cycloid. Prove that if slightly disturbed it will leave the curve at the level of the centre of the generating circle.

10. A particle is moving on a smooth curve, under gravity, and its velocity varies as the distance (measured along the arc) from the highest point. Prove that the curve must be a cycloid.

11. A particle describes a smooth curve, under gravity, in a vertical plane. If the distance travelled along the arc in time t be $a\sinh nt$, find the shape of the curve, and the initial circumstances.

12. A particle makes small oscillations of amplitude β on a smooth curve, about the lowest point. Prove that the distances of the turning points from the lowest point are

$$\pm\beta + \frac{\beta^2}{6\rho}\frac{d\rho}{ds},$$

approximately, where ρ is the radius of curvature at the lowest point.

13. A particle oscillates about the vertex of the catenary

$$s = a \tan \psi,$$

the vertex being the lowest point, with a small amplitude β. Prove that the period is

$$2\pi \sqrt{\frac{a}{g}} \cdot \left(1 + \frac{3\beta^2}{16a^2}\right),$$

approximately.

14. The period of a small oscillation of amplitude β about the lowest point of the parabola

$$y = x^2/4a,$$

where the axis of y is vertical, is

$$2\pi \sqrt{\frac{2a}{g}} \cdot \left(1 + \frac{\beta^2}{16a^2}\right),$$

approximately.

CHAPTER VI

MOTION OF A PAIR OF PARTICLES

40. Conservation of Momentum.

As a step towards the general theory of a system of isolated particles it is worth while to consider separately the case of two particles only. This will enable us to treat a number of interesting questions without any great complexity of notation.

We begin with the case of two bodies moving in the same straight line. If we assume, in accordance with Newton's Third

Fig. 38.

Law, that the mutual actions between the particles are equal and opposite, the equations of motion will be of the forms

$$m_1 \frac{du_1}{dt} = X_1 + P, \quad m_2 \frac{du_2}{dt} = X_2 - P, \quad \text{...........(1)}$$

where X_1, X_2 are the forces acting on the two particles, respectively, from without the system, and the forces $\pm P$ represent the mutual action. By addition we have

$$\frac{d}{dt}(m_1 u_1 + m_2 u_2) = X_1 + X_2, \quad \text{.................(2)}$$

shewing that the total momentum $m_1 u_1 + m_2 u_2$ increases at a rate equal to the total *external* force, and is unaffected by the mutual action.

This result can be expressed in another form. If x_1, x_2 be the coordinates of the two particles, we have

$$m_1u_1 + m_2u_2 = m_1\frac{dx_1}{dt} + m_2\frac{dx_2}{dt} = \frac{d}{dt}(m_1x_1 + m_2x_2)$$

$$= \frac{d}{dt}(m_1 + m_2)\bar{x} = (m_1 + m_2)\bar{u}, \quad (3)$$

where the symbols $\bar{x}$ and $\bar{u}$ refer to the position and the velocity of the centre of mass [*S.* 66]. The total momentum of the system is therefore the same as if the total mass were collected at the centre of mass, and endowed with the velocity of this point.

The equation (2) may accordingly be written

$$(m_1 + m_2)\frac{d\bar{u}}{dt} = X_1 + X_2. \quad \text{..................(4)}$$

This expresses that the mass-centre moves exactly as if the whole mass were concentrated there and acted on by the total external force. In particular, if there are no external forces, the mass-centre has a constant velocity. If at rest initially it will remain at rest, whatever the nature of the mutual action between the particles.

These results are easily generalized so as to apply to the case of motion in two or three dimensions. Thus for two dimensions we have

$$\left.\begin{aligned} m_1\frac{du_1}{dt} &= X_1 + P, \quad m_1\frac{dv_1}{dt} = Y_1 + Q, \\ m_2\frac{du_2}{dt} &= X_2 - P, \quad m_2\frac{dv_2}{dt} = Y_2 - Q, \end{aligned}\right\} \quad \text{............(5)}$$

where (u_1, v_1), (u_2, v_2) are the velocities of the two particles, and (X_1, Y_1), (X_2, Y_2) the external forces acting upon them, whilst the forces (P, Q), $(-P, -Q)$, represent the mutual action.

From these equations we deduce, in the same manner as before,

$$(m_1 + m_2)\frac{d\bar{u}}{dt} = X_1 + X_2, \quad (m_1 + m_2)\frac{d\bar{v}}{dt} = Y_1 + Y_2, \quad \text{...(6)}$$

if $(\bar{u}, \bar{v})$ be the velocity of the mass-centre. The interpretation is, as before, that the mass-centre moves as if it were a material particle endowed with the total mass and acted on by all the external forces. For instance, if there are no external forces, the mass-centre describes a straight line with constant velocity. Again, if two particles connected by a string (extensible or not) be projected anyhow under gravity, their mass-centre will describe a parabola.

41. Instantaneous Impulses. Impact.

For purposes of mathematical treatment a force which produces a finite change of momentum in a time too short to be appreciated is regarded as infinitely great, and the time of action as infinitely short. The effect of ordinary finite forces in this time is therefore neglected; moreover since the velocity, though it changes, remains finite, the change of *position* in the interval is also ignored. The total effect is summed up in the value of the instantaneous impulse, which replaces the time-integral of the force (cf. Art. 9).

We may apply these principles to the direct impact of spheres. Let us suppose that two spheres of masses m_1, m_2, moving in the same straight line, impinge, with the result that the velocities are suddenly changed from u_1, u_2 to u_1', u_2', respectively. Since the total impulses on the two bodies must be equal and opposite, there is no change in the total momentum, so that

$$m_1 u_1' + m_2 u_2' = m_1 u_1 + m_2 u_2. \qquad (1)$$

Some additional assumption is required in order to determine the result of the impact in any given case. If we assume that there is no loss of kinetic energy we have, in addition,

$$m_1 u_1'^2 + m_2 u_2'^2 = m_1 u_1^2 + m_2 u_2^2. \qquad (2)$$

The equations (1) and (2) may be written

$$m_2 (u_2' - u_2) = m_1 (u_1 - u_1'),$$

$$m_2 (u_2'^2 - u_2^2) = m_1 (u_1^2 - u_1'^2),$$

whence, by division,

$$u_2' + u_2 = u_1 + u_1',$$

or

$$u_2' - u_1' = -(u_2 - u_1). \qquad (3)$$

That is, the velocity of either sphere relative to the other is reversed in direction, but unaltered in magnitude. This appears to be very nearly the case with steel or glass balls.

In general, however, there is some appreciable loss of energy*. The usual empirical principle† which is assumed for dealing with such questions is that the relative velocity is reversed and at the same time diminished in a ratio e which is constant for two given bodies. This constant ratio is sometimes called the 'coefficient of restitution.' The equation (3) is then replaced by

$$u_2' - u_1' = -e(u_2 - u_1). \quad \text{.....................(4)}$$

This equation, combined with (1), determines u_1', u_2' when u_1, u_2 are given.

In the case of *oblique* impact of spheres we resolve parallel and perpendicular to the line joining the centres at the instant of impact. If the spheres are smooth, the motion of each perpendicular to this line is unaffected, and we have, with an obvious notation,

$$v_1' = v_1, \quad v_2' = v_2. \quad \text{.........................(5)}$$

The velocities in the line of centres are subject as before to the formulæ (1) and (4).

The case of impact on a fixed obstacle may be deduced by supposing m_2 to be infinite. The equation (1) then makes $u_2' = 0$ if $u_2 = 0$, and (4) therefore reduces to

$$u_1' = -eu_1. \quad \text{.........................(6)}$$

Ex. 1. If in (1) and (4) we put $u_2 = 0$ we find

$$u_2' = \frac{m_1}{m_1 + m_2}(1 + e)\,u_1. \quad \text{.........................(7)}$$

Hence when one body (m_1) impinges on another (m_2) which is at rest, the velocity communicated to m_2 is greater the greater the ratio of m_1 to m_2, but is always less than double the original velocity of m_1.

Ex. 2. Suppose that we have a very large number of particles moving in all directions within a rectangular box, whose edges a, b, c are parallel to the coordinate axes; to calculate the average impulse, per unit time, and per unit area, on the walls. We assume the coefficient of restitution (e) to be unity.

We will suppose in the first place that the particles do not interfere with one another. Consider a single particle m moving with velocity (u, v, w). The component u is unaltered by impact on the faces which are parallel to x,

* That is, of energy of *visible* motion. The energy apparently lost is spent in producing vibrations, or permanent deformation, in the colliding bodies, and takes ultimately the form of heat.

† Given by Newton in the Scholium to his Laws of Motion.

but is reversed at each impact with the faces perpendicular to x, which we will distinguish as the 'ends.' The impacts on an end therefore succeed each other at intervals $2a/u$. Hence the total impulse on an end in a time t, which we will suppose to cover a large number of impacts, will be

$$\frac{ut}{2a} \times 2mu = \frac{mu^2t}{a};$$

and the total impulse per unit area will be

$$\frac{mu^2t}{abc}. \qquad \text{...(8)}$$

If we have a large number n of such particles, moving with the same velocity q, but in all directions indifferently, then since

$$u^2+v^2+w^2=q^2,$$

the average value of u^2 will be $\frac{1}{3}q^2$, and the expression (8) is replaced by

$$\frac{1}{3}\frac{nmq^2t}{abc}. \qquad \text{.....................................(9)}$$

If we have n_1 particles of mass m_1 moving with velocity q_1, n_2 of mass m_2 moving with velocity q_2, and so on, the average impulse per unit area and per unit time will be

$$\frac{1}{3}\frac{M\overline{q}^2}{abc}, \qquad \text{.....................................(10)}$$

where

$$M=n_1m_1+n_2m_2+\ldots, \qquad \text{..............................(11)}$$

and

$$\overline{q}^2=\frac{n_1m_1q_1^2+n_2m_2q_2^2+\ldots}{n_1m_1+n_2m_2+\ldots}, \qquad \text{........................(12)}$$

i.e. M is the total mass of the particles, and $\overline{q}^2$ is the mean square of their velocities.

If the particles are so numerous as to form an apparently continuous medium, we have $M=\rho abc$, where ρ is the density, and the pressure-intensity due to the medium is

$$p=\tfrac{1}{3}\rho\overline{q}^2. \qquad \text{....................................(13)}$$

This result is not affected by encounters between the particles, if the changes of velocity are assumed to be instantaneous; for an encounter does not affect the average kinetic energy, which is involved in (13), and there is no loss (or gain) of time.

The above is the explanation of the pressure of a gas, and of Boyle's law, on the Kinetic Theory*. The comparison with the gaseous laws [*S.* 115] shews that $\overline{q}^2$ must be supposed to be proportional to the absolute temperature.

For air at freezing point, under a pressure of one atmosphere, we find, putting

$$p=76\times 13{\cdot}6\times 981, \quad \rho={\cdot}00129,$$

$$\overline{q}=\sqrt{\left(\frac{3p}{\rho}\right)}=486 \text{ m./sec.}$$

* The calculation is due in principle to Joule (1851).

42. Kinetic Energy.

Taking first the case of rectilinear motion, let us write

$$x_1 = \bar{x} + \xi_1, \quad x_2 = \bar{x} + \xi_2, \quad \text{.................}(1)$$

so that ξ_1, ξ_2 are the coordinates of the two particles relative to the mass-centre. We have then [*S.* 66]

$$m_1\xi_1 + m_2\xi_2 = 0. \quad \text{.......................}(2)$$

Differentiating (1) we have

$$u_1 = \frac{dx_1}{dt} = \bar{u} + \dot{\xi}_1, \quad u_2 = \frac{dx_2}{dt} = \bar{u} + \dot{\xi}_2, \quad \text{...........}(3)$$

and therefore

$$\tfrac{1}{2} m_1 u_1^2 + \tfrac{1}{2} m_2 u_2^2 = \tfrac{1}{2} m_1 (\bar{u} + \dot{\xi}_1)^2 + \tfrac{1}{2} m_2 (\bar{u} + \dot{\xi}_2)^2$$
$$= \tfrac{1}{2}(m_1 + m_2)\bar{u}^2 + \tfrac{1}{2} m_1 \dot{\xi}_1^2 + \tfrac{1}{2} m_2 \dot{\xi}_2^2, \quad \text{...........}(4)$$

since $\qquad \bar{u}(m_1\dot{\xi}_1 + m_2\dot{\xi}_2) = \bar{u}\dfrac{d}{dt}(m_1\xi_1 + m_2\xi_2) = 0,$

by (2).

The kinetic energy of the system is thus expressed as the sum of two parts, viz. (1) the kinetic energy

$$\tfrac{1}{2}(m_1 + m_2)\bar{u}^2 \quad \text{.......................}(5)$$

of the whole mass supposed moving with the velocity of the mass-centre, and (2) the part

$$\tfrac{1}{2} m_1 \dot{\xi}_1^2 + \tfrac{1}{2} m_2 \dot{\xi}_2^2, \quad \text{.......................}(6)$$

which may be termed the kinetic energy of the motion of the particles relative to the mass-centre. It appears from Art. 40 that the latter constituent alone can be affected by any mutual action between the particles.

The kinetic energy of the relative motion may be expressed in terms of the velocity of the two particles relative to one another. We have, from (3),

$$\left.\begin{aligned} \dot{\xi}_1 &= u_1 - \frac{m_1 u_1 + m_2 u_2}{m_1 + m_2} = \frac{m_2(u_1 - u_2)}{m_1 + m_2}, \\ \dot{\xi}_2 &= u_2 - \frac{m_1 u_1 + m_2 u_2}{m_1 + m_2} = \frac{m_1(u_2 - u_1)}{m_1 + m_2}, \end{aligned}\right\} \quad \text{...........}(7)$$

whence $\qquad \tfrac{1}{2} m_1 \dot{\xi}_1^2 + \tfrac{1}{2} m_2 \dot{\xi}_2^2 = \tfrac{1}{2} \dfrac{m_1 m_2}{m_1 + m_2}(u_1 - u_2)^2. \quad \text{...........}(8)$

It follows that in the direct impact of spheres the loss of kinetic energy is, on the empirical assumption of Art. 41,

$$\frac{1}{2}\frac{m_1 m_2}{m_1+m_2}(1-e^2)(u_1-u_2)^2. \quad \text{..................(9)}$$

The extension of the preceding results to the case of two- or three-dimensioned motion is easy. Thus in two dimensions the kinetic energy of the motion relative to the mass-centre is

$$\frac{1}{2}\frac{m_1 m_2}{m_1+m_2}\{(u_1-u_2)^2+(v_1-v_2)^2\}. \quad \text{...........(10)}$$

The last factor is of course the square of the velocity of either particle relative to the other.

Ex. 1. If a mass m_1 strike a mass m_2 at rest, and if the coefficient of restitution (e) be zero, the loss of energy in the impact is, by (9),

$$\frac{1}{2}\frac{m_1 m_2}{m_1+m_2}u_1^2.$$

The ratio which the energy lost bears to the original energy $\frac{1}{2}m_1 u_1^2$ is therefore

$$\frac{m_2}{m_1+m_2}.$$

If m_1 be large compared with m_2, as when a pile is driven into the ground by a heavy weight falling upon it, or when a nail is driven into wood by a hammer, this ratio is small, and almost the whole of the original energy is utilized (in overcoming the subsequent resistance). But if m_1 be small compared with m_2, the ratio is nearly equal to unity, and the energy is almost wholly spent in deformation of the surface of one or other of the bodies.

Ex. 2. In the case of two particles moving in a plane and connected by an inextensible string of length a, the relative velocity is ωa, where ω is the angular velocity of the string. The expression (10) is therefore equal to

$$\frac{1}{2}\frac{m_1 m_2}{m_1+m_2}\omega^2 a^2. \quad \text{..............................(11)}$$

If there are no external forces this must be constant, and ω is therefore constant.

43. Conservation of Energy.

In any motion whatever of the system the increment of the kinetic energy in any interval of time will of course be equal to the total work done on the particles by all the forces which are operative. In calculating this work, forces such as the reactions of smooth fixed curves or surfaces, or the tensions of inextensible strings or rods, may be left out of account [*S.* 50].

In some cases we are concerned with mutual actions which are functions only of the distance (r) between the particles. If we denote this action, reckoned positive when of the nature of an attraction, by $\phi(r)$, the total work done by it on the two particles when the distance changes from r to $r + \delta r$, is $-\phi(r)\,\delta r$; and the function

$$V = \int_a^r \phi(r)\,dr \quad \text{.........................(1)}$$

will therefore represent the work which would be required from extraneous sources in order to change the mutual distance from some standard value a to the actual value r, the operation being supposed performed with infinite slowness. This quantity V is, by a natural extension of a previous definition, called the 'internal potential energy' of the system.

Since the work done by the internal forces in any actual motion is equal to the diminution of V, we may assert that the sum

$$\tfrac{1}{2} m_1 v_1^2 + \tfrac{1}{2} m_2 v_2^2 + V, \quad \text{.....................(2)}$$

(where v_1, v_2 now denote the total velocities of the particles), is increased in any interval of time by the (positive or negative) work done by the *external* forces. In particular, if there are no external forces the sum (2) is constant.

If some or all of the external forces are due to a conservative field of force, the work which they do on the system may be reckoned as a diminution of the potential energy in relation to the field (cf. Art. 30).

Ex. In the case of two masses subject only to their mutual gravitation, we have

$$\phi(r) = \frac{\gamma m_1 m_2}{r^2}, \quad \text{.................................(3)}$$

where γ is the constant of gravitation (Art. 74). Hence, from (1),

$$V = -\frac{\gamma m_1 m_2}{r} + \text{const.} \quad \text{.................................(4)}$$

Since the part of the kinetic energy which is due to the motion of the mass-centre is constant, we have

$$\frac{1}{2}\frac{m_1 m_2}{m_1 + m_2} v^2 - \frac{\gamma m_1 m_2}{r} = \text{const.}, \quad \text{.........................(5)}$$

where v now stands for the relative velocity of the two particles. Hence

$$v^2 = \frac{2\gamma(m_1+m_2)}{r} + \text{const.}; \quad \text{.........................(6)}$$

cf. Art. 81.

44. Oscillations about Equilibrium.

The following problems relate to the small oscillations of a system of two particles about a state of equilibrium. The first is known as that of the 'double pendulum.'

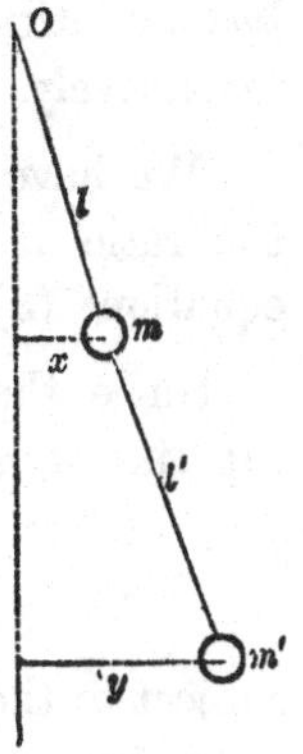

1. A mass m hangs from a fixed point O by a string of length l, and a second mass m' hangs from m by a string of length l'. The motion is supposed confined to one vertical plane.

Let x, y denote the horizontal displacements of m, m', respectively, from the vertical through O. If the inclinations of the two strings be supposed small, the tensions will have the statical values $(m+m')g$ and $m'g$, approximately, for the reason given in Art. 11. The equations of motion are therefore

Fig. 39.

$$\left.\begin{aligned} m\frac{d^2x}{dt^2} &= -(m+m')g\frac{x}{l} + m'g\frac{y-x}{l'}, \\ m'\frac{d^2y}{dt^2} &= -m'g\frac{y-x}{l'}. \end{aligned}\right\} \quad \text{.........(1)}$$

To solve these, we inquire, in the first place, whether a mode of vibration is possible in which each particle executes a simple-harmonic vibration of the same period, and with coincidence of phase. We assume, therefore, for trial,

$$x = A\cos(nt+\epsilon), \quad y = B\cos(nt+\epsilon). \quad \text{.........(2)}$$

Substituting in (1) we find that the cosines divide out, and that the equations are accordingly satisfied provided

$$\left.\begin{aligned} \left\{n^2 - \frac{(1+\mu)g}{l} - \frac{\mu g}{l'}\right\}A + \frac{\mu g}{l'}B &= 0, \\ \frac{g}{l}A + \left(n^2 - \frac{g}{l'}\right)B &= 0, \end{aligned}\right\} \quad \text{............(3)}$$

where $\mu = m'/m$.

Eliminating the ratio A/B we obtain

$$n^4 - (1+\mu)\,g\left(\frac{1}{l}+\frac{1}{l'}\right)n^2 + (1+\mu)\frac{g^2}{ll'} = 0, \quad \text{.........(4)}$$

which is a quadratic in n^2. Since the expression on the left-hand is positive when $n^2 = \infty$, negative when $n^2 = g/l$ or g/l', and positive when $n^2 = 0$, the roots are real and positive, and are moreover one greater than the greater, and the other less than the lesser of the two quantities g/l and g/l'. We denote these roots by n_1^2, n_2^2, respectively.

We have thus obtained two distinct solutions of the type (2), the ratio A/B being determined in each case by one of the equations (3), with the appropriate value of n^2 inserted.

Since the equations (1) are linear, these solutions may be superposed, and we have

$$\left.\begin{aligned} x &= A_1 \cos(n_1 t + \epsilon_1) + A_2 \cos(n_2 t + \epsilon_2), \\ y &= B_1 \cos(n_1 t + \epsilon_1) + B_2 \cos(n_2 t + \epsilon_2), \end{aligned}\right\} \quad \text{.........(5)}$$

subject to the relations

$$\frac{B_1}{A_1} = -\frac{g}{n_1^2 l' - g}, \quad \frac{B_2}{A_2} = \frac{g}{g - n_2^2 l'}. \quad \text{...............(6)}$$

The constants A_1, A_2, ϵ_1, ϵ_2 may be regarded as arbitrary and enable us to satisfy any prescribed initial conditions as to the values of x, y, $\dot{x}$, $\dot{y}$. The solution is therefore complete.

The types of motion represented by the two partial solutions, taken separately, are called the 'normal modes' of vibration. Their periods $2\pi/n_1$, $2\pi/n_2$ are fixed by the constitution of the system, whilst the amplitudes and phases depend on the initial conditions. In each mode the two particles keep step with one another, but it is to be noticed that in one mode x and y have the same sign, whilst in the other the signs are opposite; this follows from (6).

Some special cases of the problem are important. If the ratio $\mu\,(= m'/m)$ is small, the roots of (4) are g/l and g/l', approximately. In the normal mode corresponding to the former of these the upper mass m oscillates almost like the bob of a simple pendulum of length l, being only slightly affected by the influence of the smaller particle, whilst the latter behaves much like the bob of a pendulum of length l' whose point of suspension has a forced oscillation of period $2\pi\sqrt{(l/g)}$. We have, in fact, from (6)

$$\frac{y}{x}=\frac{l}{l-l'}, \qquad\qquad (7)$$

which is seen to agree with Art. 13. In the second normal mode (for which $n^2=g/l'$, approximately) the ratio x/y is small; and the upper mass is therefore comparatively at rest, whilst the lower particle oscillates much like the bob of a simple pendulum of length l'.

If on the other hand the mass of the upper particle is small compared with that of the lower one, so that μ is large, we have from (4)

$$n_1^2+n_2^2=\mu g\left(\frac{1}{l}+\frac{1}{l'}\right), \quad n_1^2n_2^2=\frac{\mu g^2}{ll'}, \qquad\qquad (8)$$

approximately. The roots are accordingly

$$\mu g\left(\frac{1}{l}+\frac{1}{l'}\right) \text{ and } \frac{g}{l+l'}, \qquad\qquad (9)$$

nearly. The former root makes

$$\frac{y}{x}=-\frac{l}{\mu(l+l')}, \qquad\qquad (10)$$

approximately. The lower mass is nearly at rest, whilst the upper one vibrates like a particle attached to a string of length $l+l'$ which is stretched between fixed points with a tension $m'g$; cf. Art. 10, Ex. 3. In the second mode we have

$$\frac{y}{x}=\frac{l+l'}{l}, \qquad\qquad (11)$$

shewing that the two masses are always nearly in a straight line with the point of suspension, m' now oscillating like the bob of a pendulum of length $l+l'$.

We have seen that the two periods can never be exactly equal, but considering the square of the difference of the roots of (4), viz.

$$(n_1^2-n_2^2)^2=(1+\mu)g^2\left\{\left(\frac{1}{l}-\frac{1}{l'}\right)^2+\mu\left(\frac{1}{l}+\frac{1}{l'}\right)^2\right\}, \quad (12)$$

we see that they will be very nearly equal if l is nearly equal to l', and the ratio μ, or m'/m, is at the same time small. The motion of each particle, being made up of two superposed simple-harmonic vibrations of nearly equal period, is practically equivalent, as explained in Art. 24, to a vibration of nearly constant period but fluctuating amplitude. Moreover if the initial circumstances be such that the amplitudes A_1, A_2 and consequently also B_1, B_2 are equal (in absolute magnitude), we have intervals of approximate rest. The energy of the motion then appears to be transferred alternately from one particle to the other, and back again, at regular intervals, the amplitude of m' being of course much the greater. The experiment is easily made, and is very striking.

2. Two equal particles m are attached symmetrically at equal distances a from the ends of a tense string of length $2(a+b)$.

If x, y denote small lateral displacements of the two particles, we have on the same principles as in Art. 10, Ex. 3,

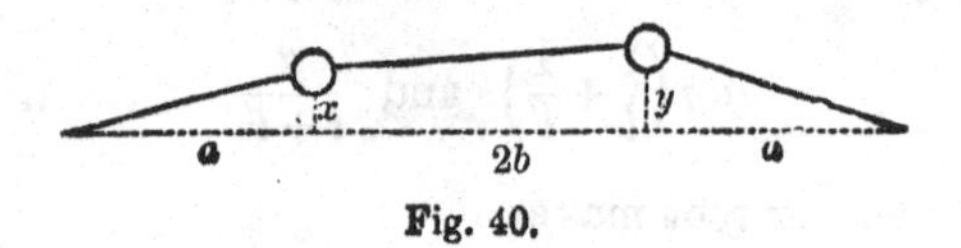

Fig. 40.

$$\left.\begin{aligned} m\frac{d^2x}{dt^2} &= -P\frac{x}{a} + P\frac{y-x}{2b}, \\ m\frac{d^2y}{dt^2} &= -P\frac{y}{a} - P\frac{y-x}{2b}, \end{aligned}\right\} \quad \text{...............(13)}$$

where P is the tension.

Assuming a trial solution

$$x = A\cos(nt+\epsilon), \quad y = B\cos(nt+\epsilon), \quad \text{.........(14)}$$

we find

$$\left.\begin{aligned} \left(n^2 - \frac{P}{m}\cdot\frac{a+2b}{2ab}\right)A + \frac{P}{2mb}B &= 0, \\ \frac{P}{2mb}A + \left(n^2 - \frac{P}{m}\cdot\frac{a+2b}{2ab}\right)B &= 0. \end{aligned}\right\} \quad \text{............(15)}$$

Hence, eliminating the ratio A/B we have

$$n^2 - \frac{P}{m}\cdot\frac{a+2b}{2ab} = \pm\frac{P}{2mb}, \quad \text{...............(16)}$$

or, denoting the two roots by n_1^2, n_2^2,

$$n_1^2 = \frac{P}{m} \cdot \frac{a+b}{ab}, \quad n_2^2 = \frac{P}{ma}. \quad \text{..............(17)}$$

The complete solution is then of the type (5).

In the normal mode corresponding to the first of the roots (17) we have $A = -B$, from (15); the configuration at the end of a

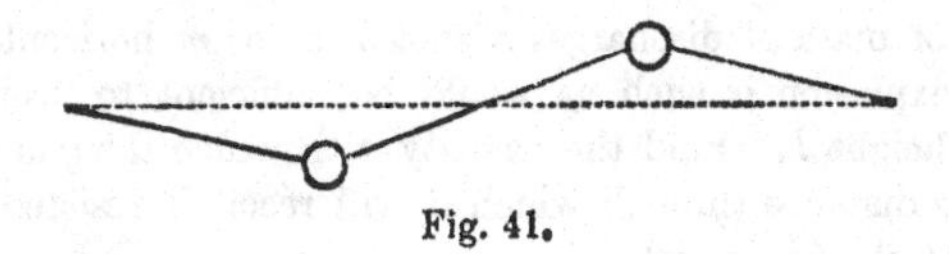

Fig. 41.

swing is therefore as shewn in Fig. 41. In the second mode we have $A = B$, the configuration being as in Fig. 42.

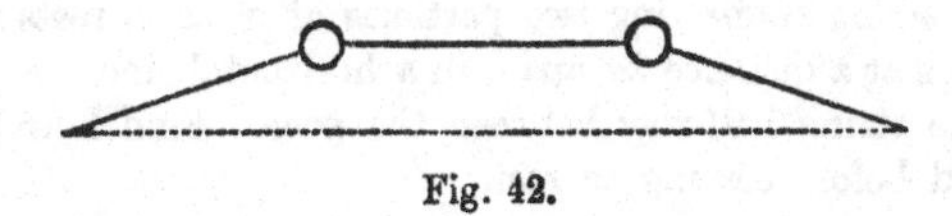

Fig. 42.

In the present problem, owing to the symmetry of the arrangement, it is easy to see beforehand that the two independent modes of vibration just described are possible, and the corresponding periods are accordingly more readily found independently.

The preceding results are particular cases of a general principle which may be here stated, although the complete proof is beyond our limits*. In any conservative dynamical system whatever, which is slightly disturbed from a state of stable equilibrium, the resulting motion may be regarded as made up by superposition of a series of independent 'normal modes' of vibration. If any one of these modes is started alone, the motion of each particle of the system is simple-harmonic, and the various particles keep step with one another, passing simultaneously through their positions of equilibrium. The number of such modes is equal to the number of degrees of freedom of the system, and their periods depend solely on its constitution. Their respective amplitudes and phases, on the other hand, are arbitrary, and depend on the nature of the initial disturbance.

* The case of a system of *two* degrees of freedom is however discussed in a general manner in Chap. XIII.

EXAMPLES. XI.

1. A mass of 5 lbs. impinges directly on a mass of 10 lbs. which is at rest, with a velocity of 12 ft./sec., and is observed to recoil with a velocity of 1 ft./sec. Find (in ft.-lbs.) the energy lost in the impact. [4·57.]

2. A weight of 3 tons falling through 6 feet drives a pile weighing 10 cwt. one inch into the ground. Find the resistance of the ground; and the time occupied by the movement of the pile. [185 tons; ·01 sec.]

3. A gun of mass M discharges a shot of mass m horizontally, and the energy of the explosion is such as would be sufficient to project the shot vertically to a height h. Find the velocity with which the gun will begin to recoil; also the distance through which it will recoil if resisted by a steady force equal to $1/n$th of its weight.

$$\left[\sqrt{\left(\frac{m^2}{M(M+m)}\cdot 2gh\right)};\ \frac{m^2}{M(M+m)}\,nh.\right]$$

4. A light string connecting two particles of mass m rests supported by two smooth pegs at a distance $2a$ apart in a horizontal line. A particle m' is attached to the string half-way between the pegs. Find how far the mass m' will descend before coming to rest.

$$\left[\frac{4mm'a}{4m^2-m'^2}.\right]$$

5. Two particles m_1, m_2 are connected by a string passing over a smooth pulley, as in Atwood's machine. Prove that if the inertia of the pulley be neglected, the mass-centre of the particles has a downward acceleration

$$\left(\frac{m_1-m_2}{m_1+m_2}\right)^2.g.$$

6. A mass M hangs by a very long string, and from it is suspended a simple pendulum of length l and mass m. Prove that the period of oscillation of m relative to M is less than if M had been fixed, in the ratio

$$1:\sqrt{\left(1+\frac{m}{M}\right)}.$$

7. If in the double pendulum of Art. 44 the inclinations of the strings l, l' to the vertical be θ, θ' respectively, prove that

$$(m+m')\,l^2\left(\frac{d\theta}{dt}\right)^2+2m'll'\frac{d\theta}{dt}\frac{d\theta'}{dt}\cos(\theta-\theta')+m'l'^2\left(\frac{d\theta'}{dt}\right)^2$$
$$=2(m+m')\,gl\cos\theta+2m'gl'\cos\theta'+C.$$

8. Prove that in the notation of Art. 44 the potential energy of the double pendulum is, for small displacements,

$$\frac{1}{2}\frac{(m+m')\,gx^2}{l}+\frac{1}{2}\frac{m'g\,(y-x)^2}{l'}.$$

9. Prove that if in the double pendulum of Art. 44, the ratio μ be small, the frequencies of the two normal modes are determined by

$$n_1^2=\frac{g}{l}\left(1+\frac{\mu l'}{l'-l}\right), \quad n_2^2=\frac{g}{l'}\left(1+\frac{\mu l}{l-l'}\right),$$

approximately, unless l and l' are nearly equal.

10. If, in the preceding example, the two strings are of equal length the frequencies are given by

$$n^2=(1\pm\sqrt{\mu})\frac{g}{l},$$

approximately.

11. Two particles m_1, m_2 are attached at P_1, P_2 to a tense string AB whose ends are fixed. Prove that the periods $(2\pi/n)$ of the two normal modes of lateral vibration are given by the equation

$$n^4-\left\{\frac{T}{m_1}\left(\frac{1}{a_1}+\frac{1}{a_2}\right)+\frac{T}{m_2}\left(\frac{1}{a_2}+\frac{1}{a_3}\right)\right\}n^2+\frac{T^2}{m_1m_2}\left(\frac{1}{a_2a_3}+\frac{1}{a_3a_1}+\frac{1}{a_1a_2}\right)=0,$$

where T is the tension, and $a_1=AP_1$, $a_2=P_1P_2$, $a_3=P_2B$.

12. Prove that in the problem of the tense string with two attached particles (Art. 44) the potential energy is

$$\tfrac{1}{2}P\left\{\frac{x^2}{a}+\frac{(x-y)^2}{2b}+\frac{y^2}{a}\right\}.$$

Prove that the total energy of a vibration is

$$P\left(\frac{A_1^2}{a}+\frac{(a+b)A_2^2}{ab}\right),$$

where A_1, A_2 have the same meanings as in Art. 44 (14).

13. Three particles each of mass m are attached to a tense string of length $4a$, at distances a, $2a$, $3a$ from one end. Prove that there are three normal modes of transverse vibration; and that the squares of the frequencies of these are as

$$2-\sqrt{2} : 2 : 2+\sqrt{2}.$$

Draw figures shewing the characters of the respective modes.

14. A mass m hangs from a fixed point by a helical spring such that the period of a small vertical vibration is $2\pi/p$. If a mass m' be suspended from m by a second spring, such that the period of a small vertical vibration of m', when m is held fixed, is $2\pi/p'$, prove that when both masses are free the periods $2\pi/n$ of the normal modes of vertical vibration of the system are given by the equation

$$n^4-\left\{p^2+\left(1+\frac{m'}{m}\right)p'^2\right\}n^2+p^2p'^2=0.$$

15. Two masses of 3 and 5 lbs. are connected by a string 6 feet long. The former being fixed, the latter is projected at right angles to the string with a velocity of 10 f. s.; find the tension in lbs.

If the former mass be now released, what is the nature of the subsequent motion, and what is the altered tension? (Neglect extraneous forces.)

[2·60 lbs.; ·976 lbs.]

16. Two particles m, m' are connected by a string of length a. The former is free to move along a smooth straight groove, and the latter is projected in a direction perpendicular to the groove with velocity v_0. Prove that the particle m will oscillate through a space $2am'/(m+m')$, and that if the ratio m'/m be small the period of an oscillation will be

$$\frac{2\pi a}{v_0}\left(1-\frac{1}{4}\frac{m'}{m}\right),$$

nearly.

17. Two gravitating masses m_1, m_2 are at rest at a distance r apart. Find their velocities when this distance has diminished to r'.

$$\left[u_1^2=\frac{2\gamma m_2^2}{m_1+m_2}\left(\frac{1}{r'}-\frac{1}{r}\right),\quad u_2^2=\frac{2\gamma m_1^2}{m_1+m_2}\left(\frac{1}{r'}-\frac{1}{r}\right).\right]$$

CHAPTER VII

DYNAMICS OF A SYSTEM OF PARTICLES

45. Linear and Angular Momentum.

When we have to deal with a system of isolated particles we may begin as in Art. 40 by forming the equations of motion of each particle separately, taking account, of course, of the mutual actions, as well as of the forces acting on the system from without. In these equations each component of internal force will appear twice over, with opposite signs, viz. in the equations of motion of the two particles between which the force in question acts.

If the system is subject to frictionless constraints, a geometrical relation is in each case supplied, which contributes to determine the unknown reaction involved in the constraint. Thus if one of the particles is restricted to lie on a smooth surface, we have the equation of the surface.

The complete solution of such problems is, as may be expected, usually difficult. For instance, the fundamental problem of Physical Astronomy, viz. the 'problem of three bodies,' which is to determine the motions of three mutually gravitating particles (e.g. the sun, the earth, and the moon), can only be solved by elaborate methods of approximation.

There are, however, two general results which hold whatever the nature of the mutual actions in the system. Before proceeding to these it is convenient to premise one or two kinematical theorems which will simplify the statements.

The momenta of the several particles of the system evidently constitute a series of *localized* vectors [*S.* 18] which, for the purposes of resolving and taking moments, may be treated by the same rules as forces in Statics. For instance, in the two-dimensional case they may be replaced by a 'linear momentum' resident in a line through any assigned point O, and an 'angular momentum' which is the sum of the moments of the momenta of the several particles with respect to O. These correspond to the 'resultant force' and 'couple' of plane Statics [*S.* 21].

The linear momentum is simply the geometric sum of the momenta of the several particles, and is therefore independent of the position of the point O. If (x, y, z) be the coordinates of a particle m, the components of linear momentum are

$$\Sigma(m\dot{x}), \quad \Sigma(m\dot{y}), \quad \Sigma(m\dot{z}), \quad \text{.................}(1)$$

where the summations include all the particles of the system.

This linear momentum is (as in the particular case of Art. 40) the same as if the total mass were concentrated at the mass-centre of the system and endowed with the velocity of that point. For if in the time δt the particle m_1 is displaced from P_1 to P_1', the particle m_2 from P_2 to P_2', and so on, we have [*S.* 64]

$$\Sigma(m\,.\,\mathrm{PP'}) = \Sigma(m)\,.\,\mathrm{GG'}, \quad \text{.................}(2)$$

where GG′ is the displacement of the mass-centre. Hence, dividing by the scalar quantity δt,

$$\Sigma\left(m\,.\,\frac{\mathrm{PP'}}{\delta t}\right) = \Sigma(m)\,.\,\frac{\mathrm{GG'}}{\delta t}\,. \quad \text{.................}(3)$$

In the limit the vector $m\,.\,\mathrm{PP'}/\delta t$ is the momentum of a particle m, and the vector $\Sigma(m)\,.\,\mathrm{GG'}/\delta t$ is the momentum of a mass $\Sigma(m)$ supposed moving with the mass-centre G. Hence the theorem.

Analytically, we have

$$\Sigma\left(m\frac{dx}{dt}\right) = \frac{d}{dt}\Sigma(mx) = \frac{d}{dt}\{\Sigma(m)\,\bar{x}\} = \Sigma(m)\,.\,\frac{d\bar{x}}{dt}, \quad \text{...}(4)$$

and similarly for the other components.

In the condensed notation of Vector Analysis, if $\mathbf{r}$ be the position-vector of a particle m relative to any origin, and ρ its position-vector relative to the mass-centre, so that

$$\mathbf{r} = \bar{\mathbf{r}} + \rho, \quad \text{........................}(5)$$

where $\bar{\mathbf{r}}$ refers to the mass-centre, we have [*S.* 64]

$$\Sigma(m\rho) = 0, \qquad\qquad (6)$$

and

$$\Sigma(m\mathbf{r}) = \Sigma(m) \,.\, \bar{\mathbf{r}}. \qquad\qquad (7)$$

Hence, differentiating with respect to t, we have

$$\Sigma(m\mathbf{v}) = \Sigma(m) \,.\, \bar{\mathbf{v}}, \qquad\qquad (8)$$

where $\mathbf{v}\,(=\dot{\mathbf{r}})$ is the velocity of m, and $\bar{\mathbf{v}}\,(=d\bar{\mathbf{r}}/dt)$ is the velocity of the mass-centre.

Some kinematical theorems relating to angular momentum are deferred for the present (see Arts. 61, 62).

46. Kinetic Energy.

If from a fixed origin O we draw a series of vectors OV_1, OV_2,... to represent the velocities of the particles m_1, m_2,... on any given scale, and if we construct the vector

$$OK = \frac{\Sigma(m\,.\,OV)}{\Sigma(m)}, \qquad\qquad (1)$$

it will represent the velocity of the mass-centre. This is of course merely a different statement of the theorem of the preceding Art. It is evident [*S.* 64] that K will coincide with the mass-centre of a system of particles m_1, m_2,... situate at V_1, V_2,..., respectively, in the auxiliary diagram now imagined.

This property leads at once to some important theorems as to the kinetic energy of the system. By Lagrange's 'First Theorem' [*S.* 74] we have

$$\tfrac{1}{2}\Sigma(m\,.\,OV^2) = \tfrac{1}{2}\Sigma(m)\,.\,OK^2 + \tfrac{1}{2}\Sigma(m\,.\,KV^2); \qquad (2)$$

i.e. the kinetic energy of the system is equal to the kinetic energy of the whole mass, supposed concentrated at the mass-centre G, and moving with the velocity of this point, together with the kinetic energy of the motion relative to G. The latter constituent may be called the 'internal kinetic energy' of the system.

A particular case of this theorem has been given in Art. 42, and the analytical proof there given is easily generalized. Thus using rectangular coordinates, and writing

$$x = \bar{x} + \xi, \quad y = \bar{y} + \eta, \quad z = \bar{z} + \zeta \qquad\qquad (3)$$

where $\bar{x}$, $\bar{y}$, $\bar{z}$ refer to the mass-centre, we have

$$\tfrac{1}{2}\Sigma m(\dot{x}^2+\dot{y}^2+\dot{z}^2)=\tfrac{1}{2}\Sigma m\left\{\left(\frac{d\bar{x}}{dt}+\dot{\xi}\right)^2+\left(\frac{d\bar{y}}{dt}+\dot{\eta}\right)^2+\left(\frac{d\bar{z}}{dt}+\dot{\zeta}\right)^2\right\}$$

$$=\tfrac{1}{2}\Sigma(m)\,.\left\{\left(\frac{d\bar{x}}{dt}\right)^2+\left(\frac{d\bar{y}}{dt}\right)^2+\left(\frac{d\bar{z}}{dt}\right)^2\right\}$$

$$+\tfrac{1}{2}\Sigma m(\dot{\xi}^2+\dot{\eta}^2+\dot{\zeta}^2),\ \ldots\ldots(4)$$

since

$$\Sigma\left(m\dot{\xi}\frac{d\bar{x}}{dt}\right)=\frac{d\bar{x}}{dt}\Sigma(m\dot{\xi})=\frac{d\bar{x}}{dt}\frac{d}{dt}(\Sigma m\xi)=0,\ \ldots\ldots(5)$$

and so on.

There is also an interesting analogue to Lagrange's 'Second Theorem' [*S.* 74], viz.

$$\tfrac{1}{2}\Sigma(m\,.\,KV^2)=\frac{\tfrac{1}{2}\Sigma(mm'\,.\,VV'^2)}{\Sigma(m)},\ \ldots\ldots\ldots\ldots(6)$$

where in the summation in the numerator each pair of particles is taken once only. This formula gives the internal kinetic energy of the system in terms of the masses and *relative* velocities of the several pairs of particles*. For the special case of two particles a proof has already been given in Art. 42.

If, continuing the vector analysis of Art. 45, we write

$$\mathbf{v}=\dot{\boldsymbol{\rho}},\ \ldots\ldots\ldots\ldots\ldots\ldots\ldots\ldots\ldots\ldots\ldots(7)$$

i.e. $\mathbf{v}$ is the velocity of a particle m relative to the mass-centre, we have

$$\tfrac{1}{2}\Sigma(m\mathbf{v}^2)=\tfrac{1}{2}\Sigma m(\bar{\mathbf{v}}+\mathbf{v})^2=\tfrac{1}{2}\Sigma(m)\,.\,\bar{\mathbf{v}}^2+\tfrac{1}{2}\Sigma(m\mathbf{v}^2),\ldots\ldots(8)$$

since

$$\Sigma(m\bar{\mathbf{v}}\mathbf{v})=\bar{\mathbf{v}}\Sigma(m\mathbf{v})=\bar{\mathbf{v}}\Sigma(m\dot{\boldsymbol{\rho}})=0,\ \ldots\ldots\ldots\ldots(9)$$

by Art. 45 (6). Since the scalar square of a vector is equal to the square of the absolute magnitude of the vector, the formula (8) is equivalent to (2).

Again we may shew, almost exactly as in the case of *Statics*, Art. 74 (13), that

$$\tfrac{1}{2}\Sigma(m\mathbf{v}^2)=\frac{\tfrac{1}{2}\Sigma mm'(\mathbf{v}-\mathbf{v}')^2}{\Sigma(m)},\ \ldots\ldots\ldots\ldots\ldots(10)$$

in agreement with (6).

* The theorem, and the method of proof, are due to A. F. Möbius, *Mechanik des Himmels*, 1843.

47. Principle of Linear Momentum.

We proceed to the proof of the two dynamical theorems referred to in Art. 45.

The first of these is known as the 'Principle of Linear Momentum.' If there are no external forces on the system the total linear momentum, i.e. the vector sum of the momenta of the several particles, is constant in every respect. For consider any two particles P, Q, and let F be the mutual action between them, reckoned positive when attractive. In the infinitesimal time δt an impulse $F\delta t$ will be given to P in the direction PQ, whilst an equal and opposite impulse is given to Q in the direction QP. These impulses produce equal and opposite momenta in the directions PQ, QP, respectively, and therefore leave the geometric sum of the momenta of the system unaltered. Similarly for any other pair of particles.

Since, as we have seen, the total momentum is the same as that of the whole mass supposed moving with the velocity of the mass-centre, it follows that the mass-centre describes a straight line with constant velocity. For instance, the mass-centre of the solar system moves in this way, so far as the system is free from action from without.

If there are external forces acting on the system, the total momentum is modified in the time δt by the (geometric) addition of the sum of the impulses of these external forces. The mass-centre will therefore move exactly as if the whole mass were concentrated there, and acted on by the external forces, supposed applied parallel to their actual directions. Thus in the case of a system of particles subject to ordinary gravity, and to any mutual actions whatever, the mass-centre will describe a parabola.

48. Principle of Angular Momentum.

The remaining theorem is called the 'Principle of Angular Momentum.'

If there are no external forces, the total angular momentum, i.e. the sum of the moments* of the momenta of the several

* The moment of a localized vector about an axis normal to any plane in which the vector lies has been already defined [*S.* 20]. To find the moment about any axis whatever, we resolve the vector into two orthogonal components, of which

particles, with respect to any fixed axis, is constant. For in the time δt the mutual action between any two particles P, Q produces equal and opposite momenta in the line PQ, and these will have equal and opposite moments about the axis in question.

If external forces are operative, the total angular momentum about the fixed axis is increased by the sum of the moments of the external impulses about this axis.

Some further developments of this theorem are given in Chap. IX., Arts. 61, 62.

Ex. In the case of a particle subject only to a force whose direction passes always through a fixed point O, the angular momentum about this point (i.e. about an axis through this point normal to the plane of the motion) will be constant. If m be the mass of the particle, v its velocity, and p the perpendicular from O on the tangent to its path, this angular momentum is mvp. Hence v will vary inversely as p.

This is verified in the case of elliptic harmonic motion (Art. 28) where v was found to vary as the semi-diameter of the ellipse parallel to the direction of motion.

49. Motion of a Chain.

Books on Dynamics usually include a number of problems on the motion of chains. Though hardly important in themselves, such problems furnish excellent illustrations of dynamical principles.

Take first the case of a uniform chain sliding over the edge of a table, the portion on the table being supposed straight and at right angles to the edge. Let l be the total length, μ the line-density (i.e. the mass per unit length), u the velocity of the chain when the length of the vertical portion is x. If T be the tension at the edge, we have, considering the vertical portion,

$$\mu x \frac{du}{dt} = g\mu x - T, \quad \text{.........................(1)}$$

one is parallel to the axis in question, and the other is in a plane perpendicular to it. The moment of the latter component is the moment required. Some convention as to sign is of course necessary.

It easily follows that the sum of the moments of two intersecting localized vectors about any axis is equal to the moment of their resultant.

and, considering the horizontal portion,

$$\mu (l - x) \frac{du}{dt} = T. \quad \text{..........................(2)}$$

By addition we find

$$\frac{du}{dt} = \frac{gx}{l}. \quad \text{..........................(3)}$$

Since $u = dx/dt$, this makes

$$\frac{d^2x}{dt^2} = n^2 x, \quad \text{..........................(4)}$$

where $n^2 = g/l$; and the general solution is

$$x = Ae^{nt} + Be^{-nt}, \quad \text{..........................(5)}$$

where the constants A, B depend on the initial conditions.

The formula (3) follows also from the principle of energy. Since the vertical portion has a mass μx, and since its mass-centre is at a depth $\frac{1}{2}x$ below the level of the table, the potential energy is less than when the whole chain is at this level by $\frac{1}{2}g\mu x^2$ [S. 50]. Hence

$$\tfrac{1}{2}\mu l u^2 = \tfrac{1}{2}g\mu x^2 + \text{const.}, \quad \text{..................(6)}$$

the constant depending on the length which was initially overhanging. Differentiating with respect to x we find

$$u \frac{du}{dx} = \frac{gx}{l}, \quad \text{..........................(7)}$$

in agreement with (3).

Let us now vary the question by supposing that the portion of the chain which is on the table forms a loose heap close to the edge. In a short time δt an additional mass $\mu u \delta t$ is set in motion with velocity u. Hence, considering the momentum generated, we have

$$\mu u \delta t \,.\, u = T\delta t, \quad \text{..........................(8)}$$

or

$$T = \mu u^2. \quad \text{..........................(9)}$$

This equation takes the place of (2). Substituting in (1), and writing $u\,du/dx$ for du/dt, we have

$$xu \frac{du}{dx} + u^2 = gx. \quad \text{......................(10)}$$

This may be put in the form

$$2x^2u\frac{du}{dx}+2xu^2=2gx^2, \quad \text{.................(11)}$$

which is immediately integrable; thus

$$x^2u^2=\tfrac{2}{3}gx^3+\text{const.} \quad \text{....................(12)}$$

This cannot as a rule be integrated further without using elliptic functions. But if the chain be just started from rest at the edge of the table, so that u is infinitesimal for $x=0$, the constant vanishes, and we have

$$u^2=\tfrac{2}{3}gx, \quad \text{...........................(13)}$$

shewing that the acceleration is constant and equal to $\frac{1}{3}g$.

It is to be noticed that the equation of energy, if applied to this form of the problem, leads to an erroneous result unless we take account of the energy continually lost in the succession of infinitesimal impacts which take place at the edge, as one element of the chain after another is set in motion. It was found in Art. 42 that when the coefficient (e) of restitution vanishes, as it is assumed to do in the present case, the energy lost in an impact between two masses m, m' is

$$\frac{1}{2}\frac{mm'}{m+m'}u^2, \quad \text{.......................(14)}$$

if u is the relative velocity just before the impact. In the present case we have $m=\mu x$, $m'=\mu u\,\delta t$, so that the loss of energy in time δt is $\frac{1}{2}\mu u^3\,\delta t$, ultimately. Hence, equating the loss of potential energy to the actual kinetic energy acquired *plus* the energy lost, we have

$$\begin{aligned}\tfrac{1}{2}g\mu x^2&=\tfrac{1}{2}\mu xu^2+\tfrac{1}{2}\mu\int u^3dt\\&=\tfrac{1}{2}\mu xu^2+\tfrac{1}{2}\mu\int u^2dx. \quad \text{..............(15)}\end{aligned}$$

If we differentiate this with respect to x we reproduce the equation (10).

50. Steady Motion of a Chain.

If T be the tension at any point of an inextensible chain moving in a given curve, the tensions on the two ends of a linear element δs contribute a force δT along the tangent, and a force $T\delta s/\rho$, where ρ is the radius of curvature, along the normal [*S.* 80].

Hence a uniform chain of line-density μ, which is subject to no external force, and is therefore free from constraint of any kind, can run with constant velocity v in the form of *any* given curve, provided

$$\delta T = 0, \quad \frac{\mu \delta s v^2}{\rho} = \frac{T \delta s}{\rho}, \quad \text{......................(1)}$$

i.e. provided

$$T = \mu v^2. \quad \text{..............................(2)}$$

This explains the observed tendency of chains running at high speed to preserve any form which they happen to have, in spite of the action of external forces such as gravity.

The formula (2) gives what is known as the 'centrifugal tension' in the case of a revolving circular chain, or even of a hoop, but we now learn that there is no restriction to the circular form.

As a particular case, we may imagine the chain to be infinitely long, and to be straight except for a finite portion, which has some other form. If we now superpose a velocity v in the direction opposite to that in which the chain is running in the straight portions, the deformation will travel along unchanged in the form of a *wave* on a chain which is otherwise at rest. This is Tait's* proof that the velocity of a wave of transverse displacement on a chain or cord stretched with a tension T is

$$v = \sqrt{\frac{T}{\mu}}. \quad \text{..............................(3)}$$

If external forces act, we may suppose them resolved, as regards an element δs, into components $\mathfrak{T}\delta s$ along the tangent and $\mathfrak{N}\delta s$ along the normal. The equations of motion of a mass-element $\mu\delta s$ will then be

$$\mu \delta s \frac{dv}{dt} = \delta T + \mathfrak{T}\delta s, \quad \mu \delta s \frac{v^2}{\rho} = \frac{T \delta s}{\rho} + \mathfrak{N}\delta s, \quad \text{.........(4)}$$

or

$$\mu \frac{dv}{dt} = \frac{dT}{ds} + \mathfrak{T}, \quad \frac{\mu v^2}{\rho} = \frac{T}{\rho} + \mathfrak{N}. \quad \text{............(5)}$$

If we put $v = 0$ we get the conditions which must be satisfied in order that the chain may be in equilibrium in the given form,

* P. G. Tait (1831-1901), Professor of Natural Philosophy at Edinburgh (1860-1901).

under the same system of external forces. Hence if the curve has a form which is one of possible equilibrium under the given forces, the equations (5) are satisfied by

$$v = \text{const.}, \quad \text{.........................(6)}$$

provided

$$T = T' + \mu v^2, \quad \text{.........................(7)}$$

where T' is the statical tension. For instance, a chain hanging in the form of a catenary between smooth pulleys can retain the same form when running with velocity v, provided the terminal tensions be increased by μv^2.

In the case of a chain constrained to move along a given smooth curve, the normal force $\mathfrak{N}\delta s$ will include a term $(R\delta s)$ contributed by the pressure of the curve. If the chain be subject to gravity, we have, then,

$$\mathfrak{T} = -\mu g \sin\psi, \quad \mathfrak{N} = -\mu g \cos\psi + R, \quad \text{.................(8)}$$

where ψ denotes the inclination of the tangent to the horizontal. Also, if the axis of y be drawn vertically upwards, that of x being horizontal, we have $\sin\psi = dy/ds$, and the former of equations (5) takes the form

$$\mu \frac{dv}{dt} = \frac{dT}{ds} - \mu g \frac{dy}{ds}. \quad \text{.........................(9)}$$

Hence, integrating over a finite length l of the chain,

$$\mu l \frac{dv}{dt} = T_2 - T_1 - \mu g (y_2 - y_1), \quad \text{...................(10)}$$

where the suffixes relate to the two ends. If both ends are free we have $T_1 = T_2 = 0$, and the acceleration will depend only on the difference of level of the ends.

This question, again, may be treated by the principle of energy. The total energy at any instant is

$$\tfrac{1}{2}\mu l v^2 + \int_{s_1}^{s_2} \mu g y \, ds, \quad \text{.........................(11)}$$

the second term representing the potential energy. In the time δt an element $\mu g y_2 v \delta t$ is added to the integral at the upper limit, whilst an element $\mu g y_1 v \delta t$ is subtracted at the lower limit. The rate of increase of the total energy is therefore

$$\mu l v \frac{dv}{dt} + \mu g (y_2 - y_1) v. \quad \text{.........................(12)}$$

This must be equal to the rate at which work is done by the tensions on the two ends, viz.

$$T_2 v - T_1 v. \quad \text{.................................(13)}$$

Equating (12) and (13), and dividing by v, we obtain the formula (10).

51. Impulsive Motion of a Chain.

Suppose that we have a chain initially at rest in the form of a plane curve, and that it is suddenly set in motion by tangential impulses at the extremities.

We now denote by T the *impulsive* tension at any point P, i.e. the time-integral of the actual tension over the infinitely short duration of the force. Let u, v be the velocities of P immediately after the impulse, in the directions of the tangent and normal respectively. Since the tangential and normal components of the impulse on an element δs are δT and $T\delta\psi$, we have

$$\mu\delta s \,.\, u = \delta T, \quad \mu\delta s \,.\, v = T\delta\psi,$$

or

$$\mu u = \frac{dT}{ds}, \quad \mu v = T\frac{d\psi}{ds}. \quad \text{...............(1)}$$

If P, Q be two adjacent points on the chain, the velocity of Q relative to P will be made up of a component

$$(u + \delta u)\cos\delta\psi - (v + \delta v)\sin\delta\psi - u = \delta u - v\delta\psi,$$

ultimately, parallel to the tangent at P, and a component

$$(u + \delta u)\sin\delta\psi + (v + \delta v)\cos\delta\psi - v = u\delta\psi + \delta v,$$

ultimately, parallel to the normal at P. Since the length (δs) of the element PQ is constant, the former component of relative velocity must vanish, whence

$$\frac{du}{ds} - v\frac{d\psi}{ds} = 0. \quad \text{.......................(2)}$$

The expression for the relative normal velocity shews that immediately after the impulse the element PQ has, in addition to its velocity of translation, an angular velocity

$$u\frac{d\psi}{ds} + \frac{dv}{ds}. \quad \text{.........................(3)}$$

Eliminating u, v between (1) and (2) we have

$$\frac{d}{ds}\left(\frac{1}{\mu}\frac{dT}{ds}\right) = \frac{T}{\mu}\left(\frac{d\psi}{ds}\right)^2 = \frac{T}{\mu\rho^2}, \quad \text{...............(4)}$$

if ρ be the radius of curvature.

If the chain be of uniform line-density, this reduces to

$$\frac{d^2T}{ds^2}=\frac{T}{\rho^2}. \quad \text{......................(5)}$$

The solution of (4) or (5) involves two arbitrary constants, which are to be determined from the data relating to the two ends. The values of u, v are then given by (1).

Ex. If the chain form a circular arc, we have $\rho=a$, the radius. Hence, putting $s=a\theta$,

$$\frac{d^2T}{d\theta^2}=T, \quad \text{..................................(6)}$$

the solution of which is

$$T=A\cosh\theta+B\sinh\theta. \quad \text{..........................(7)}$$

If T_0 be the impulsive tension applied at one end ($\theta=0$), and if the other end ($\theta=a$) be free, we have

$$A=T_0, \quad A\cosh a+B\sinh a=0, \quad \text{....................(8)}$$

whence

$$T=T_0\,\frac{\sinh(a-\theta)}{\sinh a}. \quad \text{..............................(9)}$$

Hence

$$u=-\frac{T_0}{\mu a}\cdot\frac{\cosh(a-\theta)}{\sinh a}, \quad v=\frac{T_0}{\mu a}\cdot\frac{\sinh(a-\theta)}{\sinh a}. \quad \text{...........(10)}$$

EXAMPLES. XII.

1. A mass m is on a smooth table, and is attached to a string which passes through a small hole in the table and carries a mass M hanging vertically. If m be projected at right angles to the string with velocity v, at a distance a from the hole, prove that when m is next moving at right angles to the string its distance from the hole is the positive root of the equation

$$x^2-\frac{mv^2}{2Mg}x-\frac{mv^2}{2Mg}a=0.$$

2. If in the preceding question the distance of m from the hole oscillates between the values a, b, the kinetic energy in the extreme positions will be

$$\frac{Mgb^2}{a+b}, \quad \frac{Mga^2}{a+b},$$

respectively.

3. P, Q are two particles of masses M, m, respectively. The former describes a circle of radius a about a fixed point O, and the latter describes a circle of radius b relative to P. If θ be the angle which OP makes with a fixed direction, and χ the angle which PQ makes with OP, prove that the kinetic energy of the system is

$$\tfrac{1}{2}(A\dot\theta^2+2H\dot\theta\dot\chi+B\dot\chi^2),$$

where $A=Ma^2+m(a^2+2ab\cos\chi+b^2)$, $H=mb(a\cos\chi+b)$, $B=mb^2$.

Also prove that the angular momentum about O is

$$A\dot\theta+H\dot\chi.$$

4. A uniform chain of length $2a$ hangs in equilibrium over a smooth peg. If it be just started from rest, prove that its velocity when it is leaving the peg is $\sqrt{(ga)}$.

5. A uniform chain hangs vertically from its two ends, which are close together. If one end be released, the tension at the bight after a time t, on the stationary side, is $\frac{1}{2}\mu g^2 t^2$, where μ is the mass per unit length.

Examine the loss of mechanical energy.

6. A piece of uniform chain hangs vertically from its upper end, with the lower end just clear of a horizontal table. If the upper end be released, prove that at any instant during the fall the pressure on the table is three times the weight of the portion then on the table.

7. A chain rests across a smooth circular cylinder whose axis is horizontal, its length being equal to half the circumference. Prove that if it be slightly disturbed its velocity when a length $a\theta$ has slipped over the cylinder will be

$$\sqrt{\left[\frac{ga}{\pi}\{\theta^2+2(1-\cos\theta)\}\right]},$$

if a denote the radius of the cylinder.

8. A uniform chain whose ends are free slides on a smooth cycloid whose axis is vertical; prove that its middle point moves as if the whole mass were concentrated there.

9. A number of equal particles are at consecutive vertices of a regular polygon, being connected by inextensible strings. If one of the strings be suddenly jerked, prove that the tensions of any three successive strings are connected by the relation

$$T_{n+1}-2T_n+T_{n-1}=(T_{n+1}+2T_n+T_{n-1})\tan^2\alpha,$$

where α is half the angle subtended at the centre of the polygon by a side.

10. If a chain is set in motion by instantaneous tangential impulses T_1, T_2 at the two ends, prove that the kinetic energy generated is

$$\tfrac{1}{2}(T_1u_1+T_2u_2),$$

where u_1, u_2 are the initial tangential velocities of the two ends.

11. A uniform chain in the form of an arc of an equiangular spiral of angle α is set in motion by a tangential impulse at one end. Prove that the impulsive tension is given by the equation

$$T=As^{n_1}+Bs^{n_2},$$

where n_1, n_2 are the roots of the quadratic

$$n(n-1)=\tan^2\alpha.$$

If the chain extend to infinity in one direction prove that the direction of the initial motion of each point makes the same angle with the curve.

CHAPTER VIII

DYNAMICS OF RIGID BODIES. ROTATION ABOUT A FIXED AXIS

52. Introduction.

When we pass from the consideration of a system of discrete particles to that of continuous or apparently continuous distributions of matter, whether fluid or solid, we require some physical postulate in extension of the laws of motion which have hitherto been sufficient. These laws are in fact only definite so long as the bodies of which they are predicated can be represented by mathematical points.

As to the precise form in which this new physical assumption shall be introduced there is some liberty of choice. One plan is to assume that any portion whatever of matter may be treated as if it were constituted of mathematical points, separated by finite intervals, endowed with inertia-coefficients, and acting on one another with forces in the lines joining them, subject to the law of equality of action and reaction*. In the case of a 'rigid' body these forces are supposed to be so adjusted that the general configuration of the system is sensibly constant. On this basis we can at once predicate the principles of Linear and Angular Momentum, as developed in the preceding Chapter. These principles will be found to supply all that is generally necessary as a basis for the Dynamics of Rigid Bodies.

53. D'Alembert's Principle.

Another method is to assume a principle first stated in a general form by d'Alembert†.

* This is often referred to as 'Boscovich's hypothesis,' after R. G. Boscovich, author of a treatise on Natural Philosophy (Venice, 1758) in which this doctrine was taught.

† J. le R. d'Alembert (1717–1783). The principle appears in his *Traité de dynamique*, Paris, 1743.

The product of the mass m of a particle into its acceleration is a vector quantity which we may call the 'effective force on the particle.' By the Second Law of Motion it must be equal in every respect to the resultant of all the forces acting on m. In the case of a particle forming part of a material assemblage, these forces may be divided into two classes, viz. we have (1) the 'external forces' acting from without the assemblage, and (2) the 'internal forces' or reactions due to the remaining particles. Considering the whole assemblage, we may say, then, that the system of localized vectors which represent the effective forces is statically equivalent to the two systems of external and internal forces combined.

The assumption made by d'Alembert is that the internal forces form by themselves a system in equilibrium. It follows that the system of effective forces is as a whole statically equivalent to the system of *external* forces*. In particular, the sum of the effective forces on all the particles, resolved in any given direction, must be equal to the sum of the components of the external forces in that direction; and the sum of the moments of the effective forces about any axis must be equal to the sum of the moments of the external forces about that axis.

To express these results analytically, let (x, y, z) denote the position, relative to fixed rectangular axes, of a particle m, and (X, Y, Z) the external force on this particle. Since the components of the effective force on m are

$$m\ddot{x},\quad m\ddot{y},\quad m\ddot{z},$$

we have, resolving parallel to Oz,

$$\Sigma(m\ddot{z}) = \Sigma(Z), \qquad \text{.........................(1)}$$

and taking moments about Oz,

$$\Sigma(x \,.\, m\ddot{y} - y \,.\, m\ddot{x}) = \Sigma(xY - yX), \qquad \text{...........(2)}$$

where the summations embrace all the particles of the system [*S.* 60]. In the case of a continuous distribution of matter, the summations take the form of integrations.

* This is (virtually) the original formulation of the 'principle.'

These equations may be written

$$\frac{d}{dt}\Sigma(m\dot{z}) = \Sigma(Z), \quad\text{.......................(3)}$$

and

$$\frac{d}{dt}\Sigma(x \,.\, m\dot{y} - y \,.\, m\dot{x}) = \Sigma(xY - yX). \quad\text{.........(4)}$$

There are of course two other equations of each of these types.

Since the axis of z may have any position, the equations (3) and (4) express that the rate of increase of the total momentum in any given direction is equal to the total external force in that direction, and that the rate of increase of the angular momentum about any given axis is equal to the total moment of all the external forces about that axis.

It appears, then, that whichever form of fundamental assumption we adopt we are led immediately to the principles of linear and angular momentum as above stated. It is to be observed that no restriction to the case of *rigid* bodies is made, or implied, and that the principles in question are inferred to be of universal validity. The peculiar status of rigid bodies in dynamical theory is that these principles furnish equations equal in number to the degrees of freedom of such a body, whether in two or in three dimensions, and that they are accordingly generally sufficient for the discussion of dynamical problems in which only such bodies are involved. In other cases, as e.g. in Hydrodynamics and in the theory of Elastic Vibrations, auxiliary physical assumptions of a more special kind have to be introduced.

In the form most usually given to d'Alembert's principle it is asserted that the system of external forces is in equilibrium with that of the effective forces *reversed.* This is obviously equivalent to the previous statement. Problems of Dynamics are thus brought conveniently, but somewhat unnaturally, under the rules of Statics. A particular case will be already familiar to the student; the 'reversed effective force' on a particle m describing a circle of radius r with the constant angular velocity ω is simply the fictitious 'centrifugal force' $m\omega^2 r$ which is in equilibrium with the real forces acting on the particle.

As regards the postulates themselves, it must be recognized that both forms are open to criticism. The assumption explained in Art. 52 makes dynamical investigations depend on a particular view as to the ultimate structure of matter. On the other hand, it has been objected to d'Alembert's principle that it expresses a law of motion in terms of the rules of Statics, whereas on a more rational procedure the laws of equilibrium should appear as a simple corollary, dealing with a particular case, from the general principles of Dynamics*.

In the author's view it is best to postulate the principles of linear and angular momentum as such, regarding them as natural extensions of the Newtonian Laws of Motion, suggested, although not proved, by considerations such as those of Chap. VII. Since some assumption has in any case to be made, it seems best to make it directly in the form which is most convenient for further developments, and is at the same time independent of doubtful hypotheses.

Ex. 1. A self-propelled vehicle is driven from the hind wheels; it is required to find the maximum acceleration, having given the coefficient of friction (μ) between the wheels and the road.

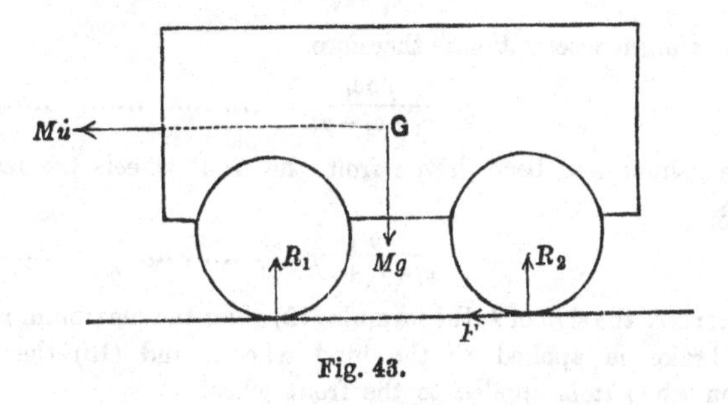

Fig. 43.

Let M be the total mass, G the centre of mass, h the height of G above the road, a the radius of the hind wheels, c_1, c_2 the distances of the vertical through G from the front and rear axles, respectively. If N be the couple which the engine exerts on the rear axle, and F the forward pull of the road on the hind wheels, preventing these from slipping backwards, we have, taking moments about the axle,

$$Fa = N, \qquad \text{.......................................(5)}$$

* Historically, d'Alembert's principle was of the greatest service as introducing a *general* method of treating dynamical questions. Previously to his time, problems of 'rigid' Dynamics had been dealt with separately, on the basis of special assumptions, which were more or less plausible, but sometimes disputed.

the rotatory inertia of the wheels being neglected. If u be the velocity of the vehicle, the momenta of the various parts are equivalent to a linear momentum Mu in a horizontal line through G, provided we neglect the inertia of the moving parts of the engine. Hence, resolving horizontally and vertically,

$$M\frac{du}{dt}=F, \qquad \text{.............................(6)}$$

and

$$R_1+R_2=Mg, \qquad \text{.............................(7)}$$

where R_1, R_2 are the vertical pressures of the ground on the front and hind pairs of wheels, respectively. Also, taking moments about G,

$$R_2c_2-R_1c_1=Fh. \qquad \text{.............................(8)}$$

Hence

$$\left.\begin{aligned} R_1&=\frac{Mgc_2}{c_1+c_2}-\frac{Fh}{c_1+c_2},\\ R_2&=\frac{Mgc_1}{c_1+c_2}+\frac{Fh}{c_1+c_2}.\end{aligned}\right\} \qquad \text{.............................(9)}$$

The effect of the propelling force is therefore to increase the pressure on the hind wheels, and to diminish that on the front wheels, by $Fh/(c_1+c_2)$.

In order that the wheels may not slip we must have

$$F \ngtr \mu R_2, \qquad \text{.............................(10)}$$

which gives

$$F \ngtr \frac{\mu Mgc_1}{c_1+c_2-\mu h}; \qquad \text{.............................(11)}$$

and the maximum acceleration is therefore

$$\frac{\mu gc_1}{c_1+c_2-\mu h}. \qquad \text{.............................(12)}$$

If the vehicle had been driven from the front wheels the result would have been

$$\frac{\mu gc_2}{c_1+c_2+\mu h}. \qquad \text{.............................(13)}$$

If we reverse the sign of h, the formula (12) gives the maximum retardation when a brake is applied to the hind wheels, and (13) the maximum retardation when it is applied to the front wheels.

As a numerical example of (12), let $c_1=8$, $c_2=4$, $h=3$, $\mu=\frac{1}{2}$. The result is $\frac{8}{45}g$, or about $5\frac{1}{2}$ ft./sec.^{2}.

Ex. 2. Let us suppose that the vehicle is propelled by a tractive force F acting in a horizontal line at a height h' above the ground. If we neglect friction at the axles, and the rotatory inertia of the wheels, there is now no tangential drag on the wheels. The equation (8) is replaced by

$$R_2c_2-R_1c_1=F(h-h'). \qquad \text{.............................(14)}$$

Combined with (7), this gives

$$\left.\begin{aligned} R_1(c_1+c_2)&=Mgc_2-F(h-h'),\\ R_2(c_1+c_2)&=Mgc_1+F(h-h').\end{aligned}\right\} \qquad \text{.............................(15)}$$

Hence if $h > h'$, the acceleration must not exceed

$$\frac{c_2}{h-h'}g, \qquad \text{.......................(16)}$$

or the value of R_1 required to satisfy (15) would be negative; i.e. the front wheels would jump off the ground. Similarly, when the motion of the vehicle is checked, the retardation must not exceed

$$\frac{c_1 g}{h-h'}, \qquad \text{.......................(17)}$$

or the hind wheels will jump. If $h < h'$, the words 'acceleration' and 'retardation,' must be interchanged in these statements, and the denominators replaced by $h'-h$. An absolutely sudden shock, which means an infinite acceleration, would cause a jump in any case unless h and h' were equal.

Ex. 3. A bar OA describes a cone of semi-angle α about the vertical through its upper end O, which is fixed; to find the requisite angular velocity ω.

This question is most simply treated by means of the fictitious centrifugal forces above referred to. If μ be the line-density at a distance x from O, the centrifugal force on an element $\mu\delta x$ will be $\mu\delta x \,.\, \omega^2 x \sin\alpha$, and its moment about O will be $\mu\delta x \,.\, \omega^2 x \sin\alpha \,.\, x\cos\alpha$. Hence, if a be the length of the bar, M its mass, and h the distance of the mass-centre from O,

$$\omega^2 \sin\alpha\cos\alpha \int_0^a \mu x^2 dx = Mgh\sin\alpha. \qquad \text{.......................(18)}$$

Hence if k be the radius of gyration about O,

$$\omega^2 = \frac{gh}{k^2\cos\alpha} = \frac{g}{l\cos\alpha}, \qquad \text{.......................(19)}$$

where l is the distance of the centre of oscillation from O.

54. Rotation about a Fixed Axis.

The position of a rigid body which is free only to turn about a fixed axis is specified by the angle θ which some plane through the axis, fixed in the body, makes with a standard position of this plane.

If in an interval δt this angle increases by $\delta\theta$, the quotient $\delta\theta/\delta t$ may be called the 'mean angular velocity' of the body in the interval. Proceeding to the limit, and writing

$$\omega = \frac{d\theta}{dt}, \qquad \text{.......................(1)}$$

ω is called the 'angular velocity' at the instant t.

Similarly, the differential coefficient

$$\frac{d\omega}{dt} \text{ or } \frac{d^2\theta}{dt^2} \quad\text{..........................(2)}$$

may be called the 'angular acceleration' of the body. If ω be regarded as a function of θ we have

$$\frac{d\omega}{dt} = \frac{d\omega}{d\theta}\frac{d\theta}{dt} = \omega\frac{d\omega}{d\theta}, \quad\text{....................(3)}$$

which is analogous to the formula (4) of Art. 2.

In the time δt a particle m of the body, whose distance from the fixed axis is r, describes a space $r\delta\theta$, and its velocity is therefore $rd\theta/dt$, or ωr. The momentum of this particle is $m\omega r$, at right angles to r and to the axis, and its moment about the axis is therefore $m\omega r . r$, or $mr^2\omega$. The total angular momentum is therefore

$$\Sigma(mr^2\omega) = \Sigma(mr^2) . \omega, \quad\text{.................(4)}$$

where the summation includes all the particles of the body.

The sum $\Sigma(mr^2)$ of the masses of the various particles multiplied by the squares of their respective distances from the axis is known as the 'moment of inertia' of the body with respect to the axis [*S.* 70]. Its value can be found in a few simple cases by integration [*S.* 71, 72]; in other cases it has to be ascertained, when required, by dynamical experiment (cf. Art. 57).

If k^2 be the 'mean square' of the distances of the particles from the axis in question [*S.* 70], i.e.

$$k^2 = \frac{\Sigma(mr^2)}{\Sigma(m)}, \quad\text{.........................(5)}$$

the linear magnitude k is called the 'radius of gyration' of the body with respect to the axis. If I be the moment of inertia, and M the total mass, we have

$$I = Mk^2; \quad\text{.........................(6)}$$

and the angular momentum may be expressed in either of the forms $I\omega$, $Mk^2\omega$.

Hence if N be the total moment of all the external forces with respect to the fixed axis of rotation, we have by the principle of angular momentum

$$\frac{d}{dt}(I\omega) = N. \quad\text{.........................(7)}$$

This may be compared with the equation of rectilinear motion of a body, viz.

$$\frac{d}{dt}(Mu) = X. \qquad \text{.........................(8)}$$

It appears that the constant I measures the inertia of the rigid body as regards rotation about the given axis, just as M measures its inertia as regards a motion of translation.

The kinetic energy of a particle m of the body is $\frac{1}{2}m(\omega r)^2$, and the total kinetic energy is therefore

$$\tfrac{1}{2}\Sigma(m\omega^2 r^2) = \tfrac{1}{2}\Sigma(mr^2)\,.\,\omega^2 = \tfrac{1}{2}Mk^2\omega^2 = \tfrac{1}{2}I\omega^2. \qquad \text{......(9)}$$

The latter form may be compared again with the expression $\frac{1}{2}Mu^2$ appropriate to the case of translation.

If we multiply the equation (7) by ω we have

$$I\omega\frac{d\omega}{dt} = N\omega = N\frac{d\theta}{dt}, \qquad \text{..................(10)}$$

or

$$\frac{d}{dt}(\tfrac{1}{2}I\omega^2) = N\frac{d\theta}{dt}. \qquad \text{....................(11)}$$

Since $N\delta\theta$ denotes [*S.* 51] the work done by the external forces when the body turns through an angle $\delta\theta$, this equation expresses that the kinetic energy is at any instant increasing at a rate equal to that at which work is being done on the body. Integrating, we infer that the increment of the kinetic energy in any interval of time is equal to the total work done by the external forces in that interval.

Ex. 1. If there are no external forces other than the constraining forces exerted by the axis, and if these have zero moment about the axis, as in the case of a perfectly smooth spindle, we have from (7)

$$\frac{d\omega}{dt} = 0, \quad \omega = \text{const.} \qquad \text{..........................(12)}$$

For instance, the angular velocity of the earth about its axis is constant, so far as the earth can be regarded as rigid, and its axis of rotation invariable.

Ex. 2. A fly-wheel, free to turn about a horizontal axis, carries a mass m suspended by a vertical string which is wrapped round an axle of radius b.

Let I be the moment of inertia of the fly-wheel about its axis, ω its angular velocity, u the downward velocity of m. Taking moments about the axis, and neglecting friction, we have

$$I\frac{d\omega}{dt} = Tb, \qquad \text{.................................(13)}$$

where T is the tension of the string. Considering the motion of m we have

$$m\frac{du}{dt}=mg-T. \qquad \text{..............................(14)}$$

Also, since when the wheel has turned through an angle $\omega\delta t$ a length $b\omega\delta t$ is added to the straight portion of the string,

$$u=b\omega. \qquad \text{..........................(15)}$$

Eliminating ω we find

$$\left(\frac{I}{b^2}+m\right)\frac{du}{dt}=mg; \qquad \text{....................(16)}$$

the acceleration of the particle m is therefore

$$\frac{du}{dt}=\frac{mb^2}{I+mb^2}g. \qquad \text{........................(17)}$$

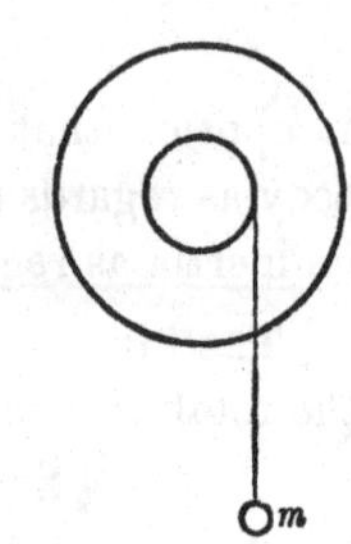

Fig. 44.

The tension is

$$T=\frac{I}{I+mb^2}mg. \qquad \text{..............................(18)}$$

55. The Compound Pendulum.

This term is applied to a rigid body of any form and constitution which is free to turn about a fixed horizontal axis, the only external forces being gravity and the pressures exerted by the axis. These pressures are supposed to have zero moment about the axis, friction being neglected.

Let θ be the angle which a plane through the axis and the mass-centre G makes with the vertical. If M be the mass of the body, and h the distance of G from the axis, the moment of the external forces about the axis, tending to increase of θ, is $-Mgh\sin\theta$. Hence if I be the moment of inertia about the axis we have, by Art. 54 (7),

$$I\frac{d^2\theta}{dt^2}=-Mgh\sin\theta. \qquad \text{......(1)}$$

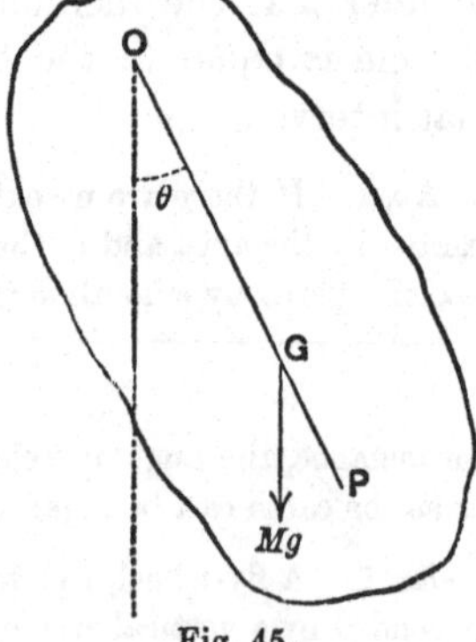

Fig. 45.

This is exactly the same as for a simple pendulum of length l (Art. 37), provided

$$l=I/Mh. \qquad \text{...........(2)}$$

If k be the radius of gyration about the axis, so that $I=Mk^2$, we have

$$l=k^2/h. \qquad \text{..........................(3)}$$

The plane of Fig. 45 is supposed to be a vertical plane through G, perpendicular to the axis and meeting it in O. If we produce OG to P, making $OP = l$, the point P is called the 'centre of oscillation.' The bob of a simple pendulum of length l, hanging from the same axis, will keep pace exactly with the point P if started with it, with the same velocity.

If κ be the radius of gyration about an axis through G parallel to the axis of suspension, we have [S. 73]

$$k^2 = \kappa^2 + h^2, \qquad (4)$$

and therefore

$$l = h + \kappa^2/h, \qquad (5)$$

or

$$GP \,.\, GO = \kappa^2. \qquad (6)$$

The symmetry of this relation shews that if the body were suspended from a parallel axis through P, the point O would become the new centre of oscillation. This is often expressed by saying that the centres of suspension and oscillation are convertible*.

For different parallel axes of suspension the period of a small oscillation will vary as $\sqrt{l}$, or $\sqrt{(GO + GP)}$. Since the product of GP and GO is fixed by (6), their sum is least when they are equal. Hence, considering any system of parallel axes, the period is least when $h = \kappa$, and therefore $l = 2\kappa$.

If the axis of a compound pendulum be tilted so as to make an angle β with the vertical, the mass-centre will oscillate in a plane making an angle β with the horizontal. The gravity Mg of the pendulum may be resolved into two components, viz. $Mg \sin\beta$ in this plane, parallel to a line of greatest slope, and $Mg \cos\beta$ normal to the plane. The latter has zero moment about the axis; and the motion is therefore the same as if the acceleration of gravity were altered from g to $g \sin\beta$. The length of the equivalent simple pendulum is therefore now equal to $k^2/(h \sin\beta)$. If β be small, as in the 'horizontal' pendulums used in seismographs, this is large, and the period correspondingly long.

The same conclusion follows also from consideration of energy. When the pendulum turns through an angle θ from its equilibrium position, the projection of OG on a line of greatest slope is $h \cos\theta$,

* The proposition is due to Huygens, *l. c. ante* p. 112.

and the depth of G below the level of O is therefore $h \cos\theta \sin\beta$. The potential energy is therefore increased by

$$Mgh(1-\cos\theta)\sin\beta.$$

Ex. When a horizontal trap-door is released, the acceleration of a point Q of it is less or greater than g, according as the distance of Q from the line of hinges is less or greater than l. Hence a weight originally resting on the door at a distance greater than l will immediately begin to fall freely.

56. Determination of g.

Methods for the determination of the value of g at any place are based on the formula

$$g = \frac{4\pi^2 l}{T^2} \quad \text{..........................(1)}$$

of Art. 11. The ideal simple pendulum there contemplated cannot of course be realized, and in practice l is the length of the simple pendulum 'equivalent' to some form of compound pendulum which is actually employed.

The period T of a complete small oscillation can be found with great accuracy by counting a large number of swings, and noting the time which they take. The practical difficulty lies chiefly in the evaluation of l. For the purpose of an accurate measurement of g two distinct plans have been followed.

In the first method a pendulum of some geometrically simple form is adopted for which the value of l can be found by calculation, using the formula (5) of Art. 55 and the values of h and k proper to the particular form. Thus in the case of a homogeneous sphere of radius a suspended by a fine wire of length λ, we have $\kappa^2 = \frac{2}{5}a^2$ [*S.* 72], and therefore

$$l = \lambda + a + \frac{2}{5}\frac{a^2}{\lambda + a}, \quad \text{......................(2)}$$

if the mass of the wire be neglected. If the mass (m) of the wire (supposed to be uniform) is taken into account we have, in the notation of Art. 55,

$$(M+m)k^2 = M(\lambda+a)^2 + \tfrac{2}{5}Ma^2 + \tfrac{1}{3}m\lambda^2, \quad \text{........(3)}$$

$$(M+m)h = M(\lambda+a) + \tfrac{1}{2}m\lambda, \quad \text{....................(4)}$$

where M is the mass of the sphere [S. 73]. Hence

$$l = \frac{k^2}{h} = \frac{M(\lambda + a)^2 + \frac{2}{5}Ma^2 + \frac{1}{3}m\lambda^2}{M(\lambda + a) + \frac{1}{2}m\lambda}. \quad \text{.............(5)}$$

Careful experiments on this plan were made by Borda and Cassini*. The theoretical value $\frac{2}{5}Ma^2$ for the moment of inertia of the sphere about a diameter rests of course on the assumption that the sphere is homogeneous. Any slight defect in this respect may be eliminated by varying the point of the sphere at which the wire is attached. Some care has also to be taken as to the mode of suspension. If the wire is *clamped* at the upper end, its stiffness comes into play, and it is not easy to say what is the precise 'point' of suspension. In Borda's experiments the wire was attached to a miniature compound pendulum furnished with a knife-edge resting on horizontal plates. The period of this miniature pendulum was adjusted so as to be as nearly equal as possible to that of the sphere and wire. The influence on the period of the latter system is then negligible, and the knife-edge may be taken as the point of suspension.

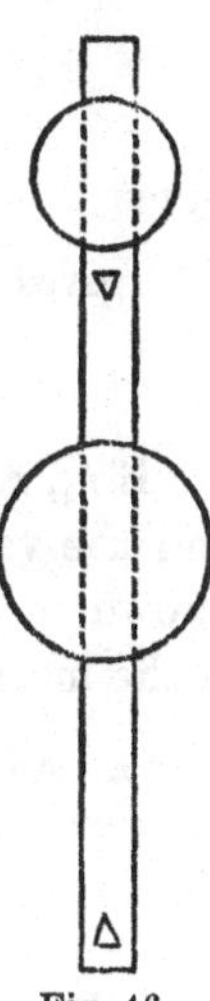

Fig. 46.

The second method above referred to is based on the principle of the convertibility of the centres of suspension and oscillation. A pendulum, whose precise form is unimportant, is constructed with two knife-edges facing one another, as nearly as possible in the same plane with the mass-centre G, and at distinctly unequal distances from this point. If it could be contrived that the period of a small oscillation should be exactly the same from whichever knife-edge the pendulum is suspended, the two edges would be in the positions of conjugate centres of suspension and oscillation, and the distance between them would give the value of l. For if

$$\frac{\kappa^2}{h_1} + h_1 = \frac{\kappa^2}{h_2} + h_2 = l, \quad \text{....................(6)}$$

we have

$$\left(\frac{\kappa^2}{h_1 h_2} - 1\right)(h_1 - h_2) = 0. \quad \text{.................(7)}$$

* *Base du système métrique*, 1810.

Hence, except in the case of $h_1 = h_2$ which has been excluded, we have*

$$\kappa^2 = h_1 h_2, \quad l = h_1 + h_2. \qquad\qquad(8)$$

With a view to the requisite adjustment, the position of one of the knife-edges is variable, or else the pendulum has a sliding weight, or both expedients may be provided.

The adjustment can however never be quite exact, and a correction is required. If T_1, T_2 be the observed periods, which are nearly, but not quite, equal, and if l_1, l_2 be the lengths of the corresponding simple pendulums, we have

$$l_1 = \frac{\kappa^2}{h_1} + h_1, \quad l_2 = \frac{\kappa^2}{h_2} + h_2, \qquad\qquad(9)$$

whence, eliminating κ,

$$h_1 l_1 - h_2 l_2 = h_1^2 - h_2^2. \qquad\qquad(10)$$

This may be written

$$\frac{1}{2}\frac{l_1 + l_2}{h_1 + h_2} + \frac{1}{2}\frac{l_1 - l_2}{h_1 - h_2} = 1. \qquad\qquad(11)$$

Since

$$l_1 = gT_1^2/4\pi^2, \quad l_2 = gT_2^2/4\pi^2, \qquad\qquad(12)$$

this gives

$$\frac{4\pi^2}{g} = \frac{\frac{1}{2}(T_1^2 + T_2^2)}{h_1 + h_2} + \frac{\frac{1}{2}(T_1^2 - T_2^2)}{h_1 - h_2}. \qquad\qquad(13)$$

If h_1, h_2 are distinctly unequal, the last term is relatively small, and the values of h_1, h_2 which occur in it need not therefore be known very accurately. The denominator $h_1 + h_2$ of the first term is the measured distance between the knife-edges.

Ex. As a numerical example, let

$$T_1 = 1{\cdot}8484 \text{ sec.}, \quad T_2 = 1{\cdot}8478 \text{ sec.},$$

$$h_1 + h_2 = 84{\cdot}88 \text{ cm.}, \quad h_1 - h_2 = 55 \text{ cm.}$$

We find

$$\frac{4\pi^2}{g} = {\cdot}040239 + {\cdot}000020,$$

whence

$$g = 980{\cdot}6 \text{ cm./sec.}^2.$$

An error even of 5 cm. in the estimated value of $h_1 - h_2$ would not affect the calculated result to the extent to which it is carried, which is all that the data warrant.

* This method had been suggested apparently by Bohnenberger, but was first employed by Captain Kater (*Phil. Trans.*, 1818).

57. Torsional Oscillations.

In various physical experiments a body is suspended by a vertical wire the upper end of which is clamped, and makes torsional oscillations about the axis of the wire. We will assume that this axis is a principal axis of inertia of the body at its mass-centre (Art. 59).

When the body is turned through an angle θ, the twist per unit length of the wire is θ/l, where l is the total length. The restoring couple called into play by the elasticity of the wire is therefore $K\theta/l$, where K is a constant, called the 'modulus of torsion of the wire.' In terms of the rigidity (μ) of the material of the wire, and the radius a of the cross-section, we have [S. 151]

$$K = \tfrac{1}{2}\pi\mu a^4. \qquad (1)$$

In the case of free vibrations we have

$$I\frac{d^2\theta}{dt^2} = -\frac{K\theta}{l}, \qquad (2)$$

where I is the moment of inertia of the suspended body about the axis of rotation*. The motion is therefore simple-harmonic, with the period

$$T = 2\pi\sqrt{\left(\frac{Il}{K}\right)}. \qquad (3)$$

This gives an experimental method of determining K, if I be known, thus

$$K = \frac{4\pi^2 Il}{T^2}. \qquad (4)$$

The value of μ can thence be inferred by (1).

If I be unknown, a second experiment may be made in which a body of regular form, e.g. a rectangular or cylindrical bar placed horizontally, or a thin cylindrical shell placed with its axis vertical, is attached symmetrically to the suspended body. If the period is thus increased to T', we have

$$K = \frac{4\pi^2 (I + I_0)\, l}{T'^2}, \qquad (5)$$

* It is here assumed that K, and therefore μ, is expressed in dynamical measure.

where I_0 is the moment of inertia of the attached body as found by weighing and calculation. Combined with (4), this gives

$$K = \frac{4\pi^2 I_0 l}{T'^2 - T^2}. \quad \text{.........................(6)}$$

We are also able to infer the value of I; thus

$$\frac{I}{I_0} = \frac{T^2}{T'^2 - T^2}. \quad \text{.........................(7)}$$

58. Bifilar Suspension.

A bar is suspended horizontally by two equal vertical strings, its mass-centre G being in the plane of these, and half-way between them.

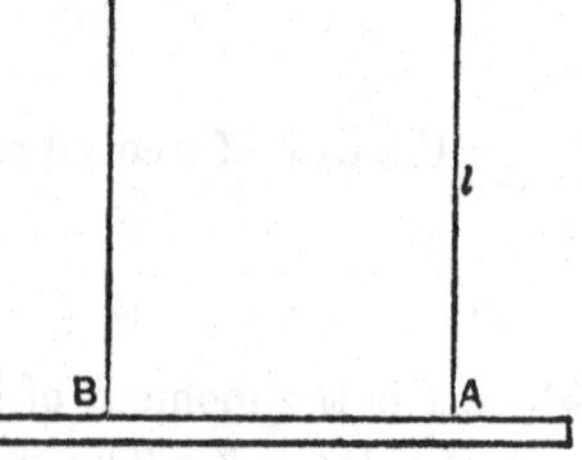

Let M be the mass of the bar, κ its radius of gyration about a vertical axis through G, $2b$ the distance between the strings when vertical, l their length.

When the bar is turned through a *small* angle θ about the vertical through G, the lower end of each string describes a small arc which may be taken to be sensibly horizontal and equal to $b\theta$; and the inclination of each string to the vertical becomes $b\theta/l$, approximately.

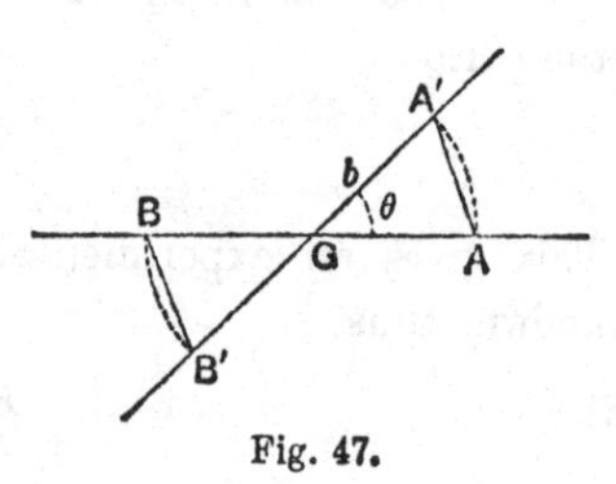

Fig. 47.

The vertical displacement of G being of the second order in θ, the tension of each string is approximately constant and equal to $\frac{1}{2}Mg$. The horizontal component of each tension is therefore $\frac{1}{2}Mgb\theta/l$, nearly. Since these components are sensibly perpendicular to the bar, they give a restoring couple $Mgb^2\theta/l$. Hence, in a free oscillation,

$$M\kappa^2 \frac{d^2\theta}{dt^2} = -Mg\frac{b^2}{l}\theta, \quad \text{.......................(1)}$$

or

$$\frac{d^2\theta}{dt^2} + \frac{gb^2}{\kappa^2 l}\theta = 0. \quad \text{.......................(2)}$$

The period is therefore

$$T = 2\pi \sqrt{\left(\frac{\kappa^2 l}{g b^2}\right)}. \quad\text{.......................(3)}$$

59. Reactions on a Fixed Axis.

In general a body rotating about a fixed axis will exert certain pressures, or reactions, on this axis.

Let us first suppose that there are no external forces on the body, other than the constraining forces at the axis, which are of course equal and opposite to the reactions under consideration. We have seen that, frictional forces being neglected, the angular velocity ω is then constant.

Instead of calculating the rates of change of angular momentum about fixed axes, it is somewhat simpler in the present case to have recourse to d'Alembert's principle, according to which the constraining forces are in equilibrium with the whole system of reversed effective forces. Since ω is constant, the reversed effective force on a particle m at a distance r from the axis will be the 'centrifugal force' $m\omega^2 r$, outwards. If we take rectangular axes of coordinates, such that Oz coincides with the axis of rotation, the three components of this centrifugal force will be

$$m\omega^2 x, \quad m\omega^2 y, \quad 0, \quad\text{.....................(1)}$$

where x, y, z are the coordinates of m. The moments of this force about the coordinate axes will therefore be

$$-m\omega^2 yz, \quad m\omega^2 xz, \quad 0, \quad\text{.................(2)}$$

respectively. The reactions in question must therefore have components

$$\omega^2 . \Sigma(mx), \quad \omega^2 . \Sigma(my), \quad 0,$$

or

$$\omega^2 . M\bar{x}, \quad \omega^2 . M\bar{y}, \quad 0, \quad\text{.....................(3)}$$

where M denotes the total mass, and the coordinates $\bar{x}$, $\bar{y}$ refer to the mass-centre. Again, the moments of the reactions about Ox, Oy, Oz must be

$$-\omega^2 . \Sigma(myz), \quad \omega^2 . \Sigma(mxz), \quad 0. \quad\text{...........(4)}$$

If we imagine the axes of x and y to revolve with the body, the coefficients of ω^2 in these expressions will be constants, depending only on the distribution of mass in the body.

It appears, then, that a body which is free to turn about a fixed *point* O, when set rotating about any given axis (Oz) through O, will not continue so to rotate unless the expressions (4) vanish. Constraining forces would be required whose moments are equal and opposite to these. Hence in order that Oz may be a possible axis of free permanent rotation, we must have

$$\Sigma (myz) = 0, \quad \Sigma (mxz) = 0. \qquad \text{.................(5)}$$

The axis in question is then said to be an axis of 'spontaneous rotation,' or a 'principal axis of inertia,' at the point O. It was in this connection that the theory of principal axes of inertia was originated*.

When the conditions (5) are fulfilled, the components of the reaction on the fixed point O are given by (3).

Since the 'products of inertia' $\Sigma (myx)$, $\Sigma (mxz)$ measure the tendency of the axis of rotation to deviate from its original direction, they are sometimes called 'deviation moments.'

Returning to the case of rotation about a fixed *axis* Oz, let us suppose that the requisite constraint is exerted by means of two smooth bearings, and consists accordingly of two forces (P_1, Q_1, 0), (P_2, Q_2, 0), respectively. Let us further suppose that the axis of x is chosen so as to pass through the mass-centre G. If the distances

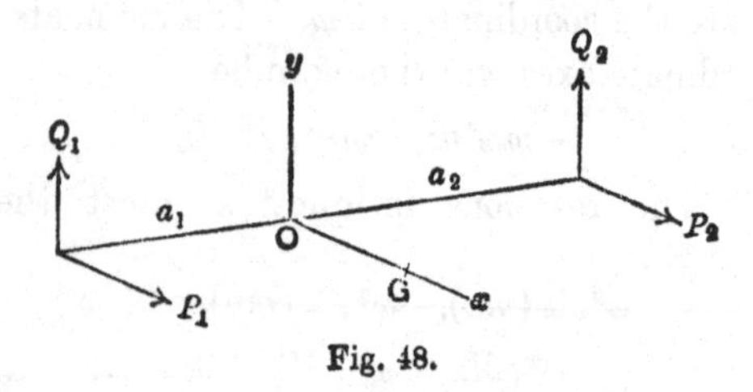

Fig. 48.

of the two bearings from O are a_1, a_2, on opposite sides, we have

$$P_1 + P_2 = -\omega^2 . M\bar{x}, \quad Q_1 + Q_2 = 0, \qquad \text{...........(6)}$$

$$-Q_1 a_1 + Q_2 a_2 = \omega^2 . \Sigma (myz), \quad P_1 a_1 - P_2 a_2 = -\omega^2 . \Sigma (mxz). \qquad \text{....(7)}$$

From these formulæ P_1, P_2, Q_1, Q_2 can be found. Since the directions of these forces revolve with the body, there is in

* By J. A. Segner (1755).

general a periodic stress on the bearings. If the period $2\pi/\omega$ coincides approximately with a natural period of elastic vibration of the supports, violent oscillations may be produced, on the principle illustrated in Art. 13. These considerations are of great importance in connection with the 'balancing' of modern high-speed machinery. It is important, not only that the mass-centre should lie in the axis of rotation, but also that the latter should be a principal axis. Under these conditions we have

$$P_1, P_2, Q_1, Q_2 = 0.$$

Ex. A uniform rectangular plate $ACBC'$ rotates about the diagonal AB, being constrained by smooth bearings at A and B.

The reactions (P) exerted by these bearings will evidently be equal and opposite, and in the plane of the rectangle. The plate is equimomental [*S.* 78] with four particles each of mass $\frac{1}{6}M$ at the middle points of the sides and a particle $\frac{1}{3}M$ at the middle point (O) of AB, M denoting the total mass. Hence, taking moments about O we have, if N be the orthogonal projection of C on AB,

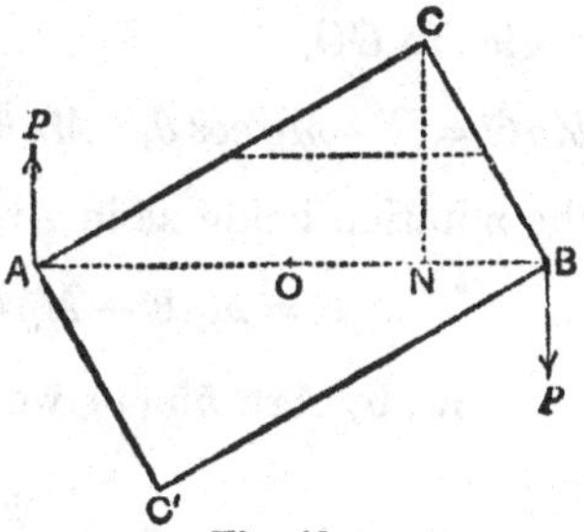

$$P \, . \, AB = 2 \times \tfrac{1}{6} M\omega^2 \, . \, \tfrac{1}{2} CN \, . \, ON.$$

Fig. 49.

Now
$$ON = \frac{AC^2 - CB^2}{2AB}, \qquad CN = \frac{AC \, . \, CB}{AB}.$$

Hence, putting $AC = a$, $BC = b$, we find

$$P = \tfrac{1}{12} M\omega^2 \, . \, \frac{ab\,(a^2 - b^2)}{(a^2 + b^2)^{\frac{3}{2}}}. \quad \text{.........................(8)}$$

This vanishes, as it ought, if $a = b$.

60. Application to the Pendulum.

The methods of the preceding Art. are easily adapted to the case where there are external forces on the body, in addition to the constraining forces of the fixed axis.

For simplicity we will take the case of the compound pendulum, and assume that the axis of suspension is a principal axis of inertia at the point O, which was defined in Art. 55 as the intersection of the axis with a plane normal to it through the mass-centre. The condition will obviously be fulfilled if this plane is a plane of symmetry of the body.

The principle of linear momentum leads easily to the required results. The reaction of the axis on the pendulum will now reduce to a single force through O. This is conveniently resolved into a component R in the direction GO, and a component S at right angles to GO, as in the figure.

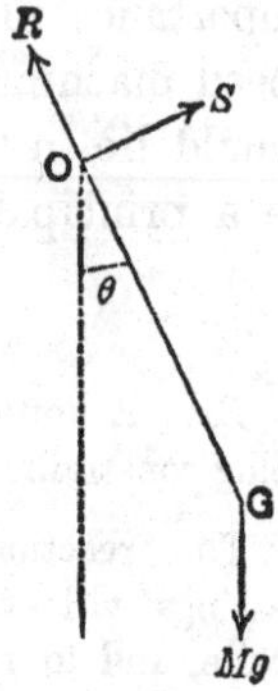

Since the motion of G is the same as if all the mass were concentrated there, and acted on by all the external forces, parallel to their actual directions, we have, resolving along and at right angles to GO,

$$Mh\dot{\theta}^2 = R - Mg\cos\theta, \quad Mh\ddot{\theta} = S - Mg\sin\theta, \text{ ...(1)}$$

Fig. 50.

the notation being as in Art. 55. Hence

$$R = Mh\dot{\theta}^2 + Mg\cos\theta, \quad S = Mh\ddot{\theta} + Mg\sin\theta. \text{ ......(2)}$$

Now, by Art. 55 (1), we have

$$\ddot{\theta} = -\frac{gh}{k^2}\sin\theta. \text{ ....(3)}$$

Again, the equation of energy is

$$\tfrac{1}{2}Mk^2\dot{\theta}^2 = Mgh\cos\theta + \text{const.}, \text{ ....(4)}$$

the right-hand side denoting the work done by gravity. Hence

$$\dot{\theta}^2 = \frac{2gh}{k^2}(\cos\theta + C), \text{ ....(5)}$$

where C is some constant. Substituting from (3) and (4) in (2), we find

$$R = Mg\cos\theta + \frac{2Mgh^2}{k^2}(\cos\theta + C), \text{ ....(6)}$$

$$S = Mg\left(1 - \frac{h^2}{k^2}\right)\sin\theta = \frac{Mg\kappa^2}{k^2}\sin\theta, \text{ ....(7)}$$

by Art. 55 (4).

The value of C will depend on the initial conditions. If the pendulum comes to rest at an inclination α, we have $C = -\cos\alpha$, and therefore

$$\frac{R}{Mg} = \cos\theta + \frac{2h}{l}(\cos\theta - \cos\alpha), \quad \text{..............(8)}$$

$$\frac{S}{Mg} = \left(1 - \frac{h}{l}\right)\sin\theta, \quad \text{..........................(9)}$$

where l is the length of the equivalent simple pendulum.

Ex. In the case of a uniform bar which just makes complete revolutions we have $\alpha = \pi$, $l = \frac{4}{3}h$, whence

$$\frac{R}{Mg} = \tfrac{1}{2}(3 + 5\cos\theta), \quad \frac{S}{Mg} = \tfrac{1}{4}\sin\theta. \quad \text{.....................(10)}$$

The component R changes sign when $\cos\theta = -\frac{3}{5}$, or $\theta = 127^\circ$, about.

EXAMPLES. XIII.

(Rotation about a Fixed Axis.)

1. A uniform circular disk 1 ft. in diameter weighs 10 lbs.; find (in ft.-lbs.) the couple which in 10 sec. would generate an angular velocity of 10 revolutions per sec. [·245.]

2. Find in kilogramme-metres the energy of a uniform solid sphere of iron, a decimetre in radius, spinning about a diameter at the rate of 5 revolutions per second, assuming that the density of iron is 7·8 and that

$g = 980$ cm./sec.^{2}. [6·57.]

3. A gyroscope is set spinning by means of a string 2 ft. long, wrapped round the axle, which is pulled with a tension of 10 lbs. Find, in revolutions per second, the angular velocity generated, assuming that the gyroscope is equivalent to a circular disk 4 in. in diameter and weighing 3 lbs. [29.]

4. A uniform circular disk of radius 10 cm., and weight 1 kg., can turn freely about its axis, which is horizontal. When started at 100 revolutions per minute it is reduced to rest in 1 min., by a frictional force applied tangentially to the rim. Assuming this force to be constant, find its magnitude in gms. [·89.]

5. If the earth contract uniformly by cooling, prove that, when the radius has diminished by the $1/n$th part, the length of the day will have diminished by the $2/n$th part.

6. The mass of a fly-wheel is 20 lbs., and a mass of 1 lb. hangs by a string wrapped round the axle, which is horizontal. This mass is observed to descend through 5 ft. from rest in 8 sec. Find the radius of gyration of the fly-wheel, having given that the radius of the axle is 2 in. [6·4 in.]

7. A fly-wheel whose moment of inertia is I has an axle of radius a, and is rotating with an angular velocity ω. As it rotates it winds up on the axle a light string which is attached to a mass M resting on the floor below. Find the ratios in which the angular velocity of the wheel, and the kinetic energy of the system, are instantaneously reduced when the string becomes tight.

8. A weight hangs from an axle of radius b, and is maintained in equilibrium by a force P applied tangentially to the circumference of a concentric wheel of radius a. Shew that if a force P' be substituted for P the weight will ascend with acceleration

$$\frac{(P'-P)\,ab}{I+Pab/g},$$

where I is the moment of inertia of the wheel and axle.

9. Two masses M_1, M_2 are connected by a string passing over a pulley of moment of inertia I and radius a, as in Atwood's machine. Prove that when the system is running freely the pressure of the pulley on its bearings is less than if the mass M_1+M_2 had been equally divided between the two sides, by the amount

$$\frac{(M_1-M_2)^2\,g}{M_1+M_2+I/a^2}.$$

10. In a machine without friction and inertia a weight P balances a weight W, both hanging by vertical cords. These weights are replaced by P', and W', which in the subsequent motion move vertically. Prove that the mass-centre of P' and W' will descend with acceleration

$$\frac{(PW'-P'W)^2}{(P^2W'+P'W^2)(P'+W')}\cdot g.$$

11. A pole is supported at its lower end, which is carried round in a horizontal circle of radius c with constant angular velocity ω. Prove that it can maintain a constant inclination a to the vertical provided

$$\omega^2\left(\frac{ac}{\sin a}-k^2\right)=\frac{ga}{\cos a},$$

where k is the radius of gyration with respect to the lower end, and a the distance of the mass-centre from this end.

EXAMPLES. XIV.

(Compound Pendulum, etc.)

1. A uniform sphere whose diameter is 10 cm. hangs by a fine string 1 metre long. Find the length of the equivalent simple pendulum.

[105·095 cm.]

2. A wheel rests with the inner face of its rim on a transverse horizontal knife edge, whose distance from the centre is 3 ft. 2 in. The period of a small oscillation about the knife edge is found to be 2·65 sec. Find the radius of gyration of the wheel about the centre. (Assume $g=32{\cdot}2$.) [2 ft. 10·2 in.]

3. Two compound pendulums of masses M, M' can swing about the same horizontal axis. The distances of the centres of gravity from this axis are h, h'; and the lengths of the equivalent simple pendulums are l, l'. Prove that if the pendulums be fastened together, they will oscillate like a simple pendulum of length

$$\frac{Mhl+M'h'l'}{Mh+M'h'}.$$

4. A bar bent into the form of an arc of a circle swings in a vertical plane about its middle point. Prove that the length of the equivalent simple pendulum is equal to the diameter of the circle.

5. A bar whose length is one metre is suspended horizontally by two equal vertical strings attached to the ends. When it swings in the direction of its length the period of a small oscillation is 3·17 secs.; and when it makes angular oscillations about the vertical through its centre (which is also the centre of mass) the period is 1·85 sec. Find its radius of gyration about the centre. [29·2 cm.]

6. Prove that in the compound pendulum (Fig. 45) the whole mass may be supposed concentrated in two particles situate at O and P, without altering the period about any axis parallel to the actual axis. What are the masses of these particles?

$$\left[\frac{M\kappa^2}{\kappa^2+h^2}, \frac{Mh^2}{\kappa^2+h^2}.\right]$$

7. A compound pendulum carries a small shelf; prove that the effect of placing a small weight on the shelf will be to lengthen or shorten the period according as the shelf is below or above the centre of oscillation.

8. Two particles m, m' are connected by a light rod of length l, and attached to a fixed point O by strings of lengths r, r', respectively. Form the equation of energy for motion in a vertical plane through O.

Prove that the system, if slightly disturbed in this plane from the position of equilibrium, will make small oscillations with a period equal to that of a simple pendulum of length

$$(mr^2+m'r'^2)/(m+m')h,$$

provided

$$h^2=\frac{mr^2+m'r'^2}{m+m'}-\frac{mm'l^2}{(m+m')^2}.$$

9. A circular hoop is suspended from fixed points by three equal vertical strings of length l, so that its plane is horizontal. Prove that the period of a small rotational oscillation about the vertical through the centre is

$$2\pi\sqrt{(l/g)}.$$

10. A horizontal bar is suspended by two equal vertical strings of length l, which are attached to it at unequal distances a, b from the centre of mass (G). Prove that the bar can perform small oscillations about a vertical axis through G, and that the length of the equivalent simple pendulum is

$$\frac{\kappa^2 l}{ab},$$

when κ is the radius of gyration about the aforesaid axis.

11. A uniform bar 3 ft. long hangs from its upper end, which is fixed. What velocity must be given to the lower end in order that the bar may just reach the position of unstable equilibrium? [24 ft./sec.]

12. A uniform bar just makes a complete revolution about one end in a vertical plane; find the pressures on the hinge (1) in the lowest position, (2) in the horizontal position. [(1) $4mg$; (2) $\frac{3}{2}mg$ horizontal, $\frac{1}{4}mg$ vertical.]

13. A compound pendulum is released from rest with the centre of gravity at the same level as the axis. At what inclination is the *horizontal* pressure on the hinge greatest? [45°.]

14. A mass m rests on a horizontal trap-door at a distance x from the line of hinges. If M be the mass of the door, and k its radius of gyration about the line of hinges, prove that when the door is released the initial pressure of the mass m on the door is changed to

$$\frac{Mmg\,(k^2 - hx)}{Mk^2 + mx^2},$$

provided $x < k^2/h$.

15. A uniform elliptic plate whose semi-axes are a, b is set in rotation about a fixed axis which coincides with a diameter. Prove that the reactions on this axis are equivalent to a couple

$$\tfrac{1}{8} M\omega^2 (a^2 - b^2) \sin 2\theta,$$

where θ is the angle which the fixed diameter makes with the axis of x.

16. A door 3 feet wide, of uniform thickness, when opened through 90° and left to itself, shuts in 2 secs.; prove that the line of hinges makes an angle of about 3° with the vertical.

CHAPTER IX

DYNAMICS OF RIGID BODIES (CONTINUED). MOTION IN TWO DIMENSIONS

61. Comparison of Angular Momenta about Parallel Axes.

If we compare the angular momenta of any material system with respect to two parallel axes, one of which passes through the mass-centre (G) of the system, we are led to the following kinematical theorem:

The angular momentum about any axis is equal to the angular momentum, about that axis, of the whole mass supposed collected at G and moving with this point, together with the angular momentum, with respect to a parallel axis through G, of the motion relative to G.

For, writing as usual

$$x = \bar{x} + \xi, \quad y = \bar{y} + \eta, \quad z = \bar{z} + \zeta, \quad \text{..............(1)}$$

for the coordinates of a particle m referred to fixed rectangular axes, the angular momentum about the axis of z is, by Art. 53,

$$\Sigma m (x\dot{y} - y\dot{x}) = \Sigma m \left\{(\bar{x} + \xi)\left(\frac{d\bar{y}}{dt} + \dot{\eta}\right) - (\bar{y} + \eta)\left(\frac{d\bar{x}}{dt} + \dot{\xi}\right)\right\}$$

$$= \Sigma(m) \,.\left(\bar{x}\frac{d\bar{y}}{dt} - \bar{y}\frac{d\bar{x}}{dt}\right) + \Sigma m (\xi\dot{\eta} - \eta\dot{\xi}), \quad \text{.........(2)}$$

the omitted terms vanishing in consequence of the relations

$$\Sigma(m\xi) = 0, \quad \Sigma(m\eta) = 0,$$

$$\Sigma(m\dot{\xi}) = 0, \quad \Sigma(m\dot{\eta}) = 0,$$

as in the theorem of Art. 46.

The first term in the last member of (2) is the angular momentum about Oz of a mass $\Sigma(m)$ moving with G, and the second term is the angular momentum, about a line through G parallel to Oz, of the motion relative to G. Since any line whatever may be taken as the axis of z, the theorem follows.

The proof in vector notation is instructive. If P be the position of any particle m of the system, the momentum $(m\mathbf{v})$ of m may be resolved into two components through P, by the formula

$$m\mathbf{v} = m\bar{\mathbf{v}} + m\boldsymbol{v}, \qquad (3)$$

where $\bar{\mathbf{v}}$ is the velocity of G, and $\boldsymbol{v}$ the velocity of m relative to G. The components $m\bar{\mathbf{v}}$ are a series of localized parallel vectors proportional to the respective masses m, and are therefore equivalent to a single localized vector $\Sigma(m).\bar{\mathbf{v}}$ through G. The moment of momentum about any axis is therefore equal to the moment of this vector, together with the sum of the moments of the vectors $m\boldsymbol{v}$ supposed localized in lines passing through the respective points P. And since $\Sigma(m\boldsymbol{v}) = 0$ the sum of the moments of the latter series of vectors about all parallel axes is the same.

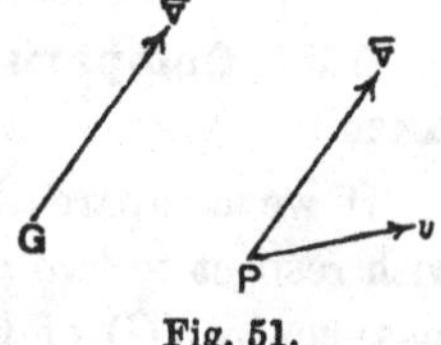

Fig. 51.

As a particular case of (2), if the axis of z be taken to pass through G, we have, putting $\bar{x} = 0$, $\bar{y} = 0$,

$$\Sigma m(x\dot{y} - y\dot{x}) = \Sigma m(\xi\dot{\eta} - \eta\dot{\xi}). \qquad (4)$$

Hence in calculating the moment of the momentum of a system about any axis through the mass-centre, it makes no difference whether we employ the actual momenta of the various particles, or whether we ignore the motion of G itself, and take account only of the momenta relative to G. The same thing follows at once from the vector proof just given.

Ex. In the compound pendulum (Art. 55) the angular momentum, about an axis through G parallel to the axis of suspension, of the motion relative to G is obviously $M\kappa^2\omega$. The moment of momentum of the whole mass, supposed concentrated at G, and moving with G, with respect to the axis of suspension would be $Mh\omega \,.\, h$. The total angular momentum about the fixed axis is therefore

$$Mh^2\omega + M\kappa^2\omega = M(h^2 + \kappa^2)\,\omega = Mk^2\omega,$$

as already found.

62. Rate of Change of Angular Momentum.

If we differentiate the equation (2) of Art. 61 with respect to t, we have

$$\frac{d}{dt}\Sigma m(x\dot{y}-y\dot{x}) = \Sigma(m).\left(\bar{x}\frac{d^2\bar{y}}{dt^2}-\bar{y}\frac{d^2\bar{x}}{dt^2}\right)+\frac{d}{dt}\Sigma m(\xi\dot{\eta}-\eta\dot{\xi}), \quad (1)$$

two terms which cancel being omitted*. An interpretation of this formula, analogous to that of the formula from which it is derived, may easily be framed; but we are chiefly concerned with the particular case where the axis Oz passes through the instantaneous position of G. On this hypothesis the formula reduces to

$$\frac{d}{dt}\Sigma m(x\dot{y}-y\dot{x}) = \frac{d}{dt}\Sigma m(\xi\dot{\eta}-\eta\dot{\xi}), \quad \text{..........(2)}$$

regarded as holding *at the instant under consideration*.

Hence, in calculating the rate of increase of the angular momentum of the system about a fixed axis through which the mass-centre G is passing at the instant under consideration, we may ignore the motion of G, and have regard only to the motion of the particles relative to G.

This theorem also can be proved without recourse to the artificial apparatus of Cartesian coordinates; but for simplicity we will consider only the case of motion in two dimensions, where the path of the mass-centre is a plane curve. Let G, G' denote the positions of the mass-centre at the instants t, $t+\delta t$, respectively, and let $\bar{\mathbf{v}}$, $\bar{\mathbf{v}}+\delta\bar{\mathbf{v}}$ be the corresponding velocities of this point. Let ν be the angular momentum about G at time t, $\nu+\delta\nu$ that about G' at time $t+\delta t$. We have seen (Art. 61) that

* This formula, like (4) of Art. 46, and (2) of Art. 61, is a particular case of the following theorem:

If $$\phi(x, y, z, \dot{x}, \dot{y}, \dot{z}, \ddot{x}, \ddot{y}, \ddot{z})$$

be any homogeneous function of the second degree in the variables indicated, and if we make the substitution (1) of Art. 61, where $x, y, z, \xi, \eta, \zeta$ refer to any particle m of a material system, then

$$\Sigma m\phi(x, y, z, \dot{x}, \dot{y}, \dot{z}, \ddot{x}, \ddot{y}, \ddot{z})$$
$$= \Sigma(m).\phi\left(\bar{x}, \bar{y}, \bar{z}, \frac{d\bar{x}}{dt}, \frac{d\bar{y}}{dt}, \frac{d\bar{z}}{dt}, \frac{d^2\bar{x}}{dt^2}, \frac{d^2\bar{y}}{dt^2}, \frac{d^2\bar{z}}{dt^2}\right)$$
$$+\Sigma m\phi(\xi, \eta, \zeta, \dot{\xi}, \dot{\eta}, \dot{\zeta}, \ddot{\xi}, \ddot{\eta}, \ddot{\zeta}).$$

Cf. *Statics*, Art. 74.

in calculating ν or $\nu+\delta\nu$ it is immaterial whether we employ the actual momenta of the particles of the system, or the momenta

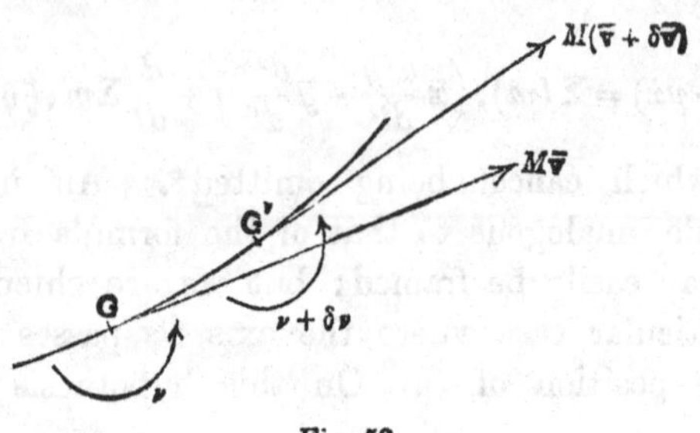

Fig. 52.

of the motion relative to the mass-centre. Now the angular momentum about G at time $t+\delta t$ is by the theorem of Art. 61 equal to $\nu+\delta\nu$, together with the moment about G of the linear momentum $\Sigma(m).(\bar{v}+\delta\bar{v})$ in a line through G' tangential to the path of the mass-centre. Since the distance of G from this line is of the second order of small quantities, this latter moment is ultimately negligible, and the increment in time δt of the angular momentum about the position G is simply $\delta\nu$.

Some important dynamical conclusions can at once be drawn. We have seen that if a mechanical system of any kind be free from external force, the mass-centre will describe a straight line with constant velocity. We now learn, in addition, that not only is the angular momentum about any fixed axis constant, but that the angular momentum of the motion relative to the mass-centre, with respect to any axis which passes through this point and moves with it (remaining constant in direction), is also constant.

For instance, the masses and *relative* velocities of the various members of the solar system determine the angular momentum of the system with respect to any axis through the mass-centre, and this angular momentum is constant, whether the system as a whole is in motion or not.

Again, if I be the moment of inertia of the earth with respect to its axis of rotation, and ω its angular velocity, the product $I\omega$ is constant, and unaffected by the earth's motion of translation. Possible retarding forces, and changes in the direction of the axis, are of course here neglected. If in consequence of physical changes, sudden or gradual, the moment of inertia and the angular velocity were altered to I' and ω', respectively, we should have

$$I'\omega' = I\omega. \qquad (3)$$

Thus a uniform contraction by cooling would involve a diminution of I and a consequent increase of ω, i.e. the length of the day would be diminished.

Ex. Two particles m_1, m_2 connected by a string of length a are in motion in one plane.

If ω be the angular velocity of the string, the angular momentum, about the mass-centre G, of the motion relative to G is

$$(m_1 r_1^2 + m_2 r_2^2)\,\omega, \qquad (4)$$

where r_1, r_2 are the distances of the two particles from G. Hence if there are no external forces, or if the external forces (as in the case of ordinary gravity) have zero moment about G, the angular velocity ω is constant (cf. Art. 42).

Again, if the external forces produce the same acceleration in all the particles (as in the case of gravity), they will not affect the relative motion. Hence, if T be the tension of the string,

$$T = m_1 \omega^2 r_1 = \frac{m_1 m_2}{m_1 + m_2}\,\omega^2 a. \qquad (5)$$

63. Application to Rigid Bodies.

We now contemplate more particularly the case of a rigid system, under the restriction that its motion is in two dimensions, i.e. the path of every particle of it is parallel to a fixed plane. It will be understood from the results of Art. 59 that a body will not in general move permanently in this manner unless the line through the mass-centre normal to the aforesaid plane is a principal axis of inertia of the body, or unless special constraining forces are introduced.

A rigid body movable in two dimensions has three degrees of freedom, and requires therefore three coordinates to specify its position [*S.* 13]. It is usually most convenient to employ the Cartesian coordinates (x, y) of the mass-centre G, and the angle θ through which the body has been turned from some standard position.

If M be the mass of the body, the components of linear momentum will be $M\dot{x}$, $M\dot{y}$, by Art. 45, and the angular momentum about G will be $I\dot{\theta}$, where I is the moment of inertia about an axis through G normal to the plane of motion. The latter expression follows from Art. 54, since in calculating the angular momentum about G we need only take account of the relative motion.

Hence if the external forces be supposed reduced, by the methods of Statics, to a force (X, Y) at G, and a couple N, we have

$$\frac{d}{dt}M\dot{x} = X, \quad \frac{d}{dt}M\dot{y} = Y, \quad \text{.................(1)}$$

$$\frac{d}{dt}I\dot{\theta} = N. \quad \text{..........................(2)}$$

These equations shew that the motions of translation and rotation are independent of one another; a principle first laid down by Euler (1749) in connection with the motion of ships.

We are now able to solve at once a number of interesting problems.

Ex. 1. In the small oscillations of a ship about a longitudinal axis through its mass-centre, we have

$$N = -Mgc\theta, \quad \text{..............................(3)}$$

where c is the metacentric height [*S.* 102]. Hence, putting $I = M\kappa^2$, we have

$$\kappa^2\ddot{\theta} = -gc\theta. \quad \text{..................................(4)}$$

The period of rolling is therefore that of a simple pendulum of length κ^2/c.

Ex. 2. A solid of revolution rolls down a plane of inclination α, with its axis horizontal. It is assumed that the reaction at the point of contact reduces to a single force; in other words, that there is no frictional *couple*.

Let a be the radius of that circle of the body which rolls in contact with the plane, and κ the radius of gyration about the axis of symmetry. Let u be the velocity of the mass-centre (G) parallel to the plane, the positive direction being downwards, and ω the angular velocity of the solid. Since the solid is turning about the point of contact as an instantaneous centre [*S.* 15], we have

$$u = \omega a. \quad \text{..(5)}$$

Resolving parallel and perpendicular to the plane, we have

$$M\frac{du}{dt} = Mg\sin\alpha - F, \quad 0 = Mg\cos\alpha - R; \quad \text{..............(6)}$$

where R, F are the normal and tangential components of the reaction of the plane on the body. Also, taking moments about G,

$$M\kappa^2 \frac{d\omega}{dt} = Fa. \qquad \text{............(7)}$$

Eliminating F and ω we find

$$\frac{du}{dt} = \frac{a^2}{\kappa^2 + a^2} g \sin \alpha. \qquad \text{.........(8)}$$

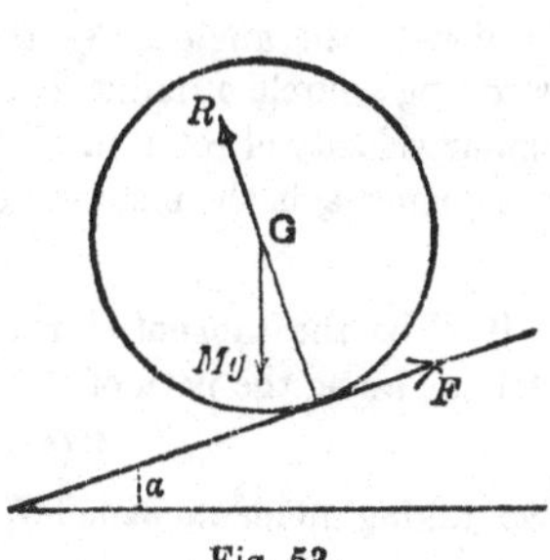

Fig. 53.

The acceleration of G is therefore less than in the case of frictionless sliding, in the ratio $a^2/(\kappa^2+a^2)$. For a homogeneous sphere ($\kappa^2 = \frac{2}{5}a^2$) this ratio is $\frac{5}{7}$; for a uniform circular disk or solid cylinder ($\kappa^2 = \frac{1}{2}a^2$) it is $\frac{2}{3}$; for a circular hoop or thin cylindrical shell ($\kappa^2 = a^2$) it is $\frac{1}{2}$.

Again, from (6) and (8),

$$F = \frac{\kappa^2}{\kappa^2+a^2} Mg \sin \alpha, \quad R = Mg \cos \alpha. \qquad \text{.....................(9)}$$

If we assume the usual law of sliding friction, viz. that F cannot exceed μR, where μ is the coefficient of friction, the condition that there should be no slipping is

$$\tan \alpha \ngtr \left(1 + \frac{a^2}{\kappa^2}\right) \mu. \qquad \text{..............................(10)}$$

Ex. 3. A movable cylinder rolls inside a fixed hollow cylinder, the sections being circular, and the axes of symmetry parallel and horizontal.

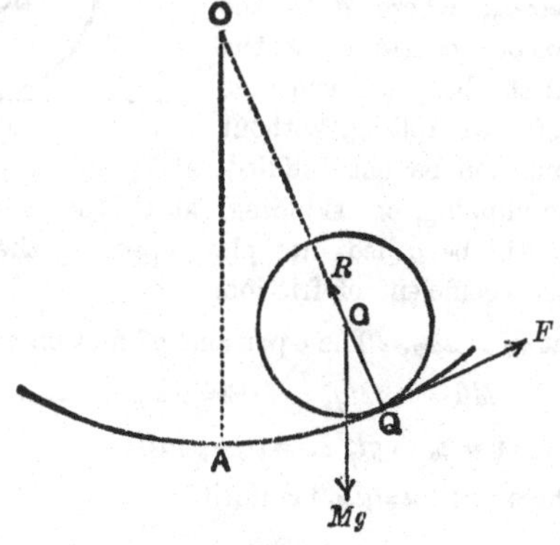

Fig. 54.

The figure represents a section by a plane perpendicular to the lengths. The point O represents the axis of the fixed cylinder, and OA is a vertical radius, whilst OQ is the radius through the mass-centre G of the moving cylinder.

We write $OA = OQ = b$, $GQ = a$, so that a is the radius of the rolling body. If θ denote the angle AOQ, the velocity of G is $(b-a)\dot{\theta}$, since this point is describing a circle of radius $b-a$ with the angular velocity $\dot{\theta}$. But if ω be the angular velocity of rotation of the moving body, the same velocity is expressed by ωa, since Q is the instantaneous centre. Hence

$$\omega a = (b-a)\dot{\theta}. \quad \text{.................................(11)}$$

If F be the tangential reaction at Q, as shewn in the figure, we have, resolving along the path of G,

$$M(b-a)\ddot{\theta} = -Mg\sin\theta + F. \quad \text{......................(12)}$$

Also, taking moments about G,

$$M\kappa^2 \frac{d\omega}{dt} = -Fa. \quad \text{.............................(13)}$$

Eliminating F and ω, we find

$$\left(1 + \frac{\kappa^2}{a^2}\right)(b-a)\ddot{\theta} = -g\sin\theta. \quad \text{...................(14)}$$

The motion of G is therefore exactly that of the bob of a simple pendulum of length

$$l = \left(1 + \frac{\kappa^2}{a^2}\right)(b-a). \quad \text{..........................(15)}$$

Ex. 4. A sphere (or a disk, hoop, etc. whose plane is vertical) is projected so as to roll and slide along a horizontal plane.

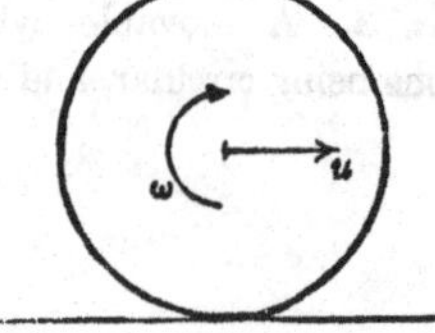

Fig. 55.

Let the initial velocity of the centre be u_0, and the initial angular velocity about the centre ω_0. If $u_0 = a\omega_0$, where a is the radius, the instantaneous centre is at the point of contact, and the body will continue to have a motion of pure rolling, without slipping. If this condition be not fulfilled, there will at first be slipping, or 'skidding,' at the point of contact, and a frictional force μMg will be called into play, opposing the relative motion there, if μ denote the coefficient of friction.

First suppose that $u_0 > a\omega_0$. The equations of motion are, at first,

$$M\dot{u} = -\mu Mg, \quad M\kappa^2\dot{\omega} = \mu Mga. \quad \text{..................(16)}$$

Hence

$$u = u_0 - \mu gt, \quad \omega = \omega_0 + \mu gat/\kappa^2. \quad \text{..................(17)}$$

These formulæ will hold until $u = a\omega$, i.e. until

$$t = \frac{u_0 - a\omega_0}{\mu g(1 + a^2/\kappa^2)}, \quad \text{.............................(18)}$$

when

$$u = \frac{a^2 u_0 + \kappa^2 a\omega_0}{a^2 + \kappa^2}. \quad \text{.............................(19)}$$

The body then proceeds to roll with a constant velocity equal to this. The loss of kinetic energy is

$$\tfrac{1}{2}M(u_0^2+\kappa^2\omega_0^2)-\tfrac{1}{2}M(a^2+\kappa^2)\omega^2=\tfrac{1}{2}\frac{M\kappa^2}{a^2+\kappa^2}(u_0-a\omega_0)^2. \quad \text{......(20)}$$

It will be observed that the results (19) and (20) are independent of μ.

The results for the case of $u_0 < a\omega_0$ will be found by changing the sign of μ in the above formulæ.

64. Equation of Energy.

It was shewn in Art. 46 that the kinetic energy of any material system is equal to the kinetic energy of the whole mass, supposed concentrated at the mass-centre and moving with this point, together with the kinetic energy of the motion relative to the mass-centre. Hence, in the case of a rigid body moving in two dimensions, if (u, v) be the velocity of the mass-centre, and ω the angular velocity, the kinetic energy is

$$\tfrac{1}{2}M(u^2+v^2)+\tfrac{1}{2}I\omega^2, \quad \text{..................(1)}$$

where M is the mass of the body, and I is its moment of inertia about an axis through the mass-centre perpendicular to the plane of motion.

Now, writing the equations of motion (Art. 63 (1), (2)) in the forms

$$\left.\begin{aligned} M\frac{du}{dt}=X,\quad M\frac{dv}{dt}=Y,\\ I\frac{d\omega}{dt}=N,\end{aligned}\right\} \quad \text{..................(2)}$$

we have
$$M\left(u\frac{du}{dt}+v\frac{dv}{dt}\right)+I\omega\frac{d\omega}{dt}=Xu+Yv+N\omega, \quad \text{......(3)}$$

or
$$\frac{d}{dt}\{\tfrac{1}{2}M(u^2+v^2)+\tfrac{1}{2}I\omega^2\}=X\frac{dx}{dt}+Y\frac{dy}{dt}+N\frac{d\theta}{dt}. \quad \text{......(4)}$$

The expression

$$X\delta x+Y\delta y+N\delta\theta$$

denotes the work done by the external forces in an infinitesimal displacement of the body [*S.* 62]. The equation (4) therefore expresses that the kinetic energy is at each instant increasing at a rate equal to that at which work is being done on the body. The total increment of kinetic energy in any finite interval of time is therefore equal to the work done by the external forces in that interval.

In applying this result we may of course ignore from the outset all forces, such as the reactions of fixed smooth surfaces, the tensions of inextensible strings, etc., which do no work [*S.* 52].

Moreover, if the body is subject to external forces which do on the whole no work in a cyclical process, by which the body is brought back to its original position after any series of displacements, we may introduce the conception of potential energy, as in Art. 30. The work done by these forces in any displacement is equivalent to the diminution of the potential energy.

An expression for the kinetic energy of a body moving in two dimensions can often be written down at once from the fact that the body is rotating about some point as instantaneous centre [*S.* 15]. If I be the moment of inertia about this point, and ω the angular velocity, the kinetic energy is $\frac{1}{2}I\omega^2$. Since the instantaneous centre is not in general a fixed point, the value of I, as well as of ω, may change as the motion proceeds.

Ex. 1. In the compound pendulum (Art. 55) the kinetic energy is $\frac{1}{2}Mk^2\dot{\theta}^2$, nd the potential energy is $-Mgh\cos\theta+\text{const.}$ Hence

$$\tfrac{1}{2}Mk^2\dot{\theta}^2-Mgh\cos\theta=\text{const.} \quad\text{.......................(5)}$$

Ex. 2. In the case of a solid of revolution rolling on an inclined plane, since the moment of inertia about the point of contact is $M(\kappa^2+a^2)$, the kinetic energy is $\frac{1}{2}M(\kappa^2+a^2)\omega^2$. Hence if x denote distance travelled down the plane we have

$$\tfrac{1}{2}M(\kappa^2+a^2)\omega^2=Mgx\sin\alpha+\text{const.}, \quad\text{....................(6)}$$

or, putting

$$\omega=u/a,$$

$$u^2=\frac{2ga^2x\sin\alpha}{\kappa^2+a^2}+\text{const.} \quad\text{..............................(7)}$$

The acceleration of the mass-centre is therefore

$$u\frac{du}{dx}=\frac{a^2}{\kappa^2+a^2}g\sin\alpha, \quad\text{.........................(8)}$$

as already found.

Ex. 3. A rod AB slides with its lower end on a smooth horizontal plane. Since there is no horizontal force on the rod, the mass-centre G has no horizontal acceleration. Its horizontal velocity is therefore constant, and may be ignored.

Let M be the mass, a the distance of G from the lower end A, κ the radius of gyration about G. If z be the altitude GN of G above the horizontal plane, and θ the inclination of the rod to the vertical, the equation of energy is

$$\tfrac{1}{2}M\dot{z}^2+\tfrac{1}{2}M\kappa^2\dot{\theta}^2+Mgz=\text{const.} \quad \text{.......................(9)}$$

Since
$$z=a\cos\theta, \quad \text{.................................(10)}$$

this gives
$$\tfrac{1}{2}(\kappa^2+a^2\sin^2\theta)\dot{\theta}^2+ga\cos\theta=\text{const.} \quad \text{................(11)}$$

The constant depends on the initial conditions; thus if the rod is just started from rest in the vertical position, we have $\dot{\theta}=0$ for $\theta=0$, and therefore

$$\tfrac{1}{2}(\kappa^2+a^2\sin^2\theta)\dot{\theta}^2=ga(1-\cos\theta). \quad (12)$$

This gives the angular velocity in any subsequent position.

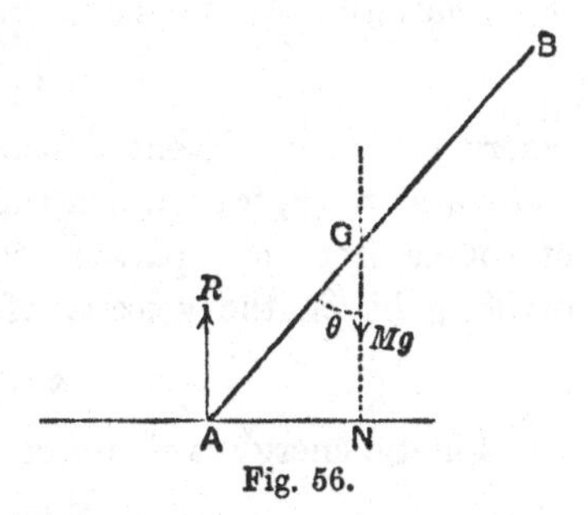

Fig. 56.

To find the vertical pressure R exerted by the plane we have, taking moments about G,

$$M\kappa^2\ddot{\theta}=Ra\sin\theta. \quad \text{......(13)}$$

Now, differentiating (12) with respect to t, and dividing out by $\dot{\theta}$, we obtain

$$(\kappa^2+a^2\sin^2\theta)\ddot{\theta}+a^2\sin\theta\cos\theta\dot{\theta}^2=ga\sin\theta. \quad \text{...........(14)}$$

By elimination of $\dot{\theta}$ and $\ddot{\theta}$ between (12), (13), and (14), the value of R can be found.

For instance, to find the pressure just before the rod becomes horizontal, we have, putting $\theta=\tfrac{1}{2}\pi$,

$$\ddot{\theta}=\frac{ga}{\kappa^2+a^2}, \qquad R=\frac{M\kappa^2}{a}\ddot{\theta}=\frac{\kappa^2}{\kappa^2+a^2}Mg. \quad \text{.................(15)}$$

65. General Theory of a System with One Degree of Freedom.

Many of the preceding examples have this feature in common, that the rigid body considered has in each case only one degree of freedom, i.e. the various positions which it can assume can all be specified by attributing the proper values to a single variable element, or 'coordinate,' as it may be called in a generalized sense. Accordingly the equation of energy, when this applies, is sufficient for determining the motion, if the initial conditions are given.

This method can obviously be extended to any conservative system which is subject to constraints such that there remains just one degree of freedom. It is convenient therefore to study the question in a general manner.

In consequence of the constraints, any particle m of the system can only move backwards or forwards along a definite path. If θ be the variable which specifies the configuration of the system, and if δs be the displacement of m along this path consequent on a variation $\delta\theta$, we have

$$\delta s = \alpha\,\delta\theta, \qquad (1)$$

where α is a coefficient depending in general on the particular configuration (θ) from which the displacement is made, and varying of course from one particle to another of the system. Hence, dividing by δt, the velocity of m is

$$v = \dot{s} = \alpha\dot{\theta}. \qquad (2)$$

The kinetic energy is therefore

$$\tfrac{1}{2}\Sigma\,(mv^2) = \tfrac{1}{2}A\dot{\theta}^2 \qquad (3)$$

where
$$A = \Sigma\,(m\alpha^2), \qquad (4)$$

the summation embracing all the particles of the system.

The coefficient A is called the 'coefficient of inertia' of the system; it is essentially positive. It is in general a function of θ, and therefore different in different configurations.

For instance, in the case of a rigid body moving in two dimensions, if θ denote the angle through which it has turned from some standard position, the coefficient α in (1) will be the distance of the particle m from the instantaneous axis of rotation, and A is accordingly the (usually variable) moment of inertia about this line.

As a further illustration we may take the system composed of the piston, connecting rod, and flywheel of a steam-engine. Let I be the moment of inertia of the flywheel, M the mass of the piston; for simplicity we neglect the inertia of the connecting rod. If the angle θ be defined as in the figure, the velocity of the piston is $OR \,.\, \dot{\theta}$ [S. 15]; and the kinetic energy is accordingly

$$\tfrac{1}{2}(I + M \,.\, OR^2)\,\dot{\theta}^2.$$

The inertia-coefficient is therefore $I + M \,.\, OR^2$. To an observer making experiments on the flywheel it would appear to have a variable moment of inertia of this amount.

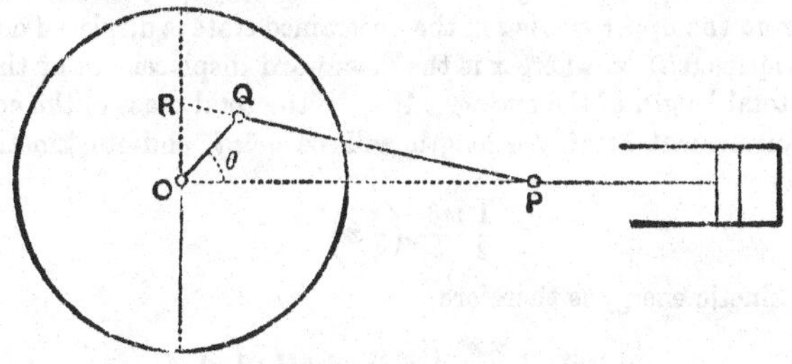

Fig. 57.

The configuration of the system may however also be specified by means of the distance OP. Denoting this by x we have

$$\dot{x} = -OR \,.\, \dot{\theta},$$

and the expression for the kinetic energy is now

$$\frac{1}{2}\left(M + \frac{I}{OR^2}\right)\dot{x}^2. \quad \text{.........................(5)}$$

From this point of view the inertia-coefficient is $M + I/OR^2$; this represents the inertia of the piston as modified by the flywheel. At the 'dead points,' where $\theta = 0$ or π, this apparent inertia* is infinite.

Some further illustrations are appended.

Ex. 1. Take the case of a waggon of mass M having n equal wheels, each of mass m.

Let the radius of each wheel be a, and the radius of gyration about the centre κ. When the waggon is moving with velocity u, the angular velocity of a wheel will be u/a, and the kinetic energy of each will therefore be

$$\tfrac{1}{2}m\left(u^2 + \frac{\kappa^2}{a^2}u^2\right).$$

The total kinetic energy is accordingly

$$\frac{1}{2}\left\{M + nm\left(1 + \frac{\kappa^2}{a^2}\right)\right\}u^2. \quad \text{.........................(6)}$$

The expression

$$M + nm\left(1 + \frac{\kappa^2}{a^2}\right)$$

measures the apparent inertia of the waggon.

* Called also by Rankine the 'reduced inertia.'

Ex. 2. The inertia of a helical spring supporting a weight which oscillates vertically (Fig. 5, p. 27) may be allowed for approximately as follows.

We assume that the spring is stretched uniformly*, so that a point whose distance from the upper end is z in the unstrained state is displaced downwards through a space $(z/l) . x$, where x is the downward displacement of the weight, and l the total length of the spring. If m be the total mass of the spring, the mass of an element δz of the length will be $m\delta z/l$, and its kinetic energy accordingly

$$\frac{1}{2}\frac{m\delta z}{l}\left(\frac{z}{l}\dot{x}\right)^2.$$

The total kinetic energy is therefore

$$\tfrac{1}{2}M\dot{x}^2 + \tfrac{1}{2}\frac{m\dot{x}^2}{l^3}\int_0^l z^2 dz = \tfrac{1}{2}(M + \tfrac{1}{3}m)\dot{x}^2, \quad \text{.................(7)}$$

where M is the suspended mass. The inertia of the spring is thus allowed for by adding one-third of its mass to that of the suspended body.

66. Oscillations about Equilibrium. Stability.

If a system of the kind above contemplated be conservative, and if V denote the potential energy, we have, when there are no extraneous forces,

$$\tfrac{1}{2}A\dot{\theta}^2 + V = \text{const.} \quad \text{....................(1)}$$

Differentiating this with respect to t, and dividing out by $\dot{\theta}$, we have

$$A\ddot{\theta} + \frac{1}{2}\frac{dA}{d\theta}\dot{\theta}^2 = -\frac{dV}{d\theta}. \quad \text{....................(2)}$$

This is the equation of motion of the system, with all reactions eliminated which on the whole do no work.

To find the configurations of equilibrium of the system we put $\dot{\theta} = 0$, $\ddot{\theta} = 0$, and obtain

$$\frac{dV}{d\theta} = 0. \quad \text{..........................(3)}$$

In words, the possible configurations of equilibrium correspond to values of θ such that the potential energy is stationary for small displacements.

To examine the nature of the equilibrium we will suppose the coordinate θ to be so modified (by the addition of a constant), that

* This condition will be fulfilled practically if the periods of free vibration of the spring itself are small compared with the period of oscillation of the suspended weight.

it vanishes in the configuration in question. The value of V for small values of θ may then be expanded in the form

$$V = V_0 + \tfrac{1}{2}a\theta^2 + \ldots, \quad \ldots\ldots(4)$$

the term of the first degree in θ being absent because, by hypothesis, (3) is satisfied for $\theta = 0$. Hence, considering a slight disturbance from equilibrium, we have

$$A\ddot{\theta} = -a\theta, \quad \ldots\ldots(5)$$

approximately, the second term in (2) being omitted as of the second order of small quantities. The coefficient A, moreover, may be taken to be constant and equal to its equilibrium value, since this involves an error of the second order only in the equation*.

If a be positive the solution of this is

$$\theta = C\cos(nt + \epsilon), \quad \ldots\ldots(6)$$

where

$$n = \sqrt{\frac{a}{A}}, \quad \ldots\ldots(7)$$

and the constants C, ϵ are arbitrary. Each particle m oscillates to and fro along its path, its displacement being given, in the notation of Art. 65, by

$$s = \alpha\theta = \alpha C\cos(nt + \epsilon), \quad \ldots\ldots(8)$$

where α is a coefficient which varies from particle to particle. The period $2\pi/n$ is fixed by the constitution of the system, and depends on the ratio which the 'coefficient of stability' a bears to the coefficient of inertia.

If a be negative, the solution of (5) is

$$\theta = Ce^{nt} + C'e^{-nt}, \quad \ldots\ldots(9)$$

where

$$n = \sqrt{(-a/A)}. \quad \ldots\ldots(10)$$

Unless the initial conditions are specially adjusted so as to make $C = 0$, the value of θ will ultimately become so great that the approximation ceases to be valid.

* In the case of a rigid body oscillating in two dimensions, a result equivalent to (5) is obtained by taking moments about the instantaneous axis as if it were fixed in the body and in space.

The equilibrium is therefore said to be stable or unstable according as a is positive or negative; i.e. according as the equilibrium value of V is a minimum or a maximum.

If extraneous forces act on the system, the work done by them in an infinitesimal displacement can be expressed in the form $\Theta\delta\theta$. The equation (1) is then replaced by

$$\frac{d}{dt}(\tfrac{1}{2}A\dot{\theta}^2 + V) = \Theta\dot{\theta}. \qquad\qquad (11)$$

Performing the differentiation, and dividing out by $\dot{\theta}$, we have

$$A\ddot{\theta} + \frac{1}{2}\frac{dA}{d\theta}\dot{\theta}^2 = -\frac{dV}{d\theta} + \Theta. \qquad\qquad (12)$$

The quantity Θ is called by analogy the generalized 'force' acting on the system. Its nature will depend on that of the coordinate θ. If θ be a line, Θ will be of the dimensions of a force in the ordinary sense of the word; if θ be an angle, Θ will be of the nature of a couple, and so on.

Ex. 1. A circular cylinder of radius a, whose mass-centre is at a distance h from the axis, rolls on a horizontal plane. This problem includes the case of a compound pendulum whose knife-edge is replaced by a cylindrical pin which rolls on horizontal supports* (Fig. 58).

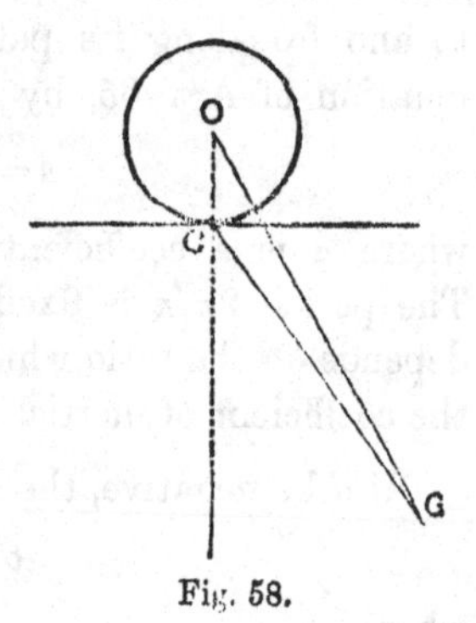

Fig. 58.

Let O represent the axis of the cylinder, C the line of contact, G the mass-centre, and let θ be the inclination of OG to the vertical OC. If κ be the radius of gyration about an axis through G parallel to the axis of the cylinder, the moment of inertia about the instantaneous axis (C) is $M(\kappa^2 + CG^2)$, and the kinetic energy is therefore

$$\tfrac{1}{2}M(\kappa^2 + CG^2)\dot{\theta}^2 = \tfrac{1}{2}M(\kappa^2 + a^2 - 2ah\cos\theta + h^2)\dot{\theta}^2. \qquad (13)$$

The equation of energy is accordingly

$$\tfrac{1}{2}M(\kappa^2 + a^2 - 2ah\cos\theta + h^2)\dot{\theta}^2 = Mgh\cos\theta + \text{const.} \qquad (14)$$

We have seen, in the general theory, that in the case of small oscillations we may put $\theta = 0$ in the value of the inertia-coefficient, and that the value of

* This problem was treated by Euler (1780).

the potential energy is only required to the second order of small quantities. Hence, putting $\cos\theta = 1 - \frac{1}{2}\theta^2$ on the right hand, we have

$$\{\kappa^2 + (h-a)^2\}\,\dot{\theta}^2 + gh\theta^2 = \text{const.} \quad \text{.....................(15)}$$

Differentiating we find

$$\{\kappa^2 + (h-a)^2\}\,\ddot{\theta} + gh\theta = 0. \quad \text{.......................(16)}$$

The length of the equivalent simple pendulum is therefore

$$l = \frac{\kappa^2 + (h-a)^2}{h}. \quad \text{..............................(17)}$$

The results evidently apply to any case of a solid of revolution rolling parallel to a vertical plane of symmetry, at right angles to the axis. In the case of a uniform solid hemisphere of radius a we have

$$\kappa^2 = \tfrac{2}{5}a^2 - (\tfrac{3}{8}a)^2, \quad h = \tfrac{3}{8}a, \text{..............................(18)}$$

whence $$l = 1{\cdot}73a.$$

Ex. 2. A cylinder, of any form of section, rocks on a horizontal plane, making small oscillations about a position of equilibrium.

In the equilibrium position the centre of gravity G is in the same vertical with the line of contact; let its height above this line be denoted by h. When the cylinder has rolled through a small angle θ from this position the vertical

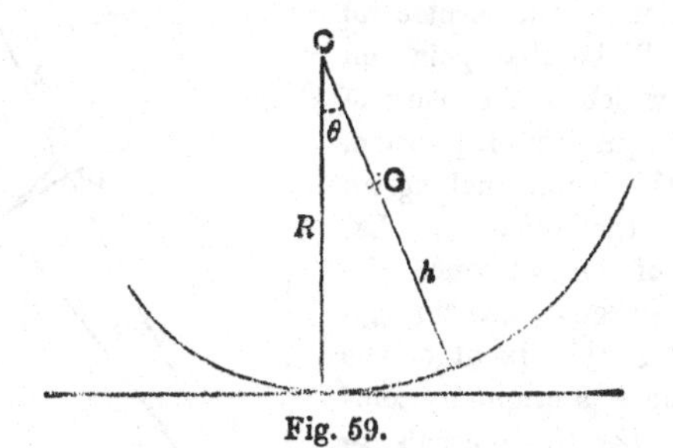

Fig. 59.

through the new line of contact, in the plane of the cross-section through G, will meet that line through G which was originally vertical in some point C; and it is known, as a matter of Infinitesimal Geometry, that the two intersecting normals will differ in length by a small quantity of the *third* order. Denoting the lengths by R, the increment of the potential energy is, to the second order,

$$Mg(R-h)(1-\cos\theta) = \tfrac{1}{2}Mg(R-h)\,\theta^2, \text{..................(19)}$$

where R may now be identified with the radius of curvature of the cross-section at the original point of contact*. The equilibrium position is

* It differs from it usually by a small quantity of the first order.

therefore stable if $h < R$ [S. 59]. Since the kinetic energy is, with sufficient approximation,

$$\tfrac{1}{2}M(\kappa^2+h^2)\dot{\theta}^2, \qquad\qquad (20)$$

where κ is the radius of gyration about a longitudinal axis through G, the length of the equivalent simple pendulum is

$$l=\frac{\kappa^2+h^2}{R-h}. \qquad\qquad (21)$$

It appears from either (17) or (21) that in the compound pendulum, if the knife-edge be slightly rounded, so that its radius of curvature is a, the value of l is changed from

$$\frac{\kappa^2+h^2}{h} \text{ to } \frac{\kappa^2+h^2}{h+a},$$

if h now denote the distance of G from the edge; i.e. it is diminished in the ratio $1-a/h$, approximately. This may be allowed for in formulæ for the determination of g by altering the observed values of T^2 (the square of the period) in the ratio $1+a/h$. The correction may easily come within the errors of observation.

Ex. 3. The more general case of one cylinder resting on another can be treated in a similar way.

The figure represents a vertical section by a plane perpendicular to the lengths, through the centre of gravity G. Let P' be that point on the lower curve which is the point of contact in the equilibrium position, the tangent at this point making an angle ψ (say) with the horizontal. Let Q be the point of contact when the upper cylinder has been turned through an angle ϕ, and P that point of the upper curve which was originally coincident with P'. Let the normals to the two curves at P and P' meet the common normal at Q in the points O and O', respectively, and let us write

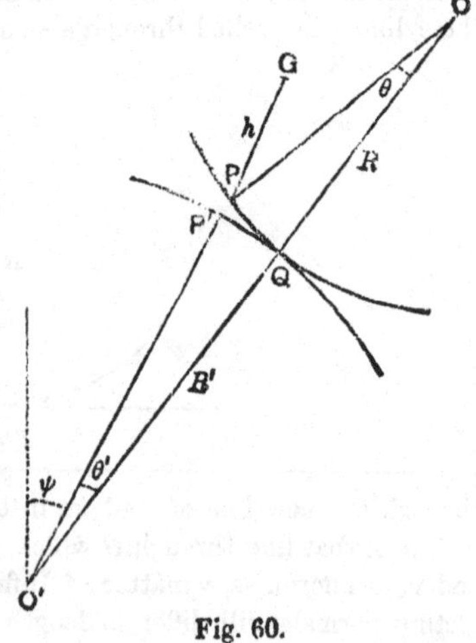

Fig. 60.

$$OP=OQ=R,\ O'P'=O'Q=R',\ PG=h, \qquad (22)$$

the error committed in treating the intersecting normals as equal being of the third order. Further, let θ, θ' denote the angles which the equal arcs PQ, $P'Q$ subtend at O, O', respectively.

Since the normals PO, $P'O'$ were originally in the same line we have

$$\theta+\theta'=\phi, \qquad\qquad (23)$$

the angle through which the upper cylinder has been turned. Hence, since

$$R\theta = R'\theta', \qquad (24)$$

with sufficient approximation, we have

$$\theta = \frac{R'\phi}{R+R'}, \quad \theta' = \frac{R\phi}{R+R'}. \qquad (25)$$

Since PG has been turned through an angle ϕ from the vertical, the altitude of G above the level of O' is

$$\begin{aligned} O'O\cos(\psi+\theta') - OP\cos(\psi+\phi) + PG\cos\phi \\ = (R+R')\{(1-\tfrac{1}{2}\theta'^2)\cos\psi - \theta'\sin\psi\} - R\{(1-\tfrac{1}{2}\phi^2)\cos\psi - \phi\sin\psi\} \\ + h(1-\tfrac{1}{2}\phi^2). \end{aligned} \qquad (26)$$

The terms of the first order in θ', ϕ cancel in virtue of (25). The increment of the potential energy in consequence of the displacement is thus found to be

$$\tfrac{1}{2}Mg \, . \, \frac{RR'\cos\psi - h(R+R')}{R+R'} \, . \, \phi^2, \qquad (27)$$

where R, R' may now be identified with the radii of curvature of the two curves at P', P, respectively.

The expression (27) is positive, and the equilibrium position is therefore stable, if

$$\frac{\cos\psi}{h} > \frac{1}{R} + \frac{1}{R'}, \qquad (28)$$

which is the usual formula [*S.* 59].

The kinetic energy is ultimately

$$\tfrac{1}{2}M(\kappa^2 + h^2)\dot{\phi}^2, \qquad (29)$$

where κ is the radius of gyration about a longitudinal line through G. The length of the equivalent simple pendulum is therefore

$$l = \frac{(\kappa^2+h^2)\left(\dfrac{1}{R}+\dfrac{1}{R'}\right)}{\cos\psi - h\left(\dfrac{1}{R}+\dfrac{1}{R'}\right)}. \qquad (30)$$

If as in Fig. 61 we describe a circle of diameter c, such that

$$\frac{1}{c} = \frac{1}{R} + \frac{1}{R'}, \qquad (31)$$

i.e. its curvature is twice the sum of the curvatures of the given curves, to touch the lower curve at the original point of contact, and if PG in its original position meet this circle in H, we have $PH = c\cos\psi$. It appears then from (28) that the equilibrium is stable or unstable, according as G is below or above H. In the theory of Roulettes this circle is called the 'circle of inflexions,' the paths of points carried by the rolling curve being convex or concave to P according as they lie inside or outside this circle, whilst points on the circle itself are at points of inflexions on their respective paths. Again, we have from (30)

$$l = \frac{\kappa^2 + h^2}{HG}. \quad \text{.............................(32)}$$

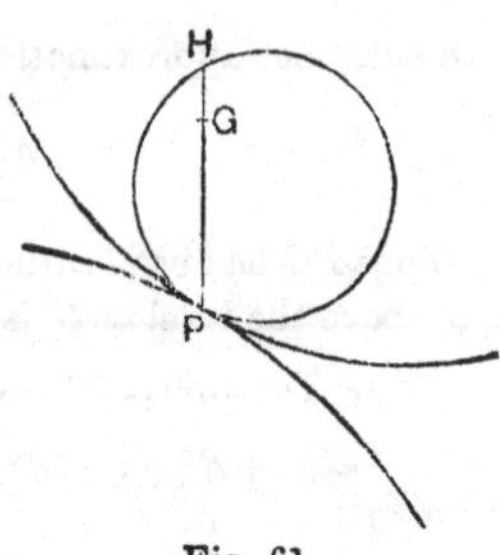

Fig. 61.

The symbols R, R', h have been taken to be positive in the case shewn in the figures. Other cases are included in the results if the proper changes of sign are made.

The whole investigation is applicable to any case of a rigid body having one degree of freedom of motion, parallel to a vertical plane, provided gravity be the only force which does work. The two curves of the figure are then the two pole-curves (i.e. loci of the instantaneous centre in the body and in space) which by a theorem of kinematics roll in contact with one another in any motion of the body [*S.* 16, 59].

67. Forced Oscillations of a Pendulum. Seismographs.

The forced oscillations of a compound pendulum (Fig. 45), due to a prescribed horizontal motion of the support O, may be treated as follows.

The oscillations being supposed small, the vertical component of the reaction at O will be Mg, approximately, since the vertical displacement of the mass-centre G is of the second order. Hence if the horizontal reaction of the support on the pendulum be X, we have, taking moments about G,

$$M\kappa^2 \frac{d^2\theta}{dt^2} = -Mgh\theta - Xh, \quad \text{.................(1)}$$

the rest of the notation being as in Art. 55. If ξ be the horizontal displacement of O, the horizontal displacement of G will be $\xi + h\theta$. Hence

$$M\frac{d^2}{dt^2}(\xi + h\theta) = X. \quad \text{....................(2)}$$

Eliminating X, we find

$$l\frac{d^2\theta}{dt^2} + g\theta = -\frac{d^2\xi}{dt^2}, \quad \text{....................(3)}$$

where $l = h + \kappa^2/h$. If we put

$$x = \xi + l\theta, \quad \text{.........................(4)}$$

so that x denotes the horizontal displacement of the centre of oscillation, the equation becomes

$$\frac{d^2x}{dt^2}+\frac{g}{l}x=\frac{g}{l}\xi, \qquad\qquad(5)$$

as in the case of a simple pendulum of length l. (Art. 13 (13).)

If we put

$$n^2=g/l, \qquad\qquad(6)$$

so that $2\pi/n$ is the period of a free oscillation, the forced oscillation due to a simple-harmonic vibration

$$\xi=C\cos pt \qquad\qquad(7)$$

of the point of support is

$$\theta=\frac{p^2\xi}{(n^2-p^2)\,l}, \qquad\qquad(8)$$

or

$$x=\frac{\xi}{1-p^2/n^2}. \qquad\qquad(9)$$

The formulæ will obviously apply to the 'horizontal pendulum' of Art. 55 provided we replace g by $g\sin\beta$, where β is the inclination of the axis of suspension to the vertical, and accordingly write

$$n^2=(g\sin\beta)/l \qquad\qquad(10)$$

in place of (6).

It appears from (9) that if p is large compared with n, x will be small compared with ξ. Hence, in the case of imposed vibrations which are rapid compared with the natural vibrations of the pendulum, the centre of oscillation will (so far as the forced oscillations are concerned) remain sensibly at rest.

These remarks have a bearing on the theory of seismographs. A seismograph is an instrument whose object is to record, as accurately as may be, some one component of the motion of the earth's surface due to earthquakes or other causes. To effect this with perfect accuracy it would be necessary to have as a basis of reference some body which did not itself participate in the motion, so far as the component in question is concerned, and which would therefore be, relatively, in neutral equilibrium. This is of course impracticable, some degree of stability being essential, but if the restoring force called into play by a relative displacement be slight, and the period of free oscillation consequently long, the body will be only slightly affected by vibrations which are comparatively rapid. A simple pendulum of sufficient length, whose motion is restricted to one vertical plane, would fulfil this condition, so far as regards either horizontal component of the displacement; but for greater convenience some form of 'horizontal' pendulum is employed, whose axis of suspension makes a very small angle with the vertical. In this way we may obtain within a moderate compass the equivalent of a simple pendulum some 300 or 400 feet long. The instrument may consist, for example, of a light rod or 'boom' AB, ending at A in a conical point which bears against a fixed surface, and carrying a weight W which is attached by a

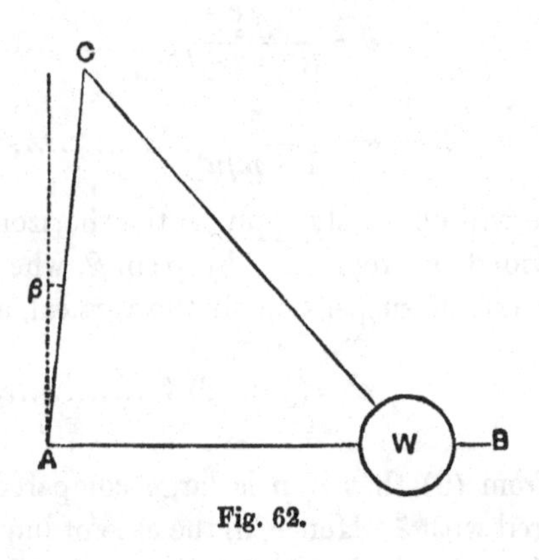

Fig. 62.

fine wire to a point C nearly, but not quite, in the same vertical with A. The axis of rotation is then AC. What is actually observed is of course the displacement of some point P of the 'boom' relatively to the framework of the apparatus. This is magnified by optical or mechanical means, and recorded on a uniformly running band of paper, so that a space-time curve is described. Unfortunately, the scale of the record varies with the frequency of the vibration, the relative displacement of P being

$$l'\theta = -\frac{1}{1 - n^2/p^2} \cdot \frac{l'}{l} \cdot \xi, \qquad \text{.....................(11)}$$

if l' be the distance of P from A. For sufficiently rapid vibrations, the first factor in (11) is sensibly equal to unity, and the scale accordingly constant, but for smaller values of the ratio p/n the record is distorted in a varying degree.

Conversely, to ascertain the true displacement (ξ) of the ground, the relative displacement of P must be multiplied by

$$-\left(1-\frac{n^2}{p^2}\right)\frac{l}{l'}.$$

The method is of course only applicable to such portions of the record as consist of a series of approximately simple-harmonic oscillations, as however is often the case near the 'maximum phase' of an earthquake.

A complete seismological observatory contains two instruments more or less of the above type, one for the N.-S. and the other for the E.-W. component of the motion. For the vertical component

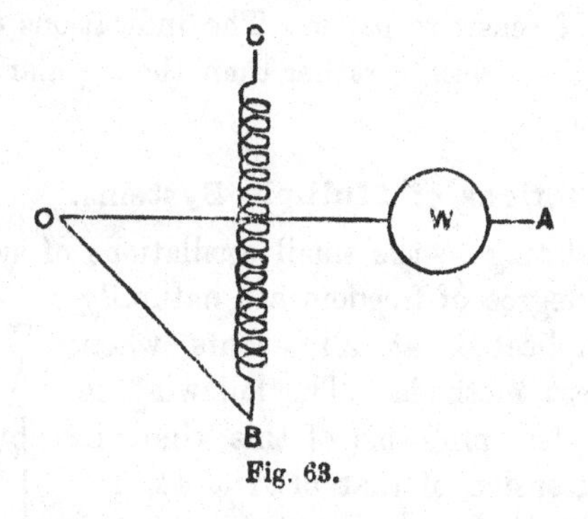

Fig. 63.

some other contrivance is necessary. In the form devised by Ewing, a rigid frame, the essential part of which is represented by AOB in the figure, can turn about a horizontal axis at O. The arm OA, which is sensibly horizontal, carries a weight W, and a point B on the arm OB is connected by a helical spring to a fixed point C. If the weight be slightly depressed from its equilibrium position the moment of gravity about O is scarcely altered; but the tension of the spring is increased, whilst its leverage about O is diminished. The circumstances are adjusted so that the former influence shall slightly prevail over the latter; the restoring force is accordingly small, and the period of a free oscillation long. The dynamical characteristics of the arrangement are therefore essentially the same as in the previous case, and its behaviour under a forced vertical oscillation of the axis O is governed by the same principles*.

So far, the *forced* oscillations have alone been referred to. Free oscillations are of course also set up, as explained in Art. 13, and tend to confuse the record, unless special damping arrangements are introduced. The operation of these will be considered in Chap. XII (Art. 95).

It may be added that in the most recent forms of seismograph†, whether horizontal or vertical, the registration is on a different principle. The pendulum carries coils of fine wire in which electric currents are induced as they swing between the poles of fixed magnets. These currents are led through a dead-beat mirror galvanometer whose mirror reflects a spot of light on to the running band of sensitive paper. The indications depend therefore on the angular *velocity*, rather than the angular displacement, of the pendulum.

68. Oscillations of Multiple Systems.

Problems relating to the small oscillations of systems having more than one degree of freedom are naturally somewhat complicated, at all events when treated by direct methods. The following is one of the simplest problems of this kind; it is a natural extension of that of Art. 44, 1.

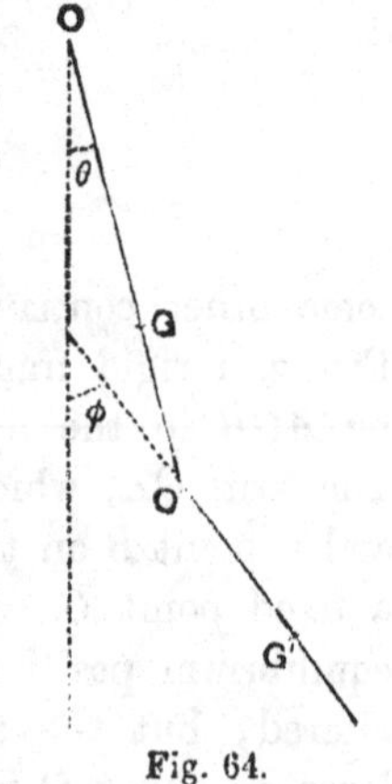

Fig. 64.

A body of mass M can turn freely about a horizontal axis O, and a second body of mass m, is suspended from it by a parallel axis O'. It is assumed for simplicity that the mass-centre G of the upper body lies in the plane of the axes O, O'. The figure represents a projection on a vertical plane perpendicular to these axes. We write

$$OG = h, \quad OO' = a, \quad O'G' = b, \quad \text{.....(1)}$$

* The general theory of the forced oscillations of a system of two degrees of freedom is given in Art. 111.

† Viz. those devised by Prince Galitzin.

where G' is the mass-centre of the lower body, and we denote by k, κ the radii of gyration of the upper body about O, and of the lower body about G', respectively. The inclinations of OG and $O'G'$ to the vertical are denoted by θ and ϕ.

In the case of small oscillations about the equilibrium position, in which G' is in the same vertical plane with the axes O, O', the vertical component of the reaction of m on M will be mg, approximately. Denoting the horizontal component by X, as indicated in Fig. 65, and taking moments about O for the upper body, we have, to the first order,

$$Mk^2\ddot{\theta} = -Mgh\theta + Xa - mga\theta. \quad \text{....(2)}$$

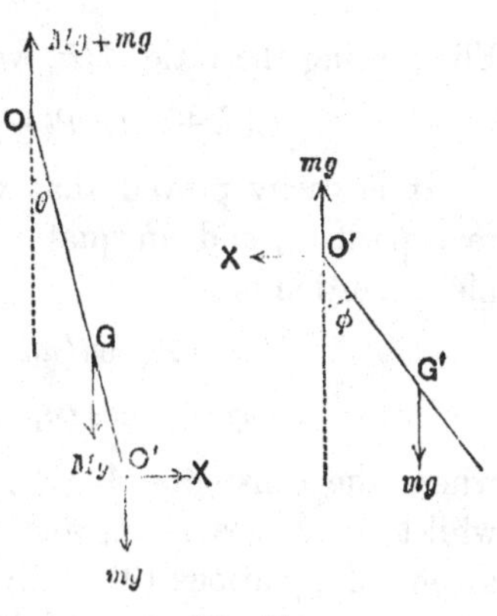

Fig. 65.

Again, since the horizontal displacement of G' from the vertical plane through O is $a\theta + b\phi$, we have

$$m(a\ddot{\theta} + b\ddot{\phi}) = -X. \quad \text{.....................(3)}$$

Also, taking moments about G', for the lower body,

$$m\kappa^2\ddot{\phi} = Xb - mgb\phi. \quad \text{.....................(4)}$$

Eliminating X, we obtain the equations

$$\left.\begin{aligned} (Mk^2 + ma^2)\,\ddot{\theta} + (Mh + ma)\,g\theta + mab\ddot{\phi} = 0, \\ ab\ddot{\theta} + (\kappa^2 + b^2)\,\ddot{\phi} + gb\phi = 0. \end{aligned}\right\} \quad \text{......(5)}$$

It is convenient to write

$$\frac{Mk^2 + ma^2}{Mh + ma} = l, \quad \frac{\kappa^2 + b^2}{b} = l'; \quad \text{...............(6)}$$

i.e. l is the length of the simple pendulum equivalent to the upper body when a particle m is attached to it at O, whilst l' relates to the oscillations of the lower body when the axis O' is fixed. The equations (5) may then be written

$$\left.\begin{aligned} (Mh + ma)(l\ddot{\theta} + g\theta) + mab\ddot{\phi} = 0, \\ a\ddot{\theta} + l'\ddot{\phi} + g\phi = 0. \end{aligned}\right\} \quad \text{...........(7)}$$

The solution follows the same course as in Art. 44. If we assume

$$\theta = A\cos(nt+\epsilon), \quad \phi = B\cos(nt+\epsilon), \quad \text{.........(8)}$$

the equations are satisfied, provided

$$\left.\begin{aligned}(Mh+ma)(n^2l-g)A + mabn^2B &= 0,\\ n^2aA + (n^2l'-g)B &= 0.\end{aligned}\right\} \quad \text{............(9)}$$

Eliminating the ratio A/B, we have

$$(Mh+ma)(n^2l-g)(n^2l'-g) - ma^2bn^4 = 0. \quad \text{.....(10)}$$

It is easily proved that the roots of this quadratic in n^2 are real, positive, and unequal. Denoting them by n_1^2, n_2^2, the complete solution is

$$\left.\begin{aligned}\theta &= A_1\cos(n_1t+\epsilon_1) + A_2\cos(n_2t+\epsilon_2),\\ \phi &= B_1\cos(n_1t+\epsilon_1) + B_2\cos(n_2t+\epsilon_2),\end{aligned}\right\} \quad \text{........(11)}$$

where the constants A_1, A_2, ϵ_1, ϵ_2 may be regarded as arbitrary, whilst the ratios B_1/A_1 and B_2/A_2 are given most simply by the second of equations (9), with the respective values of n^2 inserted. The interpretation of this solution is as in Art. 44.

If we put

$$\lambda = g/n^2, \quad \text{..........................(12)}$$

so that λ is the length of a simple pendulum having the same period as a normal mode of our system, the equation (10) may be written

$$(\lambda - l)(\lambda - l') = \frac{ma^2b}{Mh+ma}. \quad \text{..............(13)}$$

Hence one value of λ is greater than the greater, and the other is less than the smaller, of the two quantities l, l'. Since

$$\frac{B}{A} = \frac{a}{\lambda - l'}, \quad \text{.......................(14)}$$

by (9), it appears that in the slower of the two normal modes θ and ϕ have the same sign, whilst in the quicker mode the signs are opposite. Cf. Art. 44.

69. Stresses in a Moving Body.

The determination of the stresses in a moving body is an elastic problem which is usually very difficult. Questions of shearing stress and bending moment in a moving bar can however be treated by the ordinary methods of Statics [*S.* 27], provided of course that the 'effective' forces are taken into account. The following is an example.

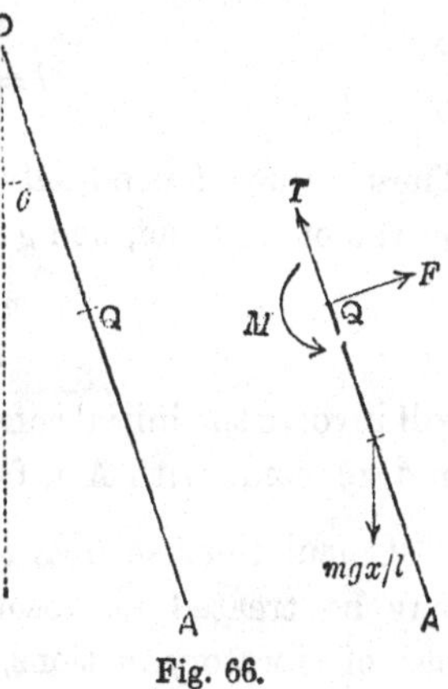

Fig. 66.

We take the case of a uniform bar of mass m and length l swinging like a pendulum about one end. The stresses across a section at a point Q at a distance x from the free end A are equivalent to a tension T, a transverse shearing stress F, and a bending moment M; we will suppose that the positive senses of these are as indicated in the figure.

The segment QA has a mass mx/l, and its centre moves on a circle of radius $l-\frac{1}{2}x$. Hence if θ be the inclination of the bar to the vertical, we have, resolving along and at right angles to the length,

$$\frac{mx}{l}(l-\tfrac{1}{2}x)\,\dot{\theta}^2 = T - \frac{mgx}{l}\cos\theta, \quad \text{..............(1)}$$

$$\frac{mx}{l}(l-\tfrac{1}{2}x)\,\ddot{\theta} = F - \frac{mgx}{l}\sin\theta. \quad \text{..............(2)}$$

It will be noticed that when $x=l$ these agree with equations (1) of Art. 60, difference of notation being allowed for. Again, the moment of inertia of AQ with respect to its centre is

$$\frac{mx}{l}\cdot\frac{1}{12}x^2;$$

hence, considering the angular motion,

$$\frac{1}{12}\frac{mx^3}{l}\ddot{\theta} = M - F\,.\,\tfrac{1}{2}x. \quad \text{.....................(3)}$$

But, putting $I=\frac{1}{3}Ml^2$, $h=\frac{1}{2}l$ in equation (1) of Art. 55*, we have

$$\ddot{\theta} = -\frac{3}{2}\frac{g}{l}\sin\theta. \quad \text{........................(4)}$$

* Or putting $M=0$, $x=l$, in (2) and (3) above, and eliminating F.

Substituting in (2) and (3) we find

$$F = \frac{mg \sin \theta}{4l^2} x (3x - 2l), \quad \ldots\ldots\ldots\ldots\ldots(5)$$

$$M = \frac{mg \sin \theta}{4l^2} x^2 (x - l). \quad \ldots\ldots\ldots\ldots\ldots(6)$$

These values depend only on the position of the bar. The tension, on the other hand, being

$$T = \frac{mgx}{l} \cos \theta + \frac{mx}{l} (l - \tfrac{1}{2}x) \dot{\theta}^2, \quad \ldots\ldots\ldots(7)$$

will involve the initial conditions. When $x = l$ we have $F = \frac{1}{4} mg \sin \theta$, in agreement with Art. 60 (7).

Calculations such as the above assume of course that the bar may be treated as absolutely rigid. This is legitimate, in the case of vibratory motions, if the period be long compared with any of the free periods of elastic vibration. When this condition is violated, the effect of elastic yielding has to be allowed for, and the results may differ greatly from those given by the preceding method.

70. Initial Reactions.

In some problems of interest a body, or a system of bodies, is released from rest in a position which is not one of equilibrium, and it is desired to know the initial accelerations of various points, or the initial reactions of the constraints.

Such questions are comparatively simple in that the initial velocities vanish, and it is therefore not necessary to integrate the equations of motion. Thus, in the case of a simple pendulum of length l released from rest at an inclination θ to the vertical, the radial acceleration $-l\dot{\theta}^2$ vanishes, and the initial tension T_0 of the string is therefore

$$T_0 = mg \cos \theta. \quad \ldots\ldots\ldots\ldots\ldots(1)$$

In some cases, however, the equation of energy is useful as an intermediate step, since on differentiation it leads to an equation involving the initial accelerations, and free from unknown reactions.

Some examples are appended.

Ex. 1. A bar, whose lower end A rests on a rough horizontal plane, is released from rest at an inclination θ to the vertical; it is required to find the horizontal and vertical components (F, R) of the initial reaction of the plane. This is really a case of the problem of Art. 60, but may be treated more simply as follows.

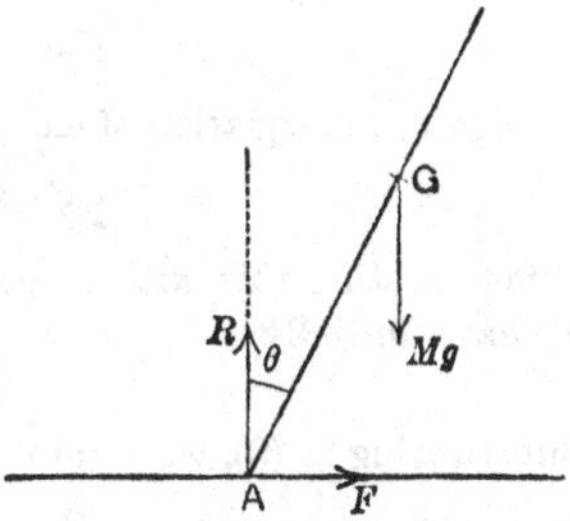

Let a be the distance of the mass-centre G from A, and κ the radius of gyration about G. Taking moments about A we have

$$M(\kappa^2+a^2)\ddot{\theta}=Mga\sin\theta. \quad ......(2)$$

Fig. 67.

Again, since the acceleration of G is $a\ddot{\theta}$ at right angles to AG we have, resolving horizontally and vertically,

$$Ma\ddot{\theta}\cos\theta=F, \quad Ma\ddot{\theta}\sin\theta=Mg-R. \quad ....(3)$$

Hence

$$\frac{F}{Mg}=\frac{a^2}{\kappa^2+a^2}\sin\theta\cos\theta, \quad \frac{R}{Mg}=\frac{\kappa^2+a^2\cos^2\theta}{\kappa^2+a^2} \quad ....(4)$$

If we assume the usual law of friction, the lower end will not begin to slip unless

$$\sin\theta\cos\theta>\mu\left(\cos^2\theta+\frac{\kappa^2}{a^2}\right), \quad ....(5)$$

where μ is the coefficient of friction. If we put $\mu=\tan\lambda$, the condition is

$$\sin(2\theta-\lambda)\not>\left(1+\frac{2\kappa^2}{a^2}\right)\sin\lambda. \quad ....(6)$$

Ex. 2. A cylindrical solid, of any form of section, free to roll on a rough horizontal plane, is released from rest in a given position.

The figure represents a section by a vertical plane through the mass-centre G, perpendicular to the length of the cylinder. Let P be the point of contact of this section, and GN the perpendicular from G to the plane. We write

$$GP=r, \quad GN=z, \quad PN=q, \quad ...(7)$$

and denote by κ the radius of gyration about a longitudinal axis through G.

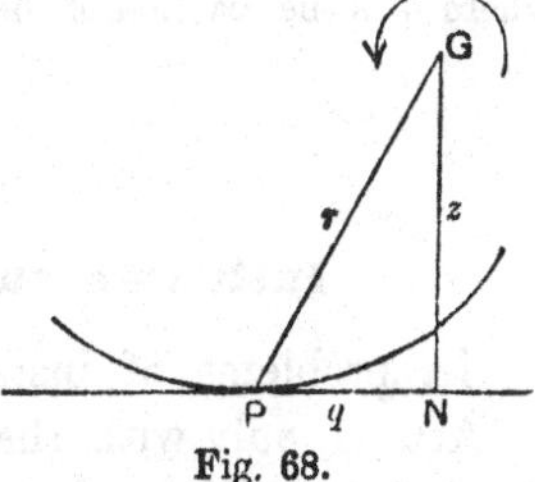

Fig. 68.

When the body turns with angular velocity ω about P, the horizontal and vertical components of the velocity (ωr) of G are

$$-\omega z, \quad \omega q,$$

respectively. Hence differentiating, and omitting terms containing $\dot{z}$ and $\dot{q}$, which vanish initially, the initial horizontal and vertical accelerations of G are

$$-\dot{\omega}z, \quad \dot{\omega}q,$$

respectively. Hence if F, R be the horizontal and vertical components of the reaction of the plane, we have

$$F = -M\dot{\omega}z, \quad R - Mg = M\dot{\omega}q. \quad \text{.......................(8)}$$

Again, the equation of energy, which holds throughout the motion, is

$$\tfrac{1}{2}M(\kappa^2 + r^2)\,\omega^2 + Mgz = \text{const.} \quad \text{.......................(9)}$$

Differentiating this with respect to t, putting $\dot{z} = \omega q$, and dividing out by ω, we have, initially,

$$(\kappa^2 + r^2)\,\dot{\omega} = -gq. \quad \text{.............................(10)}$$

Substituting in (8), we obtain

$$\frac{F}{Mg} = \frac{qz}{\kappa^2 + r^2}, \quad \frac{R}{Mg} = \frac{\kappa^2 + z^2}{\kappa^2 + r^2}. \quad \text{.......................(11)}$$

This problem includes Ex. 1 as a particular case.

Ex. 3. Let us suppose that the circumstances are the same as in Ex. 2, except that the horizontal plane is *smooth*.

Since G now moves in a vertical line, the instantaneous centre is at the intersection of the vertical through P with the horizontal through G. The equation of energy is therefore

$$\tfrac{1}{2}M(\kappa^2 + q^2)\,\omega^2 + Mgz = \text{const.} \quad \text{..........................(12)}$$

The vertical velocity of G is still given by the formula $\dot{z} = \omega q$; hence, differentiating (12), and dividing out by ω, we have, initially,

$$(\kappa^2 + q^2)\,\dot{\omega} + gq = 0. \quad \text{.............................(13)}$$

Also, since the initial vertical acceleration of G is $\dot{\omega}q$, we have

$$M\dot{\omega}q = R - Mg, \quad \text{.............................(14)}$$

where R is the reaction of the plane. Hence

$$\frac{R}{Mg} = \frac{\kappa^2}{\kappa^2 + q^2}. \quad \text{.............................(15)}$$

71. Instantaneous Impulses.

In problems of instantaneous impulse we are concerned, as in Art. 41, only with the time-integrals of the forces taken over the infinitely short duration of the impulse.

In the case of a rigid body moveable in two dimensions, let (u, v) be the velocity of the mass-centre G just before, and (u', v') its velocity just after the impulse. Also let ω, ω' be the corresponding angular velocities of the body. If ξ, η be the time-integrals of the external forces parallel to the axes of x, y, respectively, and

ν the time-integral of the moment of these forces about G, the principles of linear and angular momentum give at once

$$\left.\begin{array}{c} M(u'-u)=\xi, \quad M(v'-v)=\eta, \\ I(\omega'-\omega)=\nu, \end{array}\right\} \quad \text{..........(1)}$$

where M is the mass of the body, and I its moment of inertia about an axis through G normal to the plane xy.

These equations may also be derived from (1) and (2) of Art. 63, by integrating over the infinitely short duration of the impulse; thus

$$\left.\begin{array}{c} Mu'-Mu=\int_0^\tau X\,dt=\xi, \quad Mv'-Mv=\int_0^\tau Y\,dt=\eta, \\ I\omega'-I\omega=\int_0^\tau N\,dt=\nu. \end{array}\right\} \quad \text{...(2)}$$

Suppose, as an example, that a rigid body at rest, but free to move, is struck by an impulse ξ in a plane which is a principal plane of inertia at the mass-centre G. It is evident that G will begin to move in a line parallel to the impulse; let its initial velocity be u', and let ω' be the angular velocity communicated to the body. If we draw GP perpendicular to the line of the impulse we have

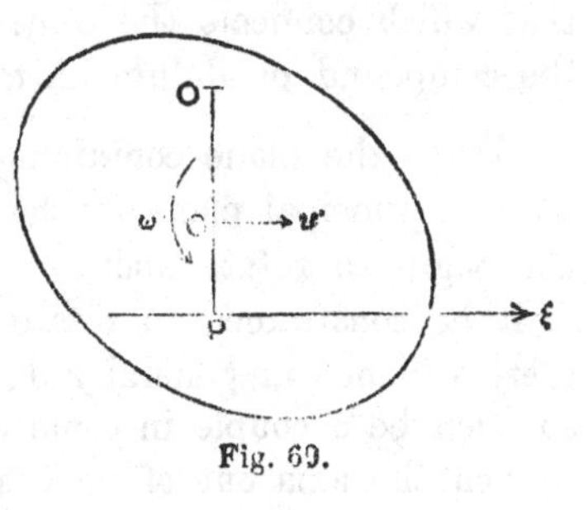

Fig. 69.

$$Mu'=\xi, \quad I\omega'=\xi.GP. \quad \text{.................(3)}$$

If PG be produced to O, the initial velocity of the point O of the body will be

$$u'-\omega'.GO=\frac{\xi}{M}\left(1-\frac{GO.GP}{\kappa^2}\right), \quad \text{...........(4)}$$

where κ is the radius of gyration about G. This will vanish, i.e. O will be the instantaneous centre of the initial rotation, provided

$$GP.GO=\kappa^2. \quad \text{.......................(5)}$$

If the body were fixed at O, but free to turn about this point, there would in general be an impulsive pressure (ξ_1, say) on this point, and an equal and opposite impulsive reaction $(-\xi_1)$ on the body. This would be determined by the equation

$$\xi - \xi_1 = Mu' = M\omega'.OG, \quad \text{.................(6)}$$

where ω' is to be found by taking moments about O; thus

$$M(\kappa^2 + OG^2)\,\omega' = \xi.OP. \quad \text{.................(7)}$$

Hence

$$\frac{\xi_1}{\xi} = 1 - \frac{M\omega'.OG}{\xi} = 1 - \frac{OP.OG}{\kappa^2 + OG^2} = \frac{\kappa^2 - GO.GP}{\kappa^2 + OG^2}. \quad \text{...(8)}$$

The impulse on the axis O will therefore vanish if the relation (5) be fulfilled, as was to be anticipated from the preceding investigation.

If, O being given, P is chosen so as to satisfy this condition, P is called the 'centre of percussion' with reference to O. It will be noticed that the relation between O and P is identical with that which connects the centres of oscillation and suspension in the compound pendulum (Art. 55).

When the plane containing the impulse and the mass-centre is not a principal plane at the latter point, the body, if free, will not begin to rotate about an axis perpendicular to this plane. If it be constrained to rotate about a given axis, the reactions there will not in general reduce to a single force; there will in addition be a couple in some plane through the axis, tending to wrench the axis out of its bearings. We do not enter into the proof of these statements, but a special case is discussed in Ex. 3 below.

Ex. 1. In the case of a uniform bar, of length $2a$, free to turn about one end, the centre of percussion is at a distance

$$a + \frac{\kappa^2}{a} = \frac{4}{3}a$$

from that end.

Again, the height above the table at which a billiard ball should be struck by a horizontal blow, in order that it may not slide, is

$$a + \frac{\kappa^2}{a} = \frac{7}{5}a,$$

where a is the radius. This should be the height of the cushions.

Ex. 2. Two similar bars AB, BC, smoothly jointed at B, are at rest in a straight line; to find the initial motion consequent on a blow F applied at C, at right angles to BC.

For simplicity we assume the mass-centres of the bars to be at their middle points. Let the length of each bar be $2a$, its radius of gyration about the centre κ, its mass M.

At the joint B there will be impulsive pressures $\pm X$, say, on the two bars. Let u_1, u_2 be the initial velocities of the centres of AB, BC, respectively, and ω_1, ω_2 the initial angular velocities, the positive senses being as shewn in the figure.

Resolving, and taking moments about the centres, we have

$$Mu_1 = -X, \quad M\kappa^2\omega_1 = -Xa, \quad \text{......................(9)}$$

$$Mu_2 = F + X, \quad M\kappa^2\omega_2 = Fa - Xa. \quad \text{..................(10)}$$

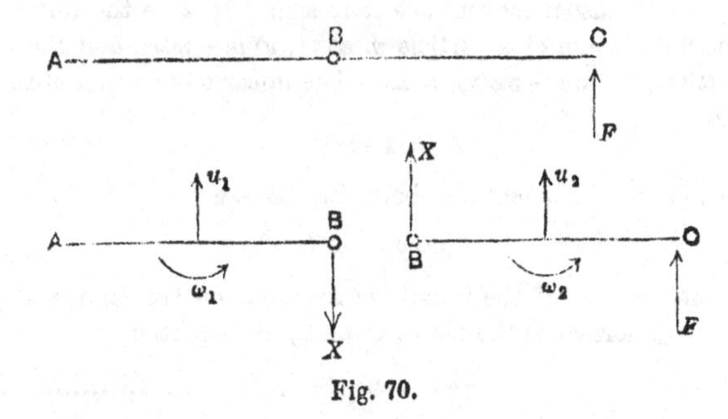

Fig. 70.

Again, the velocity of the joint B, considered as a point of the first bar, is $u_1 + \omega_1 a$, whilst considered as a point of the second bar it is $u_2 - \omega_2 a$. Hence

$$u_1 + \omega_1 a = u_2 - \omega_2 a. \quad \text{..............................(11)}$$

These five equations determine u_1, u_2, ω_1, ω_2, X. If we express the remaining unknowns in terms of X, by (9) and (10), and substitute in (11), we find

$$X = \frac{1}{2}\frac{a^2 - \kappa^2}{a^2 + \kappa^2} \cdot F. \quad \text{..............................(12)}$$

Hence

$$u_1 = -\frac{1}{2}\frac{a^2 - \kappa^2}{a^2 + \kappa^2} \cdot \frac{F}{M}, \quad a\omega_1 = -\frac{1}{2}\frac{a^2(a^2 - \kappa^2)}{\kappa^2(a^2 + \kappa^2)} \cdot \frac{F}{M}, \quad \text{......(13)}$$

$$u_2 = \frac{1}{2}\frac{3a^2 + \kappa^2}{a^2 + \kappa^2} \cdot \frac{F}{M}, \quad a\omega_2 = \frac{1}{2}\frac{a^2(a^2 + 3\kappa^2)}{\kappa^2(a^2 + \kappa^2)} \cdot \frac{F}{M}. \quad \text{.........(14)}$$

The cases $\kappa = 0$, $\kappa = a$ should be noticed, and interpreted.

If the bars be uniform, we have $\kappa^2 = \frac{1}{3}a^2$, and

$$u_1 = -\frac{1}{4}\frac{F}{M}, \quad a\omega_1 = -\frac{3}{4}\frac{F}{M}, \quad \text{.....................(15)}$$

$$u_2 = \frac{5}{4}\frac{F}{M}, \quad a\omega_2 = \frac{9}{4}\frac{F}{M}. \quad \text{.....................(16)}$$

The negative values of u_1 and ω_1 shew that the real directions of translation and rotation of AB are the opposites of those indicated by the arrows in the figure.

Ex. 3. A plane lamina, which is free to turn about a given axis in its plane, is struck at right angles by an impulse at a given point; to find the reactions at the axis.

Take rectangular axes Ox, Oy in the plane of the lamina, Oy being coincident with the axis of rotation. If (α, β) be the point at which the impulse (ζ) is delivered, the moments of the impulse about Ox, Oy will be $\zeta\beta$ and $-\zeta\alpha$, respectively, on the usual conventions as to sign. If ω be the initial angular velocity, the momentum of a particle m at (x, y) is $-m\omega x$, and the moments of this about Ox, Oy are $-m\omega xy$, $m\omega x^2$. The linear momentum of the lamina is therefore

$$-\omega\Sigma(mx),$$

and the moments of momentum about Ox, Oy are

$$-\omega\Sigma(mxy), \quad \omega\Sigma(mx^2).$$

The geometric sum of the impulsive reactions on the lamina at the axis will be a force ζ_1 normal to the plane, given by the equation

$$-\omega\Sigma(mx) = \zeta + \zeta_1. \quad \text{..........................(17)}$$

If λ_1, μ_1 be the moments of the same reactions about the coordinate axes, we have

$$-\omega\Sigma(mxy) = \lambda_1 + \zeta\beta, \quad \omega\Sigma(mx^2) = \mu_1 - \zeta\alpha. \quad \text{..............(18)}$$

Hence in order that the reactions may vanish, we must have

$$\alpha = \frac{\Sigma(mx^2)}{\Sigma(mx)}, \quad \beta = \frac{\Sigma(mxy)}{\Sigma(mx)}. \quad \text{.......................(19)}$$

The point thus determined may be called the 'centre of percussion.' When the product of inertia $\Sigma(mxy)$ vanishes, the result agrees with (5).

If we transform to new axes Ox', Oy', of which Oy' coincides with Oy, whilst Ox' makes an angle θ with it, the new coordinates of a particle m will be given by

$$x = x'\sin\theta, \quad y = y' + x'\cos\theta, \quad \text{......................(20)}$$

as is easily seen from a figure. It appears on substitution in (19) that the new coordinates α', β' of the centre of percussion are given by formulæ of the same type as before. It is therefore unnecessary to retain the accents.

There is no loss of generality if we assume the axis of x to pass through the mass-centre, so that

$$\Sigma (my)=0. \qquad\qquad(21)$$

If we now transfer the origin to this point, writing $x+h$, for x, we shall have

$$\Sigma (mx)=0, \qquad\qquad(22)$$

and we find

$$\alpha=\frac{\Sigma (mx^2)}{h\Sigma (m)}, \qquad \beta=\frac{\Sigma (mxy)}{h\Sigma (m)}, \qquad\qquad(23)$$

as the coordinates of the centre of percussion referred to axes through the mass-centre, the only restriction being that the axis of y is supposed parallel to the axis of rotation of the lamina.

The result is simplified if we make the axis of x coincide with that diameter of the 'central ellipse' of the lamina [*S.* 75, 77] which is conjugate to the axis of y. We then have $\Sigma (mxy)=0$, and

$$\alpha=\frac{a^2}{h}, \quad \beta=0, \qquad\qquad(24)$$

where

$$a^2=\frac{\Sigma (mx^2)}{\Sigma (m)}, \qquad\qquad(25)$$

i.e. a^2 is the mean square of the abscissæ of the various particles of the lamina. The geometrical meaning of (24) is that the centre of percussion is the 'anti-pole' of the axis of rotation with respect to the central ellipse.

In the case of a lamina of uniform surface-density, the centre of percussion, as determined by (19) or (24), coincides with the centre of pressure of the area when immersed so that the axis of rotation is the surface-line* [*S.* 95].

72. The Ballistic Pendulum.

The 'ballistic pendulum' is a device† for measuring the velocity of a bullet by means of the momentum communicated to the apparatus, which consists of a compound pendulum carrying a block of wood, or a box of sand, into which the bullet is fired horizontally. Owing to the resistance, the bullet is brought to relative rest before the pendulum has moved through an appreciable angle.

Let m be the mass of the bullet, v its velocity, and c the depth of its horizontal path below the axis of the pendulum. The moment of momentum of the apparatus, immediately after

* Greenhill, *Hydrostatics*, London, 1894, p. 65.

† Described by B. Robins (1707–51) in his *New Principles of Gunnery*, 1742.

the impact is therefore mvc. Hence, if ω be the angular velocity communicated to the pendulum, we have

$$mvc = Mk^2\omega, \quad\text{.........................(1)}$$

where the constants M, k refer to the pendulum as modified by the inclusion of the bullet; this correction is, however, usually unimportant. If α be the angle through which the pendulum swings before coming to rest we have, by the equation of energy,

$$\tfrac{1}{2}Mk^2\omega^2 = Mgh(1 - \cos\alpha), \quad\text{.................(2)}$$

where h denotes the distance of the mass-centre from the axis of suspension. Hence

$$v^2 = 4\left(\frac{M}{m}\right)^2 \cdot \left(\frac{k}{c}\right)^2 \cdot gh \sin^2 \tfrac{1}{2}\alpha. \quad\text{.............(3)}$$

If T be the time of a *small* oscillation of the pendulum we have

$$k^2 = \frac{gT^2h}{4\pi^2}, \quad\text{.........................(4)}$$

by Art. 55. Hence

$$v = \frac{1}{\pi} \cdot \frac{M}{m} \cdot \frac{h}{c} \cdot \sin \tfrac{1}{2}\alpha \,.\, gT. \quad\text{..................(5)}$$

The velocity v is thus expressed in terms of quantities which can be determined by observation.

The angle α is sometimes determined by means of a tape attached to the pendulum, at a point in the plane of the mass-centre and the axis of suspension, which is drawn out during the swing. If b be the distance of the point of attachment from the axis, the length thus drawn out, being equal to the chord of the arc $b\alpha$, is $2b \sin \frac{1}{2}\alpha$. Denoting this by x, we have

$$v = \frac{1}{2\pi} \cdot \frac{M}{m} \cdot \frac{h}{c} \cdot \frac{x}{b} \cdot gT. \quad\text{...................(6)}$$

In order that there may be no impulsive jar on the axis, the line of fire should pass through the centre of percussion (Art. 71).

In some varieties of the instrument the gun from which the bullet is fired forms part of the pendulum, which is set in motion by the recoil. This arrangement is less accurate, since the momentum given to the pendulum is not strictly equal and opposite to that of the bullet, but is increased on account of the momentum given to the gases generated in the explosion.

73. Effect of Impulses on Energy.

Taking the case of motion in two dimensions, let us suppose that a body is acted on by a system of instantaneous impulses applied to it at various points, and that these are equivalent to an impulsive force (ξ, η) at the mass-centre G, and an impulsive couple ν. As in Art. 71, the sudden change of motion is given by

$$\left.\begin{array}{c} M(u'-u)=\xi, \quad M(v'-v)=\eta, \\ I(\omega'-\omega)=\nu. \end{array}\right\} \qquad \text{.............(1)}$$

The increase of energy due to the impulses is therefore

$$\begin{aligned} &\{\tfrac{1}{2}M(u'^2+v'^2)+\tfrac{1}{2}I\omega'^2\}-\{\tfrac{1}{2}M(u^2+v^2)+\tfrac{1}{2}I\omega^2\} \\ &=\tfrac{1}{2}M(u'^2-u^2)+\tfrac{1}{2}M(v'^2-v^2)+\tfrac{1}{2}I(\omega'^2-\omega^2) \\ &=\xi\,.\,\tfrac{1}{2}(u+u')+\eta\,.\,\tfrac{1}{2}(v+v')+\nu\,.\,\tfrac{1}{2}(\omega+\omega'). \qquad \text{.........(2)} \end{aligned}$$

We obtain a result equivalent to this if we multiply each of the forces constituting the impulse by the arithmetic mean of the initial and final velocities of its point of application, resolved in the direction of the force. For let us imagine the same body, in the same position, to receive a displacement such that the component translations of G are $\tfrac{1}{2}(u+u')\tau$, $\tfrac{1}{2}(v+v')\tau$, and the accompanying rotation is $\tfrac{1}{2}(\omega+\omega')\tau$, where τ is some infinitesimal magnitude of the dimensions of a time. The expression in the last line of (2) would then represent the work done during this displacement by a force $(\xi/\tau, \eta/\tau)$ acting at G, and a couple ν/τ. Now if F be any one of the forces constituting the impulse in the actual problem, and q, q' the initial and final velocities of its point of application, resolved in its own direction, the resolved displacement of this point in the imagined case would be

$$\tfrac{1}{2}(q+q')\tau,$$

and the work done by a force F/τ would be $F\,.\,\tfrac{1}{2}(q+q')$. The total work done by the system of such forces would accordingly be

$$\Sigma\,\{F\,.\,\tfrac{1}{2}(q+q')\}. \qquad \text{........................(3)}$$

This result may also be obtained by calculating directly the work done by the impulsive forces. The simplest plan is to imagine these to be replaced by *constant* forces F/τ acting for a very short time τ. The changes of velocity which take place within the interval τ will then be proportional to the total impulse up to the instant considered, and therefore proportional to the time that has elapsed since the beginning of the interval. The mean velocities of the points of application, resolved in the directions of the respective forces, will accordingly be represented by $\frac{1}{2}(q+q')$, and the corresponding displacements by $\frac{1}{2}(q+q')\tau$. In this way we are led again to the expression (3) for the total work done. It will be observed that changes of velocity due to ordinary finite accelerations have been neglected; this is legitimate since the interval τ is ultimately regarded as infinitesimal.

Ex. 1. In the ballistic pendulum the impulse given to the pendulum is mv, practically, and the initial velocity of its point of application is $c\omega$. The energy given to the pendulum is therefore $\frac{1}{2}mvc\omega$, which is equal to $\frac{1}{2}Mk^2\omega^2$, by Art. 72 (1). The total loss of energy is therefore

$$\tfrac{1}{2}mv^2 - \tfrac{1}{2}Mk^2\omega^2 = \tfrac{1}{2}mv^2\left(1 - \frac{mc^2}{Mk^2}\right). \quad \text{.................(4)}$$

Ex. 2. In the problem of Art. 71, Ex. 2, the velocity given to the point C is

$$u_2 + a\omega_2 = \frac{7}{2}\frac{F}{M}, \quad \text{..............................(5)}$$

and the energy due to the impulse is therefore

$$\frac{7}{4}\frac{F^2}{M}. \quad \text{....................................(6)}$$

If the joint B had been rigid, the energy would have been F^2/M, as is easily found. This illustrates a general principle that the introduction of any constraint in a mechanical system diminishes the energy produced by given impulses. (See Art. 108.)

EXAMPLES. XV.

1. A bar rotating on a smooth horizontal plane about one end (which is fixed) suddenly snaps in two; describe the subsequent motion of each portion.

2. Having given a solid and a hollow sphere of the same size and weight, how would you ascertain which is the hollow one?

3. A vertical thread unwinds itself from a reel, the upper end of the thread being fixed. Prove that the downward acceleration of the reel is

$$\frac{a^2}{a^2+\kappa^2}\cdot g,$$

and that the tension of the thread is

$$\frac{\kappa^2}{a^2+\kappa^2}\cdot Mg,$$

where M is the mass of the reel, a its radius, and κ its radius of gyration about the axis.

4. A reel of thread rests by its slightly projecting ends on a horizontal table. The free end of the thread is brought out from the under side of the reel, and pulled horizontally with a given force F. In what direction will the centre of the reel move, and with what acceleration? (The radii of the reel at the centre and at the ends are given; also the radius of gyration.)

5. A pulley of mass M, radius a, and moment of inertia I, runs in the bight of a string one end of which is fixed, while the other passes over a fixed pulley of radius a', and moment of inertia I', and carries a mass m hanging vertically. Prove that the acceleration of m when the system is running freely, the free parts of the string being supposed vertical, is

$$\left(m-\frac{1}{2}M\right)g \div \left(m+\frac{1}{4}M+\frac{I'}{a'^2}+\frac{1}{4}\frac{I}{a^2}\right).$$

6. A circular cylinder can roll on a horizontal plane, and the latter is made to move horizontally in any manner at right angles to the axis. Prove that if the plane and cylinder were initially at rest they will come to rest simultaneously, and that the distance travelled by the centre of the cylinder will be to that travelled by the plane as $\kappa^2/(\kappa^2+a^2)$, where a is the radius of the cylinder and κ its radius of gyration.

7. A bicycle is running on the inside of a circular loop in a vertical plane. If a be the radius of each wheel, R that of the loop, v the apparent velocity of the bicycle (i.e. the velocity with which either point of contact moves along the track), find the angular velocity of the wheels, and the angular velocity of the frame. [$v(1/a-1/R)$; v/R.]

8. A circular hoop of radius a rolls in contact with a fixed circular cylinder of radius b, which it surrounds, its plane being perpendicular to the axis of the cylinder. Through what angle does the hoop turn while the point of contact makes a complete circuit of the cylinder?

Also if T be the apparent period of revolution of the hoop, find the pressure between the hoop and cylinder, neglecting gravity.

[$2\pi(a-b)/a$; $4\pi^2M(a-b)/T^2$.]

9. A uniform rod is placed like a ladder with one end against a smooth vertical wall, and the other end on a smooth horizontal plane. It is released from rest at an inclination α to the vertical. Calculate the initial pressures on the wall and plane. [$\frac{3}{4}Mg\sin\alpha\cos\alpha$; $Mg(1-\frac{3}{4}\sin^2\alpha)$.]

10. Also prove that the bar will cease to touch the wall when the upper end has fallen through one-third of its original altitude.

11. Two particles m, m' are connected by a light rod of length l, and m is free to move on a smooth horizontal plane. If the rod be slightly disturbed from the unstable position, prove that the angular velocity ω with which it reaches the horizontal position is given by

$$\omega^2 = 2g/l.$$

What is the nature of the path of m'?

12. A uniform chain of mass M and length l hangs in equilibrium over a pulley of radius a. If it be just started from rest, and does not slip relatively to the pulley, prove that the equation of motion is of the form

$$(I + Ma^2)\,\omega\frac{d\omega}{d\theta} = \frac{2Mga^2}{l}\,\theta,$$

where I is the moment of inertia of the pulley.

Find the tensions of the chain at the points where it leaves the pulley, when the latter has turned through an angle θ.

13. A uniform plank of thickness $2h$ rests across the top of a fixed circular cylinder of radius a whose axis is horizontal. Prove that if it be set in motion the equation of energy is

$$\tfrac{1}{2}(\kappa^2 + h^2 + a^2\theta^2)\,\dot{\theta}^2 + g\,\{a\theta\sin\theta - (a+h)(1-\cos\theta)\} = \text{const.},$$

on the assumption that the motion is one of pure rolling.

Hence shew that if $a > h$ the horizontal position is stable, and that the period of a small oscillation is the same as for a simple pendulum of length

$$\frac{\kappa^2 + h^2}{a - h}.$$

14. A solid ellipsoid whose semi-axes are a, b, c rests, with its shortest axis ($2c$) vertical, on a rough table. Prove that it has two normal modes of vibration, and that the lengths of the corresponding simple pendulums are

$$\frac{(a^2+6c^2)\,c}{5\,(a^2-c^2)}, \qquad \frac{(b^2+6c^2)\,c}{5\,(b^2-c^2)}.$$

Prove that the longer period belongs to that vibration which is parallel to the two shorter axes.

15. If in Kater's pendulum the knife-edges are replaced by cylindrical pins of equal radius, prove that when the period of oscillation about each is the same the length of the equivalent simple pendulum is equal to the shortest distance between the surfaces of the pins.

Prove also that if the periods are nearly but not exactly equal the formula (13) of Art. 56 will still hold very approximately, if the radius of the cylinders be small, provided h_1, h_2 denote the shortest distances of the mass-centre from the surfaces of the pins.

16. A bar AB hangs by two equal crossed strings, attached to its ends, from two points C, D at the same level, such that $AB = CD$, and it is restricted to motion in the vertical plane through CD. If $AB = 2c$, and $2b$ be the depth of AB below CD, prove that if $b > c$ the length of the equivalent simple pendulum is

$$\frac{2(\kappa^2 + b^2)b}{b^2 - c^2}.$$

17. A wheel of radius a, whose mass-centre G is at a distance c from the axis, can turn freely in a vertical plane, and carries a weight m by a string hanging tangentially on one side. Form the equation of energy in terms of I, M, θ and the other given quantities, M denoting the mass of the wheel, I its moment of inertia about the axis, and θ the inclination to the vertical of the radius through G.

When equilibrium is possible, shew that stable and unstable positions alternate; and prove that the time of a small oscillation about a stable position is

$$2\pi \sqrt{\left(\frac{I + ma^2}{Mgc \cos \alpha}\right)},$$

where α is the equilibrium value of θ.

18. Two equal particles are attached at B and C to a string $ABCD$, such that $AB = CD = l$, $BC = 2a$; and the ends A, D are fixed at the same level. When the system is in equilibrium, the portions AB, DC make angles α with the horizontal. Prove that when the system swings in the vertical plane through AD the length of the equivalent simple pendulum is

$$\frac{l \sin \alpha}{1 + l/a \,.\, \cos^3 \alpha}.$$

Obtain the corresponding result for a uniform *bar* BC suspended by equal strings AB, DC.

19. The axis of suspension of a compound pendulum is moved horizontally to and fro, its displacement at time t being ξ. Prove that the accurate equation of motion is

$$l \frac{d^2\theta}{dt^2} + g \sin \theta = -\frac{d^2\xi}{dt^2} \cos \theta,$$

where l is the length of the equivalent simple pendulum.

20. A bar hanging from a fixed point by a string of length l attached to one end makes small oscillations in a vertical plane. Prove that the periods $(2\pi/n)$ of the two normal modes of vibration are determined by the equation

$$n^4 - \frac{\kappa^2 + a^2 + al}{\kappa^2 l} gn^2 + \frac{g^2 a}{\kappa^2 l} = 0,$$

where a is the distance of the point of attachment from the mass-centre (G), and κ the radius of gyration about G.

Find approximate values of the roots when κ/l and a/l are both small; and examine the nature of the corresponding modes.

21. Prove that in the case of the double pendulum (Art. 68) the kinetic energy in any configuration is

$$\tfrac{1}{2}(A\dot{\theta}^2 + 2H\dot{\theta}\dot{\phi} + B\dot{\phi}^2),$$

where $A = Mk^2 + ma^2$, $H = mab\cos(\theta - \phi)$, $B = m(b^2 + \kappa^2)$; and that the potential energy is

$$\text{const.} - Mgh\cos\theta - mg(a\cos\theta + b\cos\phi).$$

Prove that the angular momentum about the axis O is

$$(A + H)\dot{\theta} + (H + B)\dot{\phi}.$$

22. A uniform bar of length l is made to move at right angles to its length with acceleration f by means of equal forces applied to it at the ends. Prove that the shearing force and bending moment at a distance x from one end are given by

$$F = \frac{mf}{l}(\tfrac{1}{2}l - x), \quad M = -\frac{mf}{2l}x(l - x).$$

EXAMPLES. XVI.

(Impulsive Motion.)

1. A circular hoop which is perfectly free is struck at a point of the circumference, find the initial axis of rotation (1) when the blow is tangential, (2) when it is at right angles to the plane of the ring.

2. A uniform rod at rest is struck at one end by a perpendicular impulse; prove that its kinetic energy will be greater than if the other end had been fixed, in the ratio 4 : 3.

3. A uniform semi-circular plate of radius a is struck perpendicularly to its plane at the middle point of the bounding diameter; find the distance of the axis of initial rotation from this diameter. [$\cdot 59\,a$]

4. A rigid lamina is moving in any manner in its own plane, when a point O in it is suddenly fixed. What must be the position of O in order that the lamina may be reduced to rest?

5. A square plate is spinning freely about a diagonal with angular velocity ω. Suddenly a corner not in this diagonal is fixed. Prove that the new angular velocity is $\tfrac{1}{7}\omega$.

6. A lamina free to turn in its own plane about a fixed point O is struck by an impulsive couple ν. Find the impulsive pressure at O; and prove that it cannot exceed $\tfrac{1}{2}\nu/\kappa$.

7. A bar, not necessarily uniform, is struck at right angles to its length at any point P, and the velocity produced at any other point Q is v. Prove that if the same impulse had been applied at Q, the velocity at P would have had the same value v.

8. A point P of a lamina is struck by an impulse in any direction PP' in the plane of the lamina, and the component velocity of any other point Q in a direction QQ' is v. Prove that if the same impulse had been applied at Q in the direction QQ', the velocity of P in the direction PP' would have been v.

9. A uniform rod of length $2a$ is attached at one end by a string of length l to a fixed point, about which it revolves, on a smooth horizontal plane. Find at what point an impulse must be applied to the rod to reduce it to rest.

10. A rigid lamina free to turn about a fixed point O is struck by an impulse ξ in a line meeting OG (produced) at right angles in P. Prove that the energy generated is less than if the lamina had been perfectly free by the amount

$$\frac{1}{2}\frac{(GP \,.\, GO-\kappa^2)^2}{\kappa^2(\kappa^2+OG^2)} \,.\, \frac{\xi^2}{M}.$$

11. Two equal uniform rods AB, AC, smoothly jointed at A, are at rest, forming a right angle. If a blow be delivered at C in a direction perpendicular to AC, the initial velocities of the centres of AB and AC are as $2 : 7$.

12. AB, CD are two equal parallel uniform rods, and the points A, C are connected by a string perpendicular to both. If a blow is administered at B, at right angles to AB, the initial velocity of B is 7 times that of D.

13. Four equal uniform rods, smoothly jointed together, are at rest in the form of a square. Prove that if an impulsive couple be applied to one of the rods its initial angular velocity will be one-eighth of what it would have been if it had been free.

14. Three equal uniform rods AB, BC, CD, jointed at B and C, start from rest in a horizontal line, the outer rods being in contact with smooth pegs at A and D. Prove that the initial acceleration of the middle rod is $\frac{6}{5}g$.

Why is this greater than g?

15. Two bars of masses m_1, m_2, and lengths $2a_1$, $2a_2$ are in a straight line, hinged together at a common extremity. If either bar be struck by an impulsive couple ν, the initial angular velocity of the other bar is

$$-\frac{a_1a_2}{(a_1^2+\kappa_1^2)\,m_2\kappa_2^2+(a_2^2+\kappa_2^2)\,m_1\kappa_1^2} \,.\, \nu,$$

where κ_1, κ_2 are the radii of gyration of the two bars about their centres of mass, which are supposed to be at the middle points.

16. Two equal uniform rods, hinged together, hang vertically from one extremity. If an impulsive couple ν be applied to the lower rod, the energy generated is $\frac{6}{7}\nu^2/Ma^2$, where M is the mass, and $2a$ the length, of each rod.

17. A uniform bar of length $2a$ rests symmetrically on two pegs at a distance $2c$ apart in a horizontal line. The pegs are supposed to be sufficiently rough to prevent slipping. One end of the bar is raised and released; investigate the subsequent motion on the hypothesis that there is no recoil whenever the bar strikes a peg.

Prove that if $c^2<\frac{1}{3}a^2$, the angular velocity is instantaneously reduced at each impact in the ratio $(a^2-3c^2)/(a^2+3c^2)$.

CHAPTER X

LAW OF GRAVITATION

74. Statement of the Law.

The law of gravitation propounded by Newton is to the effect that two particles of masses m, m', at a distance r apart which is great compared with the dimensions of either, attract one another with a force proportional to

$$\frac{mm'}{r^2}.$$

The steps which lead up to this induction may be stated as follows.

We have seen that the attractions of the earth on different bodies at the same place near the earth's surface are proportional to the respective masses. It is a natural assumption to make that this law of proportionality to the attracted mass is not a local peculiarity, but will hold for any position in space. Also, regarding the attraction of the earth as the resultant of the attractions of its smallest particles, the simplest assumption is that these elementary attractions follow the same law, and that consequently the attraction of a particle m' on a particle m, so far as it depends on m, is proportional to m. Since the attraction is mutual, it must also vary as m'. In this way we are led to the hypothesis that the attraction varies as the product mm', and may therefore be equated to

$$mm'\phi(r),$$

where $\phi(r)$ is some (as yet undetermined) function of the mutual distance r.

The *acceleration* produced by a mass m at a distance r will therefore be $m\phi(r)$, independent of the mass of the attracted body.

The indication as to the form of the function $\phi(r)$ is furnished by Kepler's 'Third Law' of planetary motion, which is that the squares of the periodic times of the various planets are to one another as the cubes of their mean distances from the sun. This law, if accurate, can hardly fail to hold for orbits which are exactly, and not merely approximately, circular. Hence, if r, r' be the radii of two such orbits, and T, T' the corresponding periods of revolution, the statement is

$$\frac{T^2}{T'^2} = \frac{r^3}{r'^3}. \qquad \text{.........................(1)}$$

Since the acceleration in a circular orbit is $(2\pi/T)^2 . r$, we have

$$\frac{\phi(r)}{\phi(r')} = \frac{r}{r'} . \frac{T'^2}{T^2} = \frac{r'^2}{r^2}. \qquad \text{....................(2)}$$

Hence $\phi(r)$ varies as $1/r^2$.

This argument ignores the distinction between the absolute acceleration of a planet, and its acceleration relative to the sun, which itself yields somewhat to the attraction of the planet. Owing to the great inertia of the sun compared with that of a planet, the error thus involved is very slight; we shall see presently how to allow for it (Art. 81).

If we introduce a constant γ to denote the force between two unit masses at unit distance apart, the force between two particles m, m' at a distance r will be

$$\frac{\gamma m m'}{r^2}. \qquad \text{..............................(3)}$$

This constant γ is called the 'constant of gravitation.'

75. Simple Astronomical Applications.

If the truth of the law of gravitation be assumed, some interesting astronomical applications can at once be made, on the hypothesis of circular orbits.

The only distant body which is influenced mainly by the earth's attraction is the moon. If we compare the accelerations produced by the earth on the moon and on a particle near the earth's surface, we are led indirectly to an estimate of the moon's distance, which can be compared with that found by (essentially) geometrical operations. Thus if a be the radius of the earth, and D the distance of the moon, the acceleration of the moon must be $g(a/D)^2$, on Newton's law*. But if T be the period of revolution of the moon about the earth, the acceleration in the circular orbit is $(2\pi/T)^2 . D$. Equating, we find

$$\left(\frac{a}{D}\right)^3 = \frac{4\pi^2 a}{gT^2}. \qquad \text{.........................(1)}$$

If we put

$$a = 2{\cdot}09 \times 10^7 \text{ ft.}, \quad g = 32{\cdot}2 \text{ ft./sec.}^2, \quad T = 27{\cdot}3 \times 86400 \text{ sec.},$$

we find
$$\frac{a}{D} = \frac{1}{60{\cdot}1}.$$

The ratio a/D is the sine of the moon's 'horizontal parallax,' and can be found from observation with considerable accuracy. The above numerical value agrees well with the result obtained in this way.

The above calculation was made, in a somewhat different form, by Newton, and furnished the first definite test, apart from Kepler's law, of his theory of universal gravitation.

Again, by comparing the accelerations produced by different bodies at known distances we are able to compare the masses of these bodies. In particular, we can compare the mass of any planet having a satellite with that of the sun.

Thus, denoting the mass of the sun by S, that of the earth by E, the period of revolution of the earth round the sun by T, that of the moon relative to the earth by T', the radius of the earth's orbit by D, and that of the moon's orbit by D', we have, comparing the central accelerations in the two cases,

$$\frac{E}{D'^2} : \frac{S}{D^2} = \frac{D'}{T'^2} : \frac{D}{T^2}, \qquad \text{.........................(2)}$$

whence
$$\frac{E}{S} = \left(\frac{D'}{D}\right)^3 . \left(\frac{T}{T'}\right)^2. \qquad \text{.........................(3)}$$

* It is known from the theory of Attractions that, on the law of the inverse square, any body composed of spherical layers of uniform density attracts external bodies as if its mass were concentrated at the centre.

For a rough calculation we may put

$$\frac{D}{D'} = 389, \quad \frac{T}{T'} = 13{\cdot}37,$$

whence

$$\frac{E}{S} = \frac{1}{329300}.$$

This calculation, again, is due in principle to Newton*.

76. The Problem of Two Bodies.

It is hardly necessary to say that the further investigations of Newton and his successors, in which account is taken of the attractions of the planets on one another and on the sun, have abundantly confirmed the accuracy of the law of the inverse square, by shewing that it is able to explain the actual motions of the planets in minute detail.

The first step was to consider the 'Problem of Two Bodies' without the restriction to a circular orbit. A slight simplification is made if we assume, to begin with, that the mass of one of the bodies (a planet) is so small compared with that of the other (the sun) that the acceleration produced in the latter may be neglected. We have then the problem of the motion of a particle about a fixed centre of force under a central acceleration μ/r^2, where μ denotes the acceleration at unit distance.

The particle will obviously remain in the plane containing the centre of force and the tangent to the orbit at any given instant. Since there are two degrees of freedom in this plane, we require two differential equations of motion. These may be formulated in various ways; but the simplest plan is to proceed from the two first integrals which are supplied by the principle of angular momentum and the equation of energy.

If v be the velocity of the particle, and p the perpendicular from the centre of force on the tangent to its path, we have, by Art. 48,

$$pv = h, \quad \text{.............................}(1)$$

where h is a constant.

* The sun's distance was at that time greatly under-estimated, and the ratio (1/169282) which he obtained was accordingly much too large.

Again, taking the mass of the particle as unit of mass, the kinetic energy is $\frac{1}{2}v^2$, and the potential energy is $-\mu/r+\text{const.}$, by Art. 43. Hence

$$\tfrac{1}{2}v^2-\frac{\mu}{r}=\text{const.} \quad\text{.......................(2)}$$

Combined with (1) this gives

$$\frac{h^2}{p^2}=\frac{2\mu}{r}+C, \quad\text{.........................(3)}$$

which is, virtually, a differential equation of the first order to determine the path. Any equation of this type, connecting p and r, is called a 'tangential-polar' equation; it completely determines the curve, except as to its orientation about the origin*.

The particular form (3) may be identified with the tangential-polar equation of a conic with respect to the focus as origin†. In the case of the ellipse we have

$$\frac{l}{p^2}=\frac{2}{r}-\frac{1}{a}, \quad\text{.........................(4)}$$

whilst for a branch of the hyperbola, referred to the inner focus,

$$\frac{l}{p^2}=\frac{2}{r}+\frac{1}{a}, \quad\text{.........................(5)}$$

where in each case l denotes the semi-latus-rectum, and a the semi-axis containing the focus in question. In the transition case of the parabola we have $a=\infty$, and

$$\frac{l}{p^2}=\frac{2}{r} \quad\text{.............................(6)}$$

Comparing with (3), we see that the equations are identical provided

$$l=\frac{h^2}{\mu}, \quad a=\mp\frac{\mu}{C} \quad\text{.......................(7)}$$

The orbit will therefore be an ellipse, hyperbola, or parabola according as the constant C in (3) is negative, positive, or zero, respectively.

Now if a particle were to fall from rest at infinity, so that $v=0$, $r=\infty$, initially, its velocity when at the distance r would be

* *Inf. Calc.*, Art. 143.

† *Ibid.*

$\sqrt{(2\mu/r)}$, by (2). This is called the 'velocity from infinity,' or the 'critical velocity,' corresponding to the distance r. We learn then from (7) that the orbit will be an ellipse, hyperbola, or parabola, according as the velocity at any point is less than, greater than, or equal to the velocity from infinity. In other words, if the particle is started from any position with insufficient energy to carry it to infinity it will describe an ellipse; if the energy be more than sufficient it will describe a branch of a hyperbola, approximating ultimately to motion with constant velocity along an asymptote, as the central force becomes less intense; whilst in the transition case it will describe a parabola with a velocity tending ultimately to zero.

Combining (3) with (4) and (5), we have the very convenient formula

$$v^2 = \mu\left(\frac{2}{r} \mp \frac{1}{a}\right), \quad \text{.......................(8)}$$

where the upper sign relates to an elliptic, and the lower to a hyperbolic path.

The formula (1), which holds independently of the particular law of (central) force has an important interpretation. If δs be an element of the path, $p\delta s$ is twice the area of the triangle enclosed by δs and the radii drawn from the centre of force to its extremities. Hence, denoting the area of this triangle by δA, we have

$$\delta A = \tfrac{1}{2}p\delta s = \tfrac{1}{2}pv\delta t = \tfrac{1}{2}h\delta t,$$

or

$$\frac{dA}{dt} = \tfrac{1}{2}h, \quad \text{...........(9)}$$

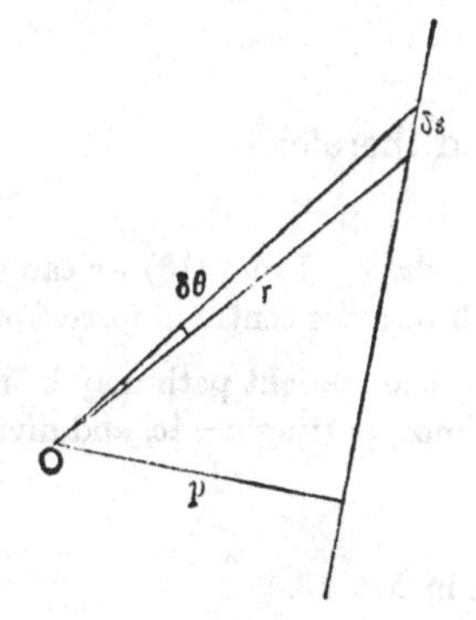

Fig. 71.

if A be the area swept over by the radius vector from some assigned epoch up to the instant t. In other words, the radius vector sweeps over equal areas in equal times, the rate per unit time being $\frac{1}{2}h$. If r, θ be polar coordinates referred to the centre of force, we may put

$$\delta A = \tfrac{1}{2}r^2\delta\theta,$$

and the equation (1) takes the form

$$r^2\frac{d\theta}{dt}=h, \qquad \text{.........................(10)}$$

which will be required presently.

The property just proved leads to an expression for the period (T) of revolution in an elliptic orbit. Since the whole area of the ellipse is swept over by the radius vector in the period, we have

$$hT=2\pi ab, \qquad \text{.........................(11)}$$

where b is the minor semi-axis. Also, from (7),

$$h=\sqrt{(\mu l)}=\sqrt{(\mu b^2/a)}. \qquad \text{.................(12)}$$

Hence $$T=\frac{2\pi a^{\frac{3}{2}}}{\sqrt{\mu}}. \qquad \text{.........................(13)}$$

It follows that in different elliptic orbits described under the same absolute acceleration (μ), the squares of the periodic times vary as the cubes of the major semi-axes*.

If n denote what is called in Astronomy the 'mean motion' in the elliptic orbit, i.e. the mean angular velocity of the radius vector about the centre of force, we have

$$nT=2\pi, \qquad \text{.........................(14)}$$

and therefore $$n=\sqrt{\frac{\mu}{a^3}}. \qquad \text{.........................(15)}$$

Ex. 1. From (13) we can deduce the time which a particle would take to fall into the centre of force from rest at any given distance c.

The straight path may be regarded as the half of an infinitely flat ellipse. Hence, putting $a=\frac{1}{2}c$, and dividing by 2, we find that the required time is

$$\frac{1}{4\sqrt{2}}\cdot\frac{2\pi c^{\frac{3}{2}}}{\sqrt{\mu}},$$

as in Art. 16.

The law of description of the path, in terms of areas swept over in the auxiliary circle of Fig. 11, is also easily inferred from the present theory.

* The major semi-axis, being the arithmetic mean of the greatest and least distances from the sun, is called in Astronomy the 'mean distance.' It must not be identified with the true mean of the distances at equal infinitesimal intervals of time. See Art. 79.

Ex. 2. To find the mean kinetic energy in an elliptic orbit, i.e. the average for equal infinitesimal intervals of time, we have

$$\int v^2 dt = \int v\,ds = h\int \frac{ds}{p} = \frac{h}{b^2}\int p' ds, \quad\text{......................(16)}$$

where p' is the perpendicular on the tangent from the second focus. The last integral, when taken round the perimeter of the ellipse, is equal to twice the area, and therefore to hT. Hence

$$\frac{1}{T}\int_0^T v^2 dt = \frac{h^2}{b^2} = \frac{\mu}{a} = n^2 a^2, \quad\text{......................(17)}$$

by (7) and (15). The mean kinetic energy is therefore the same as that of a particle revolving in a circle of radius equal to the mean distance a, with the mean angular velocity n.

77. Construction of Orbits.

The orbit of a particle which is started from a given point P with a given velocity, in a given direction, can be constructed as follows.

First, suppose that the velocity of projection is *less* than the critical velocity. The formula

$$v^2 = \mu\left(\frac{2}{r} - \frac{1}{a}\right) \quad\text{......................(1)}$$

determines the length $2a$ of the major axis, since μ, r, and v are supposed given. If a circle be described with the centre of force (S) as centre, and radius $2a$, a particle at rest anywhere on this circle will have the same energy as the given particle, if of the same mass. We may call this the 'circle of zero velocity.' It corresponds to the common directrix of the parabolic paths in Figs. 24, 25 (Art. 27).

If H be the second focus, the distance PH is given by the formula

$$PH = 2a - SP. \quad\text{......................(2)}$$

Hence H must lie on a circle, with P as centre, touching the circle of zero velocity. The direction of PH is determined by the fact that the given tangent to the path at P must be equally inclined to the two focal distances of P.

Again, let it be required to find the direction in which a particle must be projected (with the given velocity) from P, in order that it may pass through a second given point Q. The second focus must lie on a circle with Q as centre touching the

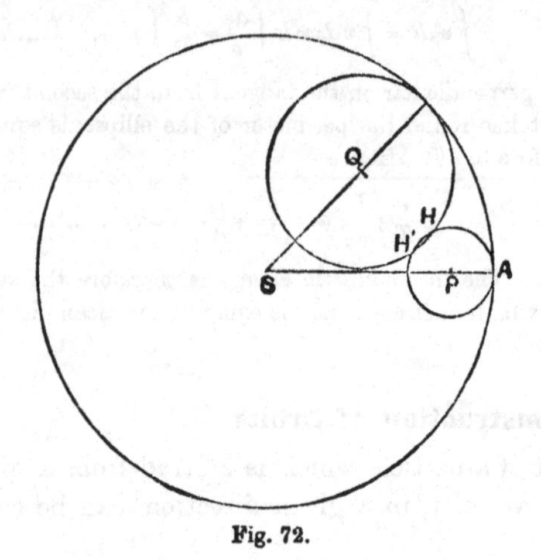

Fig. 72.

circle of zero velocity. If this circle intersect the one with P as centre there are two possible positions H, H' of the required point,

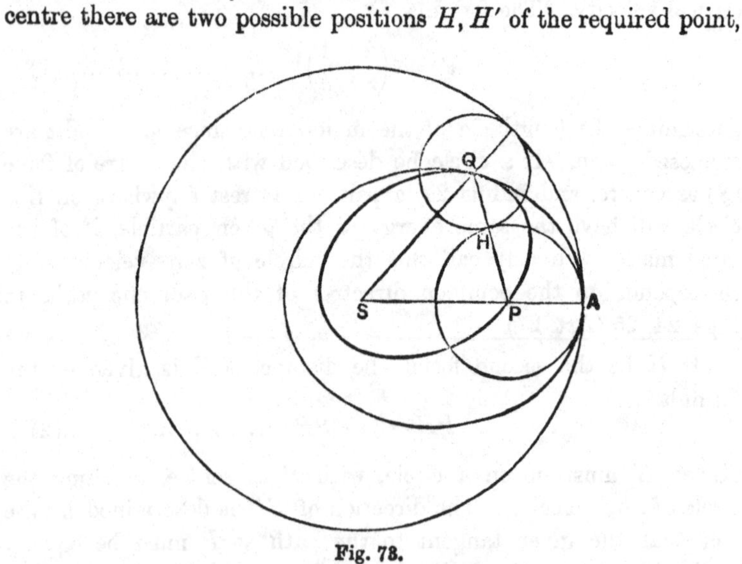

Fig. 73.

and the corresponding directions of projection will bisect the angles supplementary to SPH, SPH', respectively. If the circles do not intersect, the problem is impossible. In the extreme case, where the circles touch, as in Fig. 73, the point Q is just within range from P. It is evidently a point of ultimate intersection of consecutive paths of particles started from P with the prescribed velocity. We have then

$$SQ + QP = SQ + QH + HP = SA + AP, \quad \text{.........(3)}$$

and the envelope of the paths is therefore an ellipse, with S and P as foci, touching the circle of zero velocity at the point nearest to P.

If the velocity of projection from P *exceeds* the velocity from infinity, the formula (1) is replaced by

$$v^2 = \mu\left(\frac{2}{r} + \frac{1}{a}\right), \quad \text{........................(4)}$$

which determines the length $2a$ of the real principal axis. We now have, in place of (2),

$$HP = SP + 2a. \quad \text{........................(5)}$$

This, together with the direction of the tangent at P, determines the orbit, since the given tangent must bisect the angle between the focal distances.

If the orbit is to pass through a second given point Q, we describe about P and Q circles of radii $SP + 2a$, $SQ + 2a$, respectively. It is easily seen that these circles will always intersect, giving two real positions of the second focus. All points in the plane are in fact now within range from P, and there is accordingly no true envelope of the paths of particles started from P with the given velocity. What corresponds, geometrically, to the envelope in the previous case is an ellipse touched by the *outer* branches of the hyperbolas; this has of course no dynamical significance.

In the case where the velocity of projection is *equal* to the velocity from infinity, the direction of the axis and the focal distance SP make equal angles with the tangent at P. The parabolic orbit which is to pass through two assigned points P, Q is most easily found by drawing circles about P and Q as centres to pass through S. The two common tangents to these circles will be the directrices of the two possible paths.

Ex. In the theory of projectiles, if the variation of gravity with height be taken into account, the path *in space* relative to the earth's centre will be a conic with this point as one focus. With ordinary velocities of projection this will be the further focus, and the eccentricity will be very nearly equal to unity.

Thus, in the case of a particle released from apparent rest at a height k at a place on the equator, there is an initial horizontal velocity $\omega(c+k)$, where c is the earth's equatorial radius, and ω its angular velocity of rotation. Hence

$$h=(c+k)^2\omega. \qquad (6)$$

The equation of the conic being assumed to be

$$l/r=1-e\cos\theta, \qquad (7)$$

we have, putting $\theta=0$,
$$l=(1-e)(c+k). \qquad (8)$$

The time of describing an angle a about the earth's centre is therefore

$$t=\frac{1}{h}\int_0^a r^2 d\theta=\frac{1}{\omega}\int_0^a\left(\frac{1-e}{1-e\cos\theta}\right)^2 d\theta,$$

by (6), (7), and (8). This may be written

$$\omega t=\int_0^a \frac{d\theta}{\left(1+\frac{2e}{1-e}\sin^2\frac{1}{2}\theta\right)^2}. \qquad (9)$$

If a be the angle described when the particle reaches the ground we have putting $\theta=a$, $r=c$ in (7) and referring to (8),

$$\frac{c+k}{c}=\frac{1-e\cos a}{1-e}, \text{ or } \frac{k}{c}=\frac{2e}{1-e}\sin^2\frac{1}{2}a, \qquad (10)$$

Hence since k/c is usually a very minute fraction, (9) may be written

$$\omega t=\int_0^a\left(1+\frac{k}{c}\frac{\theta^2}{a^2}\right)^{-2} d\theta=a\left(1-\frac{2}{3}\frac{k}{c}\right), \qquad (11)$$

approximately, or
$$a=\omega t\left(1+\frac{2}{3}\frac{k}{c}\right). \qquad (12)$$

Since ωt is the angle turned through by the earth during the particle's descent, there is an easterly deviation from the point of the ground vertically beneath the starting point, of amount

$$c(a-\omega t)=\tfrac{2}{3}k\omega t. \qquad (13)$$

Since $k=\frac{1}{2}gt^2$, approximately, this may be written as $\frac{1}{3}\omega g t^3$.*

For example, if $t=5$ sec., corresponding to a fall of 400 ft., we have

$$\tfrac{2}{3}\omega t\,.\,\tfrac{1}{2}gt^2=1333\omega=\cdot 097 \text{ ft.}$$

The eccentricity e of the path is found by a comparison of (6) and (8) with the formula

$$h^2=\mu l=g'c^2 l, \qquad (14)$$

where g' is the value of true (as distinguished from apparent) gravity at the earth's surface. We find

$$1-e=\frac{\omega^2 c}{g'}\left(1+\frac{k}{c}\right)^3. \qquad (15)$$

If k/c be neglected, this makes

$$e=1-\tfrac{1}{289}=\cdot 9965, \text{ about.}$$

* I am indebted for this solution to Dr Bromwich.

78. Hodograph.

The formula (1) of Art. 76 shews that the velocity at any point P of the orbit is

$$v = \frac{h}{SY}, \quad \ldots\ldots\ldots\ldots\ldots\ldots\ldots\ldots(1)$$

where SY is the perpendicular drawn from the centre of force S to the tangent at P. In the case of the ellipse or hyperbola the locus of Y is the 'auxiliary circle'; and if YS be produced to meet this circle again in Z we have

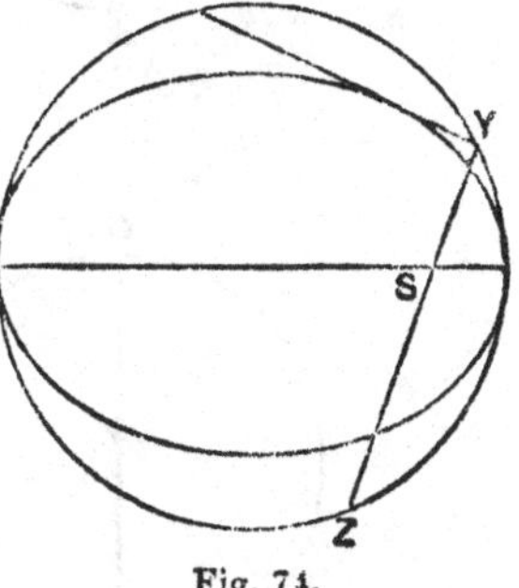

Fig. 74.

$$v = \frac{h}{SY.SZ}.SZ = \frac{h}{b^2}.SZ. \ldots(2)$$

Since the direction of this velocity is at right angles to SZ it appears that the auxiliary circle is, on a certain scale, the hodograph of the moving particle, turned through a right angle, the focus S being the pole.

In the case of the parabola, the locus of Y is the tangent at the vertex; and the hodograph will therefore be the inverse of this straight line, turned through a right angle. In other words, it is a circle through S. The following diagrams shew the hodographs in the various cases, drawn separately from the orbits, and with their true orientation.

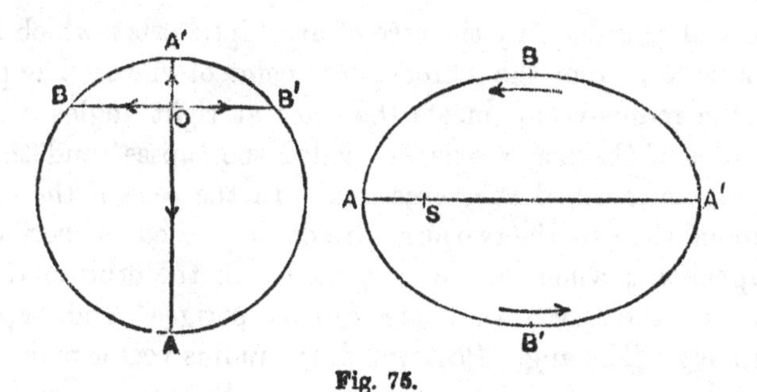

Fig. 75.

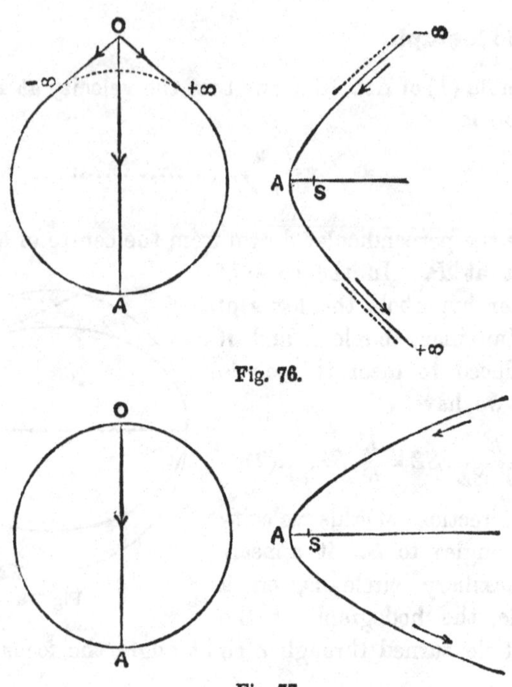

Fig. 76.

Fig. 77.

79. Formulæ for Elliptic Motion.

The equations of motion have been integrated only so far as to find the *form* of the orbit, and the law of variation of the velocity in it. For a complete solution we require to know the position in the orbit at any given time, having given the initial conditions.

We will consider only the case of an elliptic orbit, which is the most important from the astronomical point of view. The points where the radius-vector meets the orbit at right angles, viz. the extremities of the major axis, are called the 'apses,' and the line joining them is called the 'apse-line.' In the case of the earth's orbit round the sun, the two apses are distinguished as 'perihelion' and 'aphelion'; when we are concerned with the orbit of the sun relative to the earth they are called 'perigee' and 'apogee,' respectively. The angle PSA which the radius vector makes with the line drawn to the nearer apse is called the 'true anomaly.' If

Q be the point on the auxiliary circle which corresponds to P, the excentric angle of P, viz. the angle QCA, where C is the geometric centre, is called the 'excentric anomaly.'

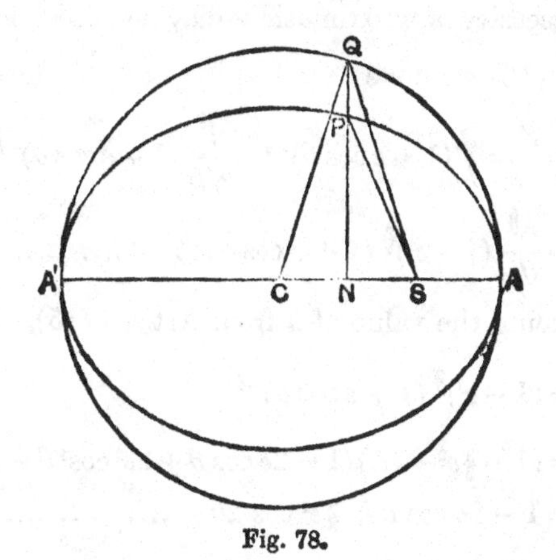

Fig. 78.

If we denote the true anomaly by θ, and the excentric angle by ϕ, we have, by known formulæ of Analytical Geometry,

$$\frac{l}{r} = 1 + e \cos \theta, \quad \text{.........................(1)}$$

where e is the excentricity of the ellipse, and

$$r = SP = a - e \,.\, CN = a(1 - e \cos \phi). \quad \text{............(2)}$$

Again, if t be the time from A to P, we have, if n denote as in Art. 76 the mean angular velocity about S,

$$\frac{nt}{2\pi} = \frac{\text{area } ASP}{\text{area of ellipse}} = \frac{\text{area } ASQ}{\text{area of circle}}$$

$$= \frac{\text{sector } ACQ - \text{triangle } SCQ}{\text{area of circle}} = \frac{\frac{1}{2}a^2\phi - \frac{1}{2}a \,.\, ae \sin \phi}{\pi a^2},$$

or

$$nt = \phi - e \sin \phi. \quad \text{.........................(3)}$$

These formulæ enable us to find the time from the apse to any given position. Thus (1) determines r in terms of θ, the angle ϕ is then found from (2), and t from (3). In Astronomy, however, it is the converse problem which is of interest, viz. to find the value of θ corresponding to any given value of t. This is known as Kepler's Problem, and has given rise to many mathematical investigations. In practice it is simplified by the fact that the excentricity e is in the case of the planetary orbits always a small quantity. If only the first few powers of e are required, the method of successive approximation may be used, as follows.

By Art. 76 (10) we have

$$\frac{dt}{d\theta}=\frac{r^2}{h}=\frac{l^2}{h}(1+e\cos\theta)^{-2}=\frac{l^{\frac{3}{2}}}{\sqrt{\mu}}(1+e\cos\theta)^{-2}$$

$$=\frac{a^{\frac{3}{2}}}{\sqrt{\mu}}(1-e^2)^{\frac{3}{2}}(1+e\cos\theta)^{-2}. \quad \text{.....................(4)}$$

Hence, introducing the value of n from Art. 76 (15),

$$\frac{ndt}{d\theta}=(1-e^2)^{\frac{3}{2}}(1+e\cos\theta)^{-2}$$

$$=(1-\tfrac{3}{2}e^2+\ldots)(1-2e\cos\theta+3e^2\cos^2\theta-\ldots)$$

$$=1-2e\cos\theta+\tfrac{3}{2}e^2\cos 2\theta-\ldots. \quad \text{..................(5)}$$

Integrating, we have

$$nt=\theta-2e\sin\theta+\tfrac{3}{4}e^2\sin 2\theta-\ldots, \quad \text{...........(6)}$$

no additive constant being necessary, since by hypothesis t vanishes with θ.

To invert this series we write it in the form

$$\theta=nt+2e\sin\theta-\tfrac{3}{4}e^2\sin 2\theta+\ldots. \quad \text{...........(7)}$$

The first approximation is

$$\theta=nt. \quad \text{.........................(8)}$$

For a second approximation we insert this value of θ in the second term on the right hand of (7), the error thus involved being of the order e^2. Thus

$$\theta=nt+2e\sin nt. \quad \text{.....................(9)}$$

For the next approximation we adopt this value of θ in the second term, and put $\theta=nt$ in the third term, the errors being now only of the order e^3. We find

$$\theta=nt+2e\sin(nt+2e\sin nt)-\tfrac{3}{4}e^2\sin 2nt$$

$$=nt+2e(\sin nt+2e\sin nt\cos nt)-\tfrac{3}{4}e^2\sin 2nt$$

$$=nt+2e\sin nt+\tfrac{5}{4}e^2\sin 2nt. \quad \text{.......................(10)}$$

To find the corresponding expression for r, we have from (3)

$$\phi = nt + e\sin\phi = nt + e\sin nt, \quad \text{.........(11)}$$

approximately. Hence

$$\frac{r}{a} = 1 - e\cos\phi = 1 - e\cos(nt + e\sin nt)$$

$$= 1 - e(\cos nt - e\sin nt \, . \sin nt)$$

$$= 1 - e\cos nt + \tfrac{1}{2}e^2(1 - \cos 2nt). \quad \text{.........(12)}$$

The mean of the values of r at equal infinitesimal intervals of time is therefore, to the present degree of approximation,

$$a(1 + \tfrac{1}{2}e^2).$$

The quantity nt which occurs in the preceding formulæ is called the 'mean anomaly'; it gives what would be the angular distance of the planet from the apse on the hypothesis of a constant rate of revolution. The difference $\theta - nt$ between the true and mean anomalies is called the 'equation of the centre.'

It may be noticed that the formula (9), where the square of e is neglected, would hold to the same degree of accuracy if we were to imagine the particle to describe a *circle* with constant velocity, with S at an excentric point, the distance of S from the centre being $2ae$, if a is the radius*. The variations in the radius vector would however not agree with (12). A better representation is obtained if we imagine a circle to be described with constant angular velocity (n) about a point H on the same diameter with S, such that $SC = CH = ae$, where C is the centre. The true orbit deviates in fact from a circle by small quantities of the order e^2, since $b^2 = a^2(1 - e^2)$. Moreover, if θ' be the angle which the radius vector HP drawn from the empty focus makes with the major axis, we easily find

$$\theta' = \theta - 2e\sin\theta = nt, \quad \text{.........(13)}$$

approximately, by (6).

80. Kepler's Three Laws.

Some sixty years before the publication of the law of gravitation by Newton, Kepler† had enunciated his celebrated three laws of planetary motion. These were not based on theory of any kind, but were intended to sum up facts of observation.

* This mode of representing the lunar motion was devised by Hipparchus (B.C. 120). The orbit, regarded as described in this way, was called an 'excentric.'

† Johann Kepler (1571-1630). The first two laws were announced in 1609, the third in 1619.

We take these laws in order, with a brief indication of the kind of evidence on which they rest.

I. The planets describe ellipses about the sun as focus.

The form of the orbit is most easily ascertained in the case of the earth, since the varying distance from the sun is indicated by the changes in the apparent diameter of the latter. The polar equation of a conic referred to the focus being

$$\frac{l}{r} = 1 + e\cos\theta, \quad \text{.......................(1)}$$

the sun's apparent diameter D, which varies inversely as the distance r, should vary as $1 + e\cos\theta$, where θ is the longitude in the orbit, measured from the nearer apse. Conversely, if a relation of this form is found to hold, the orbit must be a conic with the sun in one focus.

The excentricity e is easily deduced. If D_1, D_2 be the greatest and least values of the apparent diameter, corresponding to $\theta = 0$, $\theta = \pi$, respectively, we have

$$\frac{D_1}{D_2} = \frac{1+e}{1-e}, \quad \text{or} \quad e = \frac{D_1 - D_2}{D_1 + D_2}. \quad \text{..............(2)}$$

Taking $D_1 = 32' \, 36''$, $D_2 = 31' \, 22''$, we find $e = \frac{1}{60}$.

The same law was found to hold in the case of Mars. The verification in this instance requires more elaborate calculations, but the greater excentricity (·093) of the Martian orbit makes the test more stringent.

II. The radius vector drawn from the sun to a planet describes equal areas in equal times.

The verification is again simplest in the case of the earth. If $\delta\theta$ be the change of longitude in a given short time, e.g. a day, the product $r^2\delta\theta$ should have the same value at all times of the year; hence $\delta\theta$ should vary as D^2. This is found to be the case.

III. The squares of the periodic times of the different planets are proportional to the cubes of the respective mean distances from the sun.

The periods of revolution are easily found, to a high degree of accuracy, from observation. The relative distances can also be determined. The following table, given by Newton*, gives the periods of the planets then known, and the mean distances from the sun, as found by Kepler and Bullialdus† respectively, in terms of the mean distance of the earth as unit. The last column gives the mean distance as deduced from Kepler's law.

	Period (days)	Mean distance		
		Kepler	Bullialdus	Calculated
Mercury	87·9692	·38806	·38585	·38710
Venus	224·6176	·72400	·72398	·72333
Earth	365·2565	1·00000	1·00000	1·00000
Mars	686·9785	1·52350	1·52350	1·52369
Jupiter	4332·514	5·19650	5·22520	5·20096
Saturn	10759·275	9·51000	9·54198	9·54006

A similar comparison between the periods of the satellites of Jupiter and Saturn and their distances from the respective primaries was also given by Newton.

Now that the law of gravitation is fully established, it is usual to deduce the relative mean distances from the relative periods, on the basis of Kepler's law, subject to a slight correction to be explained in Art. 81. In the case of the planets Mercury, Venus, and Mars, whose masses are very small compared with that of the sun, the correction is almost negligible; and the calculated values in the last column agree almost exactly with those now adopted by astronomers.

It has been seen that Kepler's laws follow as simple consequences of the Newtonian law of gravitation, if we neglect the mutual influences of the various planets, and the accelerations which they produce in the central body.

Conversely it may be shewn that no other hypothesis is consistent with the laws, regarded as accurate statements of what would occur under the conditions above mentioned.

* *Principia*, lib. iii, phaenomenon iv.

† The Latinized name of Ismael Boulliau (1605–94), a French astronomer, who maintained an extensive correspondence with the scientific men of his time.

Thus, the uniform description of areas implies that the angular momentum of a planet about the sun is constant, and thence that the force on the planet is directed always towards the sun.

The elliptic form of the orbit, about the sun as focus, implies that the force in different parts of the same orbit varies inversely as the square of the radius vector. An analytical proof is given in Art. 85, but it may be noted that the result follows from the fact that the hodograph in the case of an ellipse described about a centre of force in a focus is the auxiliary circle turned through a right angle, the focus in question being the pole of the hodograph (Art. 78). The line CZ joining the centre to the point Z in Fig. 74, p. 227, is parallel to SP, and the velocity of Z is therefore $ad\theta/dt$, or ha/r^2, in the customary notation. Since this velocity represents the acceleration of P (turned through a right angle), on the same scale on which SZ represents the velocity

$$\frac{h}{b^2}\cdot SZ,$$

the acceleration is

$$\frac{h^2}{l}\cdot\frac{1}{r^2} \quad \text{..............................(3)}$$

towards S. Cf. Art. 76 (7).

Finally, the formula

$$T=\frac{2\pi a^{\frac{3}{2}}}{\sqrt{\mu}}, \quad \text{..........................(4)}$$

obtained in Art. 76, shews that if T^2 varies as a^3, the quantity μ, which denotes the acceleration at unit distance, must be the same for all the planets. This shews that the *forces* on the different planets must vary as the respective masses.

81. Correction to Kepler's Third Law.

When prolonged observations are made, it is found that Kepler's laws do not give a perfect description of the planetary motions; and the theory of universal gravitation supplies a reason why they should be departed from. The laws in question would, on this theory, hold rigorously for a system of planets which were themselves destitute of attractive power; but the accelerations actually produced by the planets in one another, and in the sun, though comparatively slight, are sufficient to produce modifications of the orbits. As some of the effects are cumulative, the changes may in time become considerable.

In the problem of two bodies, the effect of the attraction of the planet on the sun is easily allowed for. If m_1, m_2 be the masses of two gravitating bodies at a distance r apart, the particle m_1 has an acceleration towards m_2, proportional to m_2/r^2, whilst m_2 has an acceleration towards m_1 proportional to m_1/r^2. The acceleration of m_1 *relative* to m_2 is therefore proportional on the same scale to $(m_1+m_2)/r^2$. Hence the apparent acceleration (μ) at unit distance is greater than if m_2 were fixed in the ratio $(m_1+m_2)/m_2$. Hence, comparing the cases of two planets m, m' revolving round the sun, whose mass is S (say), we have, by Art. 76 (13),

$$T^2 : T'^2 = \frac{a^3}{\mu} : \frac{a'^3}{\mu'}, \qquad (1)$$

or

$$\left(\frac{a}{a'}\right)^3 = \frac{\mu}{\mu'}\left(\frac{T}{T'}\right)^2 = \frac{S+m}{S+m'}\cdot\left(\frac{T}{T'}\right)^2. \qquad (2)$$

This is the amended form of Kepler's third law.

In the case of the four inner planets the ratio m/S is very minute, being less than $\frac{1}{300000}$ even in the case of the earth. The correction to the relative distances, as deduced from the law, is therefore almost negligible. The greatest value of m/S for an outer planet is about $\frac{1}{1000}$, which is the case of Jupiter.

Although the consideration of the relative acceleration leads most immediately to the foregoing result, it is instructive to look at the matter from a different point of view. If r_1, r_2 be the distances of two particles m_1, m_2 from their mass-centre G, and r their mutual distance, we have

$$r_1 = \frac{m_2}{m_1+m_2}\,r, \quad r_2 = \frac{m_1}{m_1+m_2}\,r. \qquad (3)$$

Since the distances of the two particles from the point G and from one another are in constant ratios, the orbits which they describe relative to G and to one another will be geometrically similar. Moreover, on the principles of Art. 48, we may in considering the motion relative to G ignore the motion of G itself.

Regarding G, then, as a fixed point, the acceleration of m_2 towards it is $\gamma m_1/r^2$, or

$$\frac{\mu_2}{r_2^{\,2}}, \qquad \text{.........(4)}$$

provided $$\mu_2 = \frac{\gamma m_1^{\,3}}{(m_1+m_2)^2}. \qquad \text{.........(5)}$$

If a_2 be the mean distance in the orbit of m_2 about G, we have

$$a_2 = \frac{m_1}{m_1+m_2}\,a, \qquad \text{.........(6)}$$

where a refers to the orbit relative to m_1. Hence if T be the period, we have

$$T^2 = \frac{4\pi^2 a_2^{\,3}}{\mu_2} = \frac{4\pi^2 a^3}{\mu}, \qquad \text{.........(7)}$$

where $$\mu = \gamma\,(m_1+m_2). \qquad \text{.........(8)}$$

This leads again to the formula (2).

The same conclusion follows also from a consideration of the energy of the system. We have seen in Art. 42 that the kinetic energy of the motion relative to the mass-centre, which is the only variable part of the kinetic energy in the present case, is

$$\frac{1}{2}\frac{m_1 m_2}{m_1+m_2}\,v^2, \qquad \text{.........(9)}$$

where v is the relative velocity of the two particles. The mutual potential energy is

$$-\frac{\gamma m_1 m_2}{r}. \qquad \text{.........(10)}$$

The equation of energy may therefore be written

$$\frac{1}{2}\frac{m_1 m_2}{m_1+m_2}\,v^2 = \frac{\gamma m_1 m_2}{r} + \text{const.} \qquad \text{.........(11)}$$

This is identical with (2) of Art. 76, provided μ in that equation has the value (8).

Ex. To deduce the mean distance of Jupiter from the formula (2), we put

$$\frac{T}{T'} = 11{\cdot}8618, \quad \frac{m}{S} = \frac{1}{1047}, \quad \frac{m'}{S} = 0,$$

where the accented letters relate to the earth, the mass of the earth being negligible to the degree of accuracy aimed at. We have

$$\left(\frac{a}{a'}\right)^3 = 1{\cdot}00096 \times (11{\cdot}8618)^2,$$

whence

$$\frac{a}{a'} = 5{\cdot}2028.$$

82. Perturbations.

The study of the 'perturbations' produced in the orbit of one body by the attraction of another is the special problem of Celestial Mechanics. In a book like the present only one or two of the simpler points can be noticed.

In the first place it may be remarked that it is not the absolute acceleration due to the disturbing body which affects the orbit of a planet relative to the sun, or of a satellite relative to its primary, but rather the acceleration relative to the sun or the primary, i.e. the geometric difference of the accelerations produced in the planet and the sun, or in the satellite and the primary, respectively. It is this relative acceleration which is implied in the term 'disturbing force.'

The disturbing forces being usually very small, their effects are only gradually felt. For this reason Lagrange introduced the conception of the 'instantaneous ellipse,' i.e. the elliptic orbit which a planet would at any instant proceed to describe if the disturbing force were then to cease. The changes in this ellipse are comparatively slow.

The method is illustrated (in principle) if we examine the changes produced in the ellipse by a slight instantaneous *impulse* in its plane. Such an impulse may be resolved into two components, along and perpendicular to the direction of motion, respectively; and the effects of these may be considered separately.

Suppose, then, in the first place, that we have a small *tangential* impulse which changes the velocity from v to $v + \delta v$. The equation

$$v^2 = \mu\left(\frac{2}{r} - \frac{1}{a}\right) \quad \text{.......................(1)}$$

gives at once the change in the mean distance, since the only quantities involved in it which are instantaneously affected are v

and a. The small changes in these quantities are therefore connected by the relation

$$2v\,\delta v = \frac{\mu}{a^2}\,\delta a,$$

or
$$\frac{\delta a}{a} = \frac{2v\,\delta v}{n^2 a^2}, \quad \text{..........................(2)}$$

if n be the mean angular velocity (Art. 76).

The direction of motion is unaltered, but the distance of the unoccupied focus H from the position P of the body is changed from PH to PH', where $HH' = 2\delta a$. There is therefore in

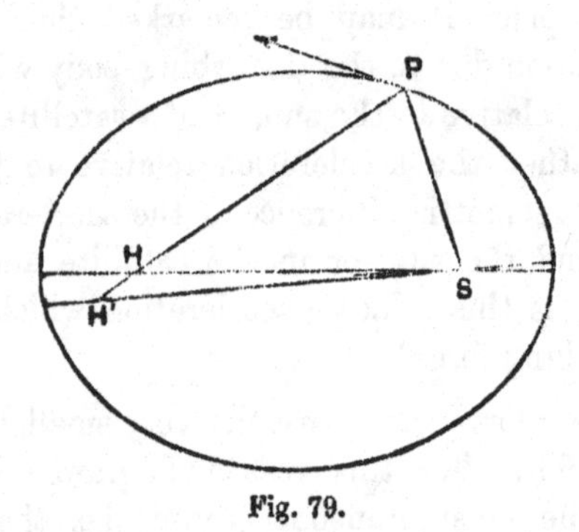

Fig. 79.

general a change in the direction of the apse-line, viz. from SH to SH'.

To find the change in the minor axis we have

$$b^2 = al = ah^2/\mu = ap^2v^2/\mu, \quad \text{..................(3)}$$

by Art. 76 (7). Since there is no instantaneous change in the value of p, we have, taking logarithmic differentials,

$$\frac{\delta b}{b} = \frac{1}{2}\frac{\delta a}{a} + \frac{\delta v}{v} = \left(1 + \frac{v^2}{n^2a^2}\right)\frac{\delta v}{v}, \quad \text{..............(4)}$$

by (2).

Since
$$n^2a^3 = \mu, \quad \text{..........................(5)}$$

the change in the mean motion n is given by

$$\frac{\delta n}{n} = -\frac{3}{2}\frac{\delta a}{a} = -3\,\frac{v\,\delta v}{n^2a^2}. \quad \text{..................(6)}$$

If K denote the mean kinetic energy in the orbit, we have, by Art. 76, Ex. 2,

$$K = \tfrac{1}{2}n^2a^2 = \frac{1}{2}\frac{\mu}{a}, \quad\quad (7)$$

and therefore

$$\delta K = -\frac{1}{2}\frac{\mu}{a^2}\delta a = -v\delta v. \quad\quad (8)$$

In the case of a nearly circular orbit, where $v = na$, nearly, we have

$$\frac{\delta a}{a} = \frac{\delta b}{b} = 2\frac{\delta v}{v}, \quad \frac{\delta n}{n} = -3\frac{\delta v}{v}, \quad \frac{\delta K}{K} = -2\frac{\delta v}{v}. \quad\quad (9)$$

For the application of these formulæ to the case of a resisting medium, see Art. 100. Another interesting illustration is furnished by the theory of the reaction of the earth's tides, supposed retarded by friction, on the moon. This will consist in the main of a small tangential acceleration f. It appears that, f being positive, the effect would be to gradually increase the size of the moon's orbit, whilst its angular motion and its mean kinetic energy would diminish. This diminution of the kinetic energy is of course more than compensated by the increase of the potential energy, the total energy (of the moon) being increased by the accelerating force.

Take next the case of a small *normal* impulse, and let δv be the velocity which it would produce in the body if the latter were at

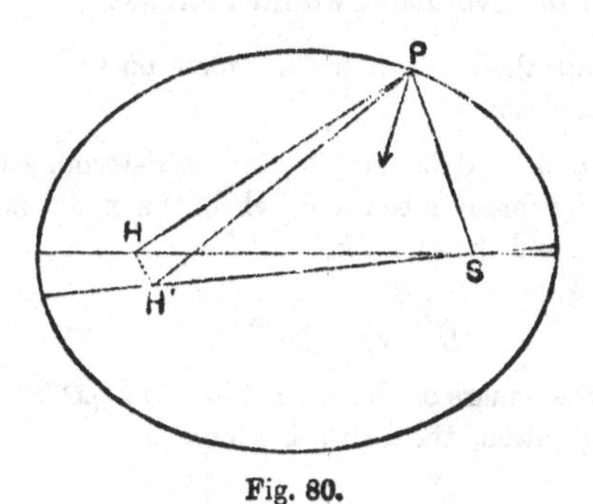

Fig. 80.

rest. The resultant velocity after the impulse is $\sqrt{\{v^2 + (\delta v)^2\}}$, and is therefore equal to v, to the first order. The formula (1) then shews that there is no instantaneous change in the value of a.

The direction of motion is, however, turned through an angle $\delta v/v$, and the second focus is therefore displaced from H to a point H', such that $PH'=PH$, and

$$HH'=PH.\frac{2\delta v}{v}.$$

The direction of the apse-line is in general changed.

The effect of a sudden slight change in the *absolute acceleration* μ is found by differentiating (1) on the supposition that μ and a are alone varied. Thus

$$\frac{\mu}{a^2}\delta a+\left(\frac{2}{r}-\frac{1}{a}\right)\delta\mu=0,$$

or

$$\frac{\delta a}{a}=-\frac{r'}{r}\frac{\delta\mu}{\mu}, \qquad (10)$$

where r' is the second focal distance. Also, since $n^2a^3=\mu$,

$$\frac{\delta n}{n}=\frac{1}{2}\frac{\delta\mu}{\mu}-\frac{3}{2}\frac{\delta a}{a}=\left(\frac{3a}{r}-1\right)\frac{\delta\mu}{\mu}. \qquad (11)$$

In the case of a nearly circular orbit these reduce to

$$\frac{\delta a}{a}=-\frac{\delta\mu}{\mu},\quad \frac{\delta n}{n}=2\frac{\delta\mu}{\mu}. \qquad (12)$$

For instance, if the sun's effective mass* were increased by the falling in of meteoric matter, the earth's orbit would contract, whilst the speed of revolution would increase.

Ex. To compare the sun's disturbing force on the moon, when greatest, with the earth's attraction.

With the notation used in Art. 75, in the determination of the earth's mass, the disturbing force of the sun, when the moon is in 'conjunction,' i.e. between the sun and the earth, is

$$\frac{\gamma S}{(D-D')^2}-\frac{\gamma S}{D^2}=\frac{2\gamma SD'}{D^3}, \qquad (13)$$

approximately, if the square of the small fraction D'/D be neglected. When the moon is in 'opposition,' the disturbing force is

$$\frac{\gamma S}{(D+D')^2}-\frac{\gamma S}{D^2}=-\frac{2\gamma SD'}{D^3}. \qquad (14)$$

* If a cloud of meteoric matter, symmetrically distributed, surround the sun, the portion included within a distance equal to the radius of the planet's orbit attracts as if its mass were already added to that of the sun. The change $\delta\mu$ must therefore be understood to be due to matter coming from *outside* the planet's orbit.

In each case the disturbing force is *outwards* from the earth. The ratio which (13) bears to the earth's attraction on the moon, viz. $\gamma E/D'^2$, is

$$2\frac{S}{E}\cdot\left(\frac{D'}{D}\right)^3. \qquad\qquad (15)$$

By Art. 75 (3) this is equal to

$$2\left(\frac{T'}{T}\right)^2=\frac{2}{(13\cdot77)^2}=\frac{1}{90}, \qquad\qquad (16)$$

nearly.

83. The Constant of Gravitation.

The constant (γ) of gravitation which occurs in the expression

$$\frac{\gamma m}{r^2} \qquad\qquad (1)$$

for the acceleration produced by a mass m in a particle at a distance r, being equal to the acceleration which the unit mass exerts at unit distance, will depend on the units of mass, length, and time which are adopted. Since $\gamma m/r^2$ is an acceleration, whose dimensions are L/T^2, the dimensions of γ will be

$$\mathsf{L}^3\mathsf{M}^{-1}\mathsf{T}^{-2}.$$

The numerical value of γ in terms of ordinary terrestrial units can of course only be found by actual measurement of the attraction of known masses. This is a matter of great delicacy, since the forces concerned are extremely minute. The best accredited result* is that

$$\gamma=6\cdot658\times10^{-8}, \qquad\qquad (2)$$

in C.G.S. units. This is the acceleration, in centimetres per second per second, which a mass of one gramme, supposed concentrated at a point, would produce at a distance of one centimetre.

A knowledge of the value of γ leads at once to that of the earth's mean density. As already stated, we may assume from the theory of Attractions that the earth attracts external particles as if its whole mass were condensed at the centre. Hence if R be the earth's radius, σ its mean density, we have

$$\frac{\gamma\,.\,\frac{4}{3}\pi R^3\,.\,\sigma}{R^2}=g,$$

or

$$\sigma=\frac{3g}{4\pi\gamma R}. \qquad\qquad (3)$$

* That of C. V. Boys, *Phil. Trans.*, 1895.

If we put $g = 981$ cm./sec.^{2}, $R = 6{\cdot}37 \times 10^8$ cm., and adopt the value (2) of γ, we have

$$\sigma = 5{\cdot}522, \quad \text{.........................(4)}$$

in grammes per cubic centimetre. It is remarkable that Newton had surmised that the mean density was about $5\frac{1}{2}$ times that of water.

For the purposes of Astronomy a knowledge of the value of γ is not essential; the science had indeed been carried to a high degree of development long before the value of γ in terrestrial measure had been ascertained with any precision. Astronomy is in fact concerned only with *relative* masses (and distances). To avoid the continual recurrence of an unknown and irrelevant constant in the formulæ, it has accordingly been customary to employ a special unit of mass. One plan is to take the mass of the sun as unit. If the earth's mean distance be taken as the unit of length, and a day as the unit of time, the value of γ is then determined by the formula

$$\frac{\gamma S}{a^2} = \left(\frac{2\pi}{T}\right)^2 a, \quad \text{.......................(5)}$$

if the ratio of the earth's mass to that of the sun be treated as negligible. Putting $S = 1$, $a = 1$, $T = 365{\cdot}25$, we get

$$\gamma = \frac{4\pi^2}{T^2} = 2{\cdot}96 \times 10^{-4}.$$

Another method is to fix the unit of mass so that $\gamma = 1$; the acceleration produced by a mass m at distance r being then m/r^2, simply. A unit so chosen is called an 'astronomical' unit. If the earth's mean distance be taken as unit of length, and the day as unit of time, the sun's mass in astronomical units will now be given by the above number, viz.

$$S = \frac{4\pi^2}{T^2} = 2{\cdot}96 \times 10^{-4}.$$

Ex. The mass m, in grammes, which would produce an acceleration of 1 cm./sec.^{2} at a distance of 1 cm. is to be found from (1), viz. we have $m\gamma = 1$, and

$$m = \frac{1}{\gamma} = 1{\cdot}502 \times 10^7.$$

The mass m, in grammes, which would attract an equal mass at a distance of 1 cm. with a force of 1 dyne is given by $m^2\gamma = 1$, or

$$m = \frac{1}{\sqrt{\gamma}} = 3876.$$

EXAMPLES. XVII.

1. What form does the law of "equal areas" assume in the case of a projectile?

2. Prove that if the earth's orbital velocity were increased by rather less than one-half it would escape from the solar system.

3. Particles are projected from the same point with the same velocity, in different directions, under a central force varying inversely as the square of the distance. Find the locus of the centres of the orbits.

4. If particles be projected from a given point with the same velocity in different directions, the lengths of the minor axes of the orbits will vary as the perpendiculars from the centre of force on the directions of projection.

5. Prove that the mean apparent diameter of the sun as seen from a planet describing an elliptic orbit is equal to the apparent diameter when the planet is at a distance equal to the major semi-axis of the orbit.

6. If θ be the sun's longitude from perigee, prove that the apparent diameter is given by

$$D = D_1 \cos^2 \tfrac{1}{2}\theta + D_2 \sin^2 \tfrac{1}{2}\theta,$$

where D_1, D_2 are the greatest and least values of D.

7. In elliptic motion about the focus the geometric mean of the velocities at the ends of any diameter is constant, and equal to the velocity at the mean distance.

8. Prove that the angular velocity about the empty focus is

$$\frac{nab}{CD^2},$$

where CD is the semi-diameter parallel to the direction of motion.

9. Prove that the velocity at any point of an elliptic orbit can be resolved into two constant components, viz. a velocity $na/\sqrt{(1-e^2)}$ at right angles to the radius vector, and a velocity $nae/\sqrt{(1-e^2)}$ at right angles to the major axis.

10. Prove that the two parts into which the earth's orbit is divided by the latus rectum are described in 178·7 and 186·5 days, respectively. ($e = \frac{1}{60}$.)

11. Assuming that the earth's orbital velocity is 30 km./sec., find that of Jupiter, whose distance from the sun is 5·20 times as great. [13·2 km./sec.]

12. Prove that the velocity acquired by a particle in falling from a great distance into the sun is $\sqrt{(2V^2/\alpha)}$, where V is the earth's orbital velocity, and α is the apparent angular radius of the sun as seen from the earth.

13. Prove that if a particle be projected in the plane of the equator, with a velocity small compared with that due to the earth's rotation, its path in space is a conic whose semi-latus-rectum is about 14 miles.

14. Prove that in a parabolic orbit described about a centre of force in the focus, the component velocity perpendicular to the axis varies inversely as the radius vector.

15. If the (parabolic) path of a comet cross the earth's orbit at the extremities of a diameter, for how many days will the path be inside the earth's orbit? [$77\frac{1}{2}$.]

16. Shew that the time spent by a comet within the earth's orbit (regarded as circular) varies as

$$(a+2c)\sqrt{(a-c)},$$

where a is the radius of the earth's orbit, and c is the comet's least distance from the sun.

Prove that this is greatest when $c=\frac{1}{2}a$.

EXAMPLES. XVIII.

(Elliptic Motion, etc.)

1. Prove from first principles that in the case of a central force varying inversely as the square of the distance the curvature of the hodograph is constant, and thence that the orbit is a conic.

2. A particle describes the hyperbolic branch

$$x=a\cosh u, \quad y=b\sinh u$$

about the inner focus; prove that the time from the apse is given by the formula

$$nt=e\sinh u-u,$$

where $n=\sqrt{(\mu/a^3)}$, and e is the eccentricity.

3. Prove that the equation of the centre (Art. 79) is a maximum for

$$\cos\theta=\frac{(1-e^2)^{\frac{3}{4}}-1}{e}.$$

Taking $e=\frac{1}{60}$, prove that the maximum is 1° 54′ 30″ about.

4. In elliptic motion about the focus, prove that the mean of the values of the radius vector at equal infinitesimal intervals of time is

$$a(1+\tfrac{1}{2}e^2),$$

accurately.

5. Prove that the Cartesian equations of motion under a central acceleration μ/r^2 are

$$\ddot{x}=-\mu x/r^3, \quad \ddot{y}=-\mu y/r^3.$$

Deduce the integral $$x\dot{y} - y\dot{x} = h,$$

and shew that $$\ddot{x} = -\frac{\mu}{h}\frac{d}{dt}\left(\frac{y}{r}\right), \quad \ddot{y} = \frac{\mu}{h}\frac{d}{dt}\left(\frac{x}{r}\right).$$

Hence shew that the equation of the path is of the form

$$r = \frac{h^2}{\mu} + Ax + By.$$

6. In elliptic motion about the focus, if the particle receive a slight impulse in the direction of the normal, the new orbit will intersect the old one at the other extremity of the chord through the empty focus.

7. It has been ascertained that the star Spica is a double star whose components revolve round one another in a period of 4·1 days, the greatest relative orbital velocity being 36 miles per second. Deduce the mean distance between the components of the star, and the total mass, as compared with that of the sun, the mean distance of the earth from the sun being taken as $92\frac{3}{4}$ million miles. [mean dist. $= 2{\cdot}03 \times 10^6$ miles; total mass $= {\cdot}083$.]

8. The effect on the minor axis of a small change in the absolute acceleration is given by

$$\frac{\delta b}{b} = -\frac{a}{r} \cdot \frac{\delta\mu}{\mu}.$$

9. Prove that the effect of a small tangential impulse on the excentricity of an orbit is given by

$$e\delta e = \frac{2b^2}{a^2}\left(\frac{a}{r} - 1\right)\frac{\delta v}{v}.$$

If θ be the longitude measured from the apse, prove that

$$\delta e = 2(\cos\theta + e)\frac{\delta v}{v}.$$

EXAMPLES. XIX.

(Constant of Gravitation.)

1. Assume that one of the minor planets is spherical and of the same mean density as the earth, and that its diameter is 100 miles. Shew that a particle projected upwards from its surface with a velocity exceeding 460 f.s. would never return.

2. Calculate the velocity in ft. per sec. with which a particle must be projected upwards from the surface of the moon in order that it may escape. (The earth's radius is 21×10^6 ft.; and the moon's mass and radius are respectively $\frac{1}{81}$ and $\frac{1}{4}$ of those of the earth.) [8150 f.s.]

3. Two equal gravitating spheres are at rest at a distance apart large compared with the radius of each. Prove that if they are abandoned to their mutual attraction the time which they will take to come together will be ·707 of what it would be if one of the spheres were fixed.

4. Prove that two equal gravitating spheres, of radius a, and density equal to the earth's mean density, starting from rest at a great distance apart, and subject only to their mutual attraction, will collide with a velocity

$$\sqrt{(\tfrac{1}{2}ga^2/R)},$$

where R is the earth's radius.

5. Two spheres of lead, each a metre in diameter, are placed a kilometre apart. Prove that if subject only to their mutual attraction they would come together in about 450 days. (Assume that the earth's radius is $6{\cdot}37 \times 10^8$ cm., that its mean density is one half that of lead, and that $g=981$ cm./sec.^{2}.)

6. Two spheres of iron, 1 metre in diameter, are placed with their centres 2 metres apart. Prove that if influenced only by their mutual attraction they would come into contact in about an hour. (Assume that the earth's radius is $6{\cdot}37 \times 10^8$ cm., and that its mean density is about $\frac{5}{8}$ that of iron.)

7. One of the satellites of Jupiter revolves in 7 d. 3 h. 40 m. and is at a distance from his centre equal to 15 times the radius of the planet. Our moon revolves in 27 d. 7 h. 40 m. and is at a distance from the earth's centre equal to 60 times the earth's radius. Find the ratio of Jupiter's mean density to that of the earth. [·214.]

8. Having given the sun's apparent semi-diameter (16′), the angle which the earth's radius subtends at the moon (57′), and the number of lunar revolutions in a year (13·4), find the ratio of the sun's mean density to that of the earth. [·252.]

9. Prove that the period of revolution of a satellite close to the surface of a spherical planet depends on the mean density of the planet, and not on its size.

What would be the period if the mean density were that of water? [3 hr. 20 min.]

10. Prove that if the dimensions of the various bodies of the solar system, and their mutual distances and velocities, were altered in any uniform ratio, the densities remaining the same, the changes of configuration in any given time would be exactly the same as at present.

What would be the effect if the densities were changed in any uniform ratio?

CHAPTER XI

CENTRAL FORCES

84. Determination of the Orbit.

The discovery of the orbits described under the law of gravitation naturally led Newton and his successors to examine the case of other laws of force, and to study also the converse problem, viz. to ascertain under what law of force directed to a given point a given orbit could be described.

As in the previous case, these questions are treated most concisely by means of the principle of angular momentum and the equation of energy. The former principle gives as before

$$pv = h, \quad \text{.............................(1)}$$

where v is the velocity, p the perpendicular from the centre of force on the tangent to the path, and h is a constant. The radius vector therefore describes equal areas in equal times, the rate per unit time being $\frac{1}{2}h$ (Art. 76).

Again if $\phi(r)$ be the acceleration towards the centre, due to the force, the potential energy, i.e. the work required to bring the particle from rest at some standard distance from the centre to the actual distance r, is

$$\int \phi(r)\, dr$$

per unit mass [S. 49]. The equation of energy is therefore

$$\tfrac{1}{2}v^2 + \int \phi(r)\, dr = \text{const.} \quad \text{..................(2)}$$

Substituting from (1) we have the tangential polar equation of the path, viz.

$$\frac{h^2}{p^2} = C - 2\int \phi(r)\, dr. \quad \text{.....................(3)}$$

This determines the shape and size of the orbit, when C is given, but not its orientation about the centre of force.

Thus in the case of

$$\phi(r) = \mu r \quad \text{.........................(4)}$$

we have

$$\frac{h^2}{p^2} = C - \mu r^2. \quad \text{.........................(5)}$$

If we compare this with the tangential polar equation of an ellipse referred to the centre, viz.*

$$\frac{a^2b^2}{p^2} = a^2 + b^2 - r^2, \quad \text{.........................(6)}$$

we see that the two equations are identical provided

$$a^2b^2 = h^2/\mu, \quad a^2 + b^2 = C/\mu. \quad \text{.................(7)}$$

Since, by (5), C is necessarily positive, the values of a^2 and b^2 determined by (7) will be real and positive, provided

$$C^2 > 4\mu h^2. \quad \text{.........................(8)}$$

Now if p, r refer to any given point on the orbit, we have from (5)

$$C^2 - 4\mu h^2 = \left(\frac{h^2}{p^2} - \mu r^2\right)^2 + 4\mu h^2\left(\frac{r^2}{p^2} - 1\right), \quad \text{........(9)}$$

which is positive, since $p < r$. We conclude that the orbit can in all cases be identified with an ellipse having its centre at the centre of force.

The period of revolution is, by (7),

$$T = \frac{2\pi ab}{h} = \frac{2\pi}{\sqrt{\mu}}, \quad \text{.....................(10)}$$

as in Art. 28. Again, from (5) and (7) we have

$$v^2 = \mu(a^2 + b^2 - r^2), \quad \text{.........................(11)}$$

shewing that the velocity at any point P varies as the semi-diameter parallel to the tangent at P.

The case of $\phi(r) = \mu/r^2$ has already been considered. In the case of a *repulsive* force varying inversely as the square of the distance we have

$$\phi(r) = -\mu/r^2, \quad \text{.......................(12)}$$

and therefore

$$\frac{h^2}{p^2} = C - \frac{2\mu}{r}. \quad \text{.......................(13)}$$

* *Inf. Calc.*, Art. 143.

Comparing this with the equation of a branch of a hyperbola referred to the *outer* focus, viz.

$$\frac{l}{p^2} = \frac{1}{a} - \frac{2}{r}, \qquad \text{.........................(14)}$$

we see that the two are identical provided

$$l = h^2/\mu, \quad a = \mu/C. \qquad \text{.....................(15)}$$

Since C is necessarily positive, these conditions can always be satisfied.

85. The Inverse Problem.

To solve the inverse problem, i.e. to ascertain under what law of force to a given centre a given orbit can be described, we have only to differentiate the formula

$$\frac{h^2}{p^2} = C - 2\int \phi(r)\, dr \qquad \text{.....................(1)}$$

with respect to r. We find

$$\phi(r) = \frac{h^2}{p^3}\frac{dp}{dr}. \qquad \text{.........................(2)}$$

The tangential-polar equation of the prescribed orbit determines p as a function of r, and the required law of force is thus ascertained.

The formula (2) may be obtained more directly as follows. Resolving along the normal to the path we have

$$\frac{v^2}{\rho} = \phi(r)\sin\phi, \qquad \text{......(3)}$$

where ρ is the radius of curvature, and ϕ the angle which the tangent makes with the radius vector. Putting $v = h/p$, $\sin\phi = p/r$, and

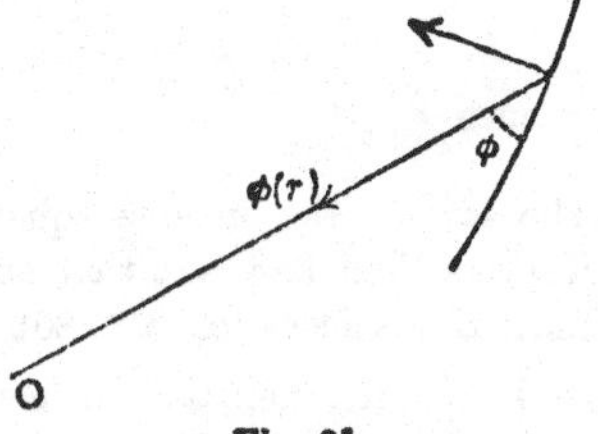

Fig. 81.

$$\rho = r\frac{dr}{dp}, \qquad \text{.........................(4)}$$

by a known formula of the Calculus, we reproduce the result (2).

Since the chord of curvature through the centre of force is $2\rho\sin\phi$, the formula (3) expresses that the velocity at any point of the orbit is equal to that which would be acquired by a particle in falling from rest through a space equal to one-fourth the chord of curvature, under a constant acceleration equal to the actual acceleration $\phi(r)$ at the point. A formula equivalent to this was employed by Newton in his geometrical investigations in the present subject.

Ex. 1. Taking the case of an ellipse, with the centre of force at the geometrical centre, we have

$$\frac{a^2b^2}{p^2}=a^2+b^2-r^2, \qquad \text{..............................(5)}$$

and the required law is, by (2),

$$\phi(r)=\frac{h^2}{a^2b^2}\cdot r. \qquad \text{..............................(6)}$$

The ellipse can therefore be described under the law

$$\phi(r)=\mu r, \qquad \text{..............................(7)}$$

provided

$$h=\sqrt{\mu}\,.\,ab. \qquad \text{..............................(8)}$$

To examine whether an orbit of the given type will *always* be described under the law (7), whatever the initial conditions, we note that these conditions imply a given point, a given tangent there, and a given value of the angular momentum h. In the present case, the latter circumstance determines the product ab, by (8). Now the problem of describing an ellipse with a given centre, so as to pass through a given point, to touch a given line there, and to have a given area, is possible and determinate. It is in fact identical with the problem of describing an ellipse having given two conjugate diameters.

Ex. 2. In the case of a conic referred to an inner focus, we have

$$\frac{l}{p^2}=\frac{2}{r}\mp\frac{1}{a}, \qquad \text{..............................(9)}$$

whence

$$\phi(r)=\frac{h^2}{l}\cdot\frac{1}{r^2}. \qquad \text{..............................(10)}$$

The law of the inverse square is therefore the only one consistent with Kepler's First Law, viz. that an undisturbed planet describes an ellipse with the sun at a focus (cf. Art. 80).

It appears, then, that the conic can be described about the focus in question under the law

$$\phi(r)=\frac{\mu}{r^2}, \qquad \text{..............................(11)}$$

provided

$$h=\sqrt{(\mu l)}. \qquad \text{..............................(12)}$$

This type of orbit is again a general one, under the law (11). For the problem of describing a conic with a given point as focus, so as to touch a given straight line at a given point, and to have a given latus rectum, is determinate. The equation (9) in fact determines a when p, r, and l are given, and thence the position of the second focus (cf. Art. 77).

Ex. 3. The tangential-polar equation of a circle referred to a point on the circumference is

$$p = r^2/c, \quad \text{.................................(13)}$$

where c is the diameter. The formula (2) therefore makes

$$\phi(r) = \frac{2h^2c^2}{r^5}, \quad \text{.................................(14)}$$

or

$$\phi(r) = \frac{\mu}{r^5}, \quad \text{.................................(15)}$$

provided

$$c^2 = \mu/2h^2. \quad \text{.................................(16)}$$

The force must therefore vary inversely as the fifth power of the distance, but the orbit in question is not a general one for this law. A circle described so as to pass through the centre of force, and through two other given (coincident) points, will not in general have the particular diameter c which is required by (16) to satisfy the initial condition as to angular momentum, unless this be specially adjusted.

Ex. 4. For the equiangular spiral

$$p = r \sin a, \quad \text{.................................(17)}$$

the law of force to the pole is

$$\phi(r) = \frac{h^2}{\sin^2 a} \cdot \frac{1}{r^3}, \quad \text{.................................(18)}$$

or

$$\phi(r) = \frac{\mu}{r^3}, \quad \text{.................................(19)}$$

if

$$\sin^2 a = h^2/\mu. \quad \text{.................................(20)}$$

The orbit is not however a general one for the law of the inverse cube. For an equiangular spiral, having a given pole, is completely determined by two coincident points on it, and its angle (a) will therefore not in general satisfy the relation (20).

A complete examination of the various orbits which can be described under the law (19) is given later (Art. 91).

86. Polar Coordinates.

Investigations relating to 'central forces' are often conducted in terms of polar coordinates, which are indeed for some purposes essential, especially in Astronomy.

Let the polar coordinates of a moving point be r, θ at the instant t, and $r + \delta r$, $\theta + \delta\theta$ at the instant $t + \delta t$. In the annexed figure $OP = r$, $OP' = r + \delta r$, and the angle POP' is $\delta\theta$. The displacement parallel to OP in the interval δt is

$$OP' \cos \delta\theta - OP = (r + \delta r) \cos \delta\theta - r = \delta r,$$

to the first order, and the component velocity at P in the direction OP is therefore

$$u = \frac{dr}{dt}. \quad \text{.........(1)}$$

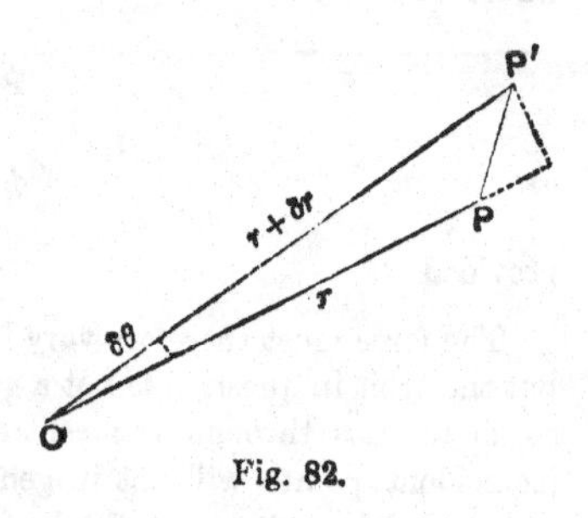

Fig. 82.

Again, the displacement perpendicular to OP is

$$OP' \sin \delta\theta = (r + \delta r) \sin \delta\theta = r\delta\theta,$$

to the first order. The component velocity at P in the direction at right angles to OP is therefore

$$v = r\frac{d\theta}{dt}. \quad \text{..........................(2)}$$

The quantities which are here denoted by u, v are called the 'radial' and 'transverse' components of the velocity, respectively. Their values might have been written down from known formulæ of the Calculus. Thus if ϕ be the angle which the tangent to the path makes with the radius vector, we have

$$\left.\begin{aligned} u &= \frac{ds}{dt}\cos\phi = \frac{ds}{dt}\frac{dr}{ds} = \frac{dr}{dt}, \\ v &= \frac{ds}{dt}\sin\phi = \frac{ds}{dt}\frac{rd\theta}{ds} = \frac{rd\theta}{dt}. \end{aligned}\right\} \quad \text{..............(3)}$$

To find the radial and transverse components of *acceleration*, let the component velocities at the instant $t + \delta t$ be $u + \delta u$, $v + \delta v$. The directions of these will of course make angles $\delta\theta$ with the directions to which u, v relate (Fig. 83). In the time δt, the velocity parallel to OP is increased by

$$(u + \delta u)\cos\delta\theta - (v + \delta v)\sin\delta\theta - u = \delta u - v\delta\theta,$$

ultimately; and the radial acceleration at time t is therefore

$$\alpha = \frac{du}{dt} - v\frac{d\theta}{dt}. \quad \text{.......................(4)}$$

Again, the velocity at right angles to OP is increased by

$$(u + \delta u) \sin \delta\theta + (v + \delta v) \cos \delta\theta - v = \delta v + u\,\delta\theta,$$

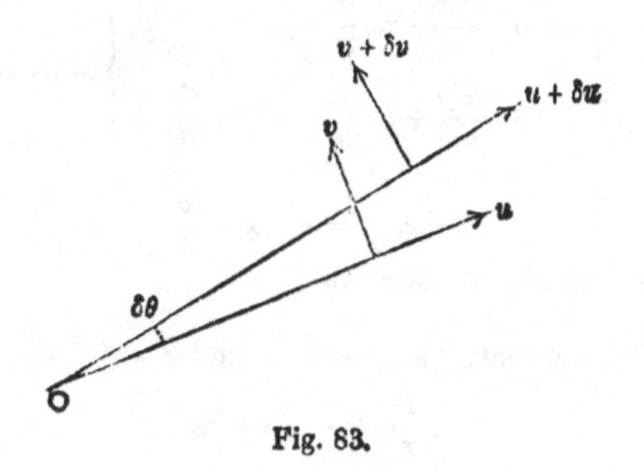

Fig. 83.

ultimately; and the transverse acceleration at time t is therefore

$$\beta = \frac{dv}{dt} + u\frac{d\theta}{dt}. \quad \text{.......................(5)}$$

Substituting from (1) and (2), we have

$$\left.\begin{aligned} \alpha &= \frac{d^2r}{dt^2} - r\left(\frac{d\theta}{dt}\right)^2, \\ \beta &= r\frac{d^2\theta}{dt^2} + 2\frac{dr}{dt}\frac{d\theta}{dt} = \frac{1}{r}\frac{d}{dt}\left(r^2\frac{d\theta}{dt}\right). \end{aligned}\right\} \text{...........(6)}$$

Hence if R, S denote the radial and transverse components, respectively, of a force acting on a particle m, the equations of motion are

$$m\left\{\frac{d^2r}{dt^2} - r\left(\frac{d\theta}{dt}\right)^2\right\} = R, \text{....................(7)}$$

$$m\frac{d}{dt}\left(r^2\frac{d\theta}{dt}\right) = Sr. \quad \text{...................(8)}$$

The latter equation might have been written down from the principle of angular momentum (Art. 48). The moment of momentum of the particle about the origin is the product of the transverse momentum mv into the radius vector r, and is therefore equal to $mr^2 d\theta/dt$, by (2). The equation (8) expresses that this increases at a rate equal to the moment Sr of the force acting on the particle.

Ex. 1. The formulæ (6) may also be deduced from the Cartesian equations of motion referred to rotating axes. When the angular velocity ω of the axes is variable, the equations (5) of Art. 33 are replaced by

$$\left.\begin{aligned}\alpha&=\frac{d^2x}{dt^2}-2\omega\frac{dy}{dt}-\omega^2x-\frac{d\omega}{dt}y,\\ \beta&=\frac{d^2y}{dt^2}+2\omega\frac{dx}{dt}-\omega^2y+\frac{d\omega}{dt}x.\end{aligned}\right\}\quad\text{.........(9)}$$

If we put

$$x=r,\quad y=0,\quad \omega=\frac{d\theta}{dt},\quad\text{.........(10)}$$

we reproduce the formulæ in question.

Ex. 2. To find the central acceleration under which the conic

$$\frac{l}{r}=1+e\cos\theta\quad\text{.........(11)}$$

can be described.

Since there is no transverse acceleration, we have

$$\frac{d}{dt}\left(r^2\frac{d\theta}{dt}\right)=0,\ \text{or}\ r^2\frac{d\theta}{dt}=h,\quad\text{.........(12)}$$

as in Art. 76. Again, differentiating (11) with respect to t,

$$\frac{l}{r^2}\frac{dr}{dt}=e\sin\theta\frac{d\theta}{dt}\quad\text{.........(13)}$$

and therefore, by (12),

$$\frac{dr}{dt}=\frac{eh}{l}\sin\theta.\quad\text{.........(14)}$$

Hence

$$\frac{d^2r}{dt^2}=\frac{eh}{l}\cos\theta\frac{d\theta}{dt}=\frac{h^2}{lr^2}\left(\frac{l}{r}-1\right),\quad\text{.........(15)}$$

by (11) and (12), and therefore

$$\frac{d^2r}{dt^2}-r\left(\frac{d\theta}{dt}\right)^2=-\frac{\mu}{r^2},\quad\text{.........(16)}$$

where

$$\mu=h^2/l,\quad\text{.........(17)}$$

in agreement with Art. 85 (10).

In the case of a hyperbolic branch referred to the outer focus as pole we have

$$\frac{l}{r}=e\cos\theta-1\quad\text{.........(18)}$$

in place of (11). A similar result follows, except that the acceleration is now outwards.

It may be noted that the equation (14) shews that the distance of a planet from the sun changes most rapidly at 90° from perihelion, the maximum rate being

$$\frac{eh}{l}=\frac{2\pi abe}{Tl}=\frac{2\pi a}{T}\frac{e}{\sqrt{(1-e^2)}},\quad\text{.........(19)}$$

where T is the period of revolution. In the case of the earth, where

$$a = 93 \times 10^6 \text{ miles}, \quad e = \tfrac{1}{60},$$

this is about 27,000 miles per day.

Ex. 3. In the case of a central orbit, we have, if the origin be at the centre of force,

$$R = -m\phi(r), \quad S = 0, \quad \text{.........................(20)}$$

and therefore

$$\ddot{r} - r\dot{\theta}^2 = -\phi(r), \quad \text{.........................(21)}$$

$$r^2\dot{\theta} = h. \quad \text{.........................(22)}$$

Hence, eliminating $\dot{\theta}$,

$$\ddot{r} - \frac{h^2}{r^3} = -\phi(r). \quad \text{.........................(23)}$$

If we multiply by $2\dot{r}$, and integrate with respect to t, we obtain

$$\dot{r}^2 + \frac{h^2}{r^2} = C - 2\int \phi(r)\, dr, \quad \text{.........................(24)}$$

which is equivalent to the equation of energy.

These formulæ may be applied to reproduce known results. Thus if

$$\phi(r) = n^2 r \quad \text{.........................(25)}$$

we have

$$\dot{r}^2 + \frac{h^2}{r^2} = C - n^2 r^2. \quad \text{.........................(26)}$$

The stationary values of r are given by $\dot{r} = 0$, or

$$n^2 r^4 - Cr^2 + h^2 = 0. \quad \text{.........................(27)}$$

It is easily seen, as in Art. 84, that the roots of this quadratic in r^2 are real and positive. Denoting them by a^2, b^2 we have

$$C = n^2(a^2 + b^2), \quad h^2 = n^2 a^2 b^2. \quad \text{.........................(28)}$$

The equation (26) may now be written

$$r^2\dot{r}^2 = n^2(a^2 - r^2)(r^2 - b^2), \quad \text{.........................(29)}$$

where a^2 is taken to be the greater of the two quantities a^2, b^2. Hence, as r diminishes from a to b,

$$\frac{r\dot{r}}{\sqrt{\{(a^2 - r^2)(r^2 - b^2)\}}} = -n. \quad \text{.........................(30)}$$

If we put

$$r^2 = a^2\cos^2\phi + b^2\sin^2\phi, \quad \text{.........................(31)}$$

we find

$$\frac{d\phi}{dt} = n, \quad \phi = nt + \epsilon. \quad \text{.........................(32)}$$

Again, from (22), we have

$$\dot{\theta} = h/r^2 = nab/r^2,$$

whence

$$\frac{d\theta}{d\phi} = \frac{ab}{a^2\cos^2\phi + b^2\sin^2\phi}, \quad \text{.........................(33)}$$

and

$$\theta = \tan^{-1}\left(\frac{b\sin\phi}{a\cos\phi}\right) \quad \text{.........................(34)}$$

if the origin of θ coincide with that of ϕ. It appears from (31) and (34) that $a\cos\phi$, $b\sin\phi$ are rectangular Cartesian coordinates of a point on the orbit, which is therefore the ellipse

$$\frac{x^2}{a^2}+\frac{y^2}{b^2}=1. \quad \text{.................................(35)}$$

Ex. 4. A particle of mass m, moving on a smooth table, is attached to a string which passes through a small hole in the table and carries a mass m' hanging vertically.

Using polar coordinates, we have, for the motion of m,

$$m(\ddot{r}-r\dot{\theta}^2)=-P, \quad r^2\dot{\theta}=h, \quad \text{......................(36)}$$

where P is the tension of the string. Since the acceleration of m' is $\ddot{r}$, upwards, we have

$$m'\ddot{r}=P-m'g. \quad \text{...............................(37)}$$

Eliminating $\ddot{r}$ and $\dot{\theta}$, we obtain

$$P=\frac{mm'}{m+m'}\left(g+\frac{h^2}{r^3}\right). \quad \text{..........................(38)}$$

Hence the particle m moves as if subject to a central force which is made up of a constant part, and of a part varying inversely as the cube of the distance. The intensity of the latter part depends however on the value of h, and therefore on the circumstances of projection.

87. Disturbed Circular Orbit.

A circular orbit is of course always possible under a central attractive force which is a function of the distance only, provided the circumstances be properly adjusted. If a be the radius of the orbit, ω the angular velocity in it, and $\phi(r)$ the central acceleration at distance r, we must have

$$\omega^2 a=\phi(a). \quad \text{........................(1)}$$

The angular momentum is therefore given by

$$h^2=\omega^2 a^4=a^3\phi(a). \quad \text{....................(2)}$$

This leads to the consideration of a *nearly* circular orbit. In particular, we may inquire whether the circular motion is stable, i.e. whether a particle slightly disturbed from revolution in a circle about the centre of force will always remain near this circle, or not.

If we eliminate $d\theta/dt$ between the equations

$$\frac{d^2r}{dt^2}-r\left(\frac{d\theta}{dt}\right)^2=-\phi(r), \quad r^2\frac{d\theta}{dt}=h, \quad \text{..........(3)}$$

we obtain as before

$$\frac{d^2r}{dt^2} - \frac{h^2}{r^3} = -\phi(r). \qquad \text{.........................(4)}$$

Writing

$$r = a + x, \qquad \text{.........................(5)}$$

we have, on the supposition that x is small,

$$\frac{d^2x}{dt^2} - \frac{h^2}{a^3}\left(1 - \frac{3x}{a}\right) = -\phi(a) - x\phi'(a), \qquad \text{...........(6)}$$

approximately. This shews that $a^3\phi(a)$ must be very nearly equal to h^2, i.e. the radius of the circle from which the particle is assumed to deviate only slightly must be connected with the angular momentum by the relation (2), very nearly. It is therefore possible, by a slight adjustment of the constant a in (5), to arrange that this condition shall be fulfilled *exactly*. On this understanding we have

$$\frac{d^2x}{dt^2} + \left\{\phi'(a) + \frac{3}{a}\phi(a)\right\}x = 0. \qquad \text{.............(7)}$$

Hence if

$$\phi'(a) + \frac{3}{a}\phi(a) > 0 \qquad \text{....................(8)}$$

the variations of x are simple-harmonic, viz. we have

$$x = C\cos(nt + \epsilon), \qquad \text{......................(9)}$$

provided

$$n^2 = \phi'(a) + \frac{3}{a}\phi(a). \qquad \text{.....................(10)}$$

The condition (8) is therefore the condition for stability.

In terms of ω we have

$$\frac{n^2}{\omega^2} = \frac{a\phi'(a)}{\phi(a)} + 3, \qquad \text{....................(11)}$$

by (1).

For instance, in the case of a force varying as some power of the distance, say

$$\phi(r) = \frac{\mu}{r^s}, \qquad \text{.........................(12)}$$

we have

$$\frac{n^2}{\omega^2} = 3 - s. \qquad \text{.........................(13)}$$

The circular form of orbit is therefore stable if $s < 3$, and unstable if $s > 3$. The case $s = 3$ must also be reckoned as unstable, since we have then

$$\frac{d^2x}{dt^2} = 0, \quad x = At + B, \quad \text{.................(14)}$$

indicating a progressive increase in the absolute value of x.

The general criterion (8) may be interpreted as expressing that for stability it is necessary that the central force should, in the neighbourhood of the circle, diminish outwards, or increase inwards, at a less rate than if it varied inversely as the cube of the distance.

It appears from (9) that the period of a complete oscillation in the length of the radius vector is $2\pi/n$. In the case of

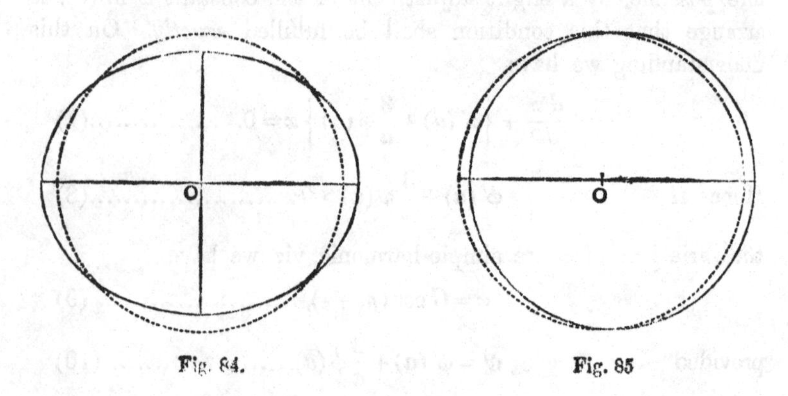

Fig. 84. Fig. 85

$\phi(r) = \mu r$, we find, putting $s = -1$ in (13), that $n = 2\omega$, and the period in question is therefore half that of a revolution. Again, if $\phi(r) = \mu/r^2$, so that $s = 2$, we have $n = \omega$, and the period is that of revolution. This agrees of course with what we know of the accurate orbits.

Ex. In the problem of Art. 86, Ex. 4, we have

$$\ddot{r} - \frac{h^2}{r^3} = -\frac{m'}{m+m'}\left(g + \frac{h^2}{r^3}\right), \quad \text{....................(15)}$$

or

$$(m+m')\ddot{r} - \frac{mh^2}{r^3} = -m'g. \quad \text{........................(16)}$$

Hence for a circular path of radius a we must have

$$mh^2 = m'ga^3. \quad \text{...............................(17)}$$

Writing $r=a+x$, and approximating, we find

$$(m+m')\ddot{x}+\frac{3m'g}{a}x=0. \quad \text{.........................(18)}$$

The circular path is therefore stable, and the period of a small oscillation about it is

$$2\pi\sqrt{\left\{\frac{m+m'}{3m'}\cdot\frac{a}{g}\right\}}. \quad \text{.........................(19)}$$

88. Apses.

A point where the radius drawn from the centre of force meets the orbit at right angles is called an 'apse,' and the radius in question is called an 'apse-line.'

If the force is always the same at the same distance, an apse-line will divide the orbit symmetrically. For if at an apse the velocity of the particle were reversed, it would retrace its previous path. Moreover, the paths described by two particles which are started from an apse with equal and opposite velocities must obviously be symmetrical.

It follows that apses will recur, if at all, at equal angular intervals. For if OA, OB be radii drawn from the centre of force to two consecutive apses, the point A' which is the image of A with respect to OB, will be the next apse, on account of the symmetry with respect to OB. The next apse will be at B', the image of B with respect to OA', and so on. The angles $AOB, BOA', A'OB', \ldots$ will all be equal; and their magnitude is called the 'apsidal' angle of the orbit. The distances OA, OB, OA', $OB', \ldots$ are called the 'apsidal distances'; the alternate ones are equal. In an elliptic orbit about the centre the apsidal angle is $\frac{1}{2}\pi$, and the apsidal distances are the semi-axes a, b; in an ellipse described about the focus the angle is π, and the apsidal distances are $a(1 \pm e)$. See Figs. 84, 85.

In any nearly circular orbit, the time-interval from a maximum to a minimum of the radius vector is π/n, where n is given by (11) of Art. 87. In this interval the radius vector describes an angle

$$\frac{\pi\omega}{n}=\pi\div\sqrt{\left\{\frac{a\phi'(a)}{\phi(a)}+3\right\}}, \quad \text{.................(1)}$$

which is accordingly the apsidal angle. In the case of a force varying as a power of the distance, say

$$\phi(r) = \frac{\mu}{r^s}, \quad \text{.........................(2)}$$

the angle is

$$\frac{\pi}{\sqrt{(3-s)}}, \quad \text{.........................(3)}$$

and accordingly independent of the size of the circle.

Conversely, we can assert that the apsidal angle cannot have the same value for all nearly circular orbits unless the force varies as some power of the distance. If the angle in question has the constant value π/m, we must have

$$\frac{a\phi'(a)}{\phi(a)} + 3 = m^2,$$

or

$$\frac{\phi'(a)}{\phi(a)} = \frac{m^2-3}{a}. \quad \text{.........................(4)}$$

Hence

$$\log \phi(a) = (m^2 - 3)\log a + \text{const.},$$

or

$$\phi(a) = \mu a^{m^2-3}. \quad \text{.........................(5)}$$

In particular the apsidal angle cannot be equal to π unless the force vary with the distance according to the Newtonian law. It follows, as was pointed out by Newton, that if the true law of gravitation deviated ever so slightly from proportionality to the inverse square of the distance, a progressive movement of the perihelion of each planet would result. For instance, if the index s in (2) had the value $2+\lambda$, where λ is small, the apsidal angle would be $(1+\frac{1}{2}\lambda)\pi$, nearly, by (3), and the nearer apse would advance in each revolution by the amount $\lambda\pi$. As a matter of fact a progressive motion of the apse is observed in the case of every planet, as well as in the case of the moon, where it is considerable; but this motion is accounted for, on the whole satisfactorily, by the disturbing effect of other bodies*.

* It appears, however, that there are certain discrepancies, especially in the case of Mercury, Venus, and Mars, which are as yet unexplained. These are very slight, amounting for instance in the case of Mercury to about 40″ per century. A slight amendment of the law of gravitation, to the extent of making $\lambda = 1{\cdot}5 \times 10^{-7}$, would account for this, but would apparently give too great a result in the cases of Venus and Mars.

In order that a nearly circular orbit may be closed, or re-entrant, after *one* revolution, the apsidal angle must be contained an even number of times in 2π. Hence the value of m in (5) must be integral. The only case which gives a force diminishing with increasing distance is that of $m = 1$. The law of the inverse square is therefore the only one under which the orbit of an undisturbed planet, when it is finite, is necessarily an oval curve. This has an application to the case of double stars. The relative orbit of the two components of a double star, when it can be sufficiently observed, is found to be an oval curve, in fact sensibly an ellipse, although the body to which the motion is referred may not be at a focus. The preceding remark leads to the conclusion that the law of gravitation holds in this case also, the apparent deviation of the 'centre of force' from the focus being explained by the fact that what is observed is not the true orbit, which is oblique to the line of sight, but its projection on the background of the sky.

89. Critical Orbits.

Subject to a certain proviso, a particle projected directly from a centre of attractive force, with a certain definite velocity depending on its position, will recede to infinity with a velocity tending asymptotically to zero. This definite velocity of projection has been called in Art. 76 the 'critical velocity' corresponding to the initial position. The orbit described when the particle is started in any other direction with the critical velocity is called a 'critical orbit.' In other words, the characteristic property of a critical orbit is that the energy of the particle is just sufficient to carry it to infinity if it be properly directed. We shall see that it does not follow that a critical orbit necessarily extends to infinity.

The equation of energy, viz.

$$v^2 = C - 2\int \phi(r)\, dr, \quad \text{.....................(1)}$$

becomes, if v is to vanish for $r = \infty$,

$$v^2 = 2\int_r^\infty \phi(r)\, dr,$$

and the proviso above referred to is that this integral must be convergent. This will be the case if

$$\lim_{r\to\infty} r^{1+\epsilon}\,\phi(r), \quad \text{.......................(2)}$$

where ϵ is any positive quantity, is finite. If the integral is not convergent, *no* velocity of projection, however great, will avail to carry the particle to infinity.

We have, then, in a critical orbit,

$$\left(\frac{dr}{dt}\right)^2 + \left(\frac{r\,d\theta}{dt}\right)^2 = 2\int_r^\infty \phi(r)\,dr, \quad \text{..............(3)}$$

with

$$r^2\frac{d\theta}{dt} = h. \quad \text{.............................(4)}$$

Hence

$$\left(\frac{dr}{dt}\right)^2 + \frac{h^2}{r^2} = 2\int_r^\infty \phi(r)\,dr. \quad \text{...................(5)}$$

The question cannot be carried further in the general case; but if

$$\phi(r) = \frac{\mu}{r^s} \quad \text{...........................(6)}$$

we have

$$\left(\frac{dr}{dt}\right)^2 + \frac{h^2}{r^2} = \frac{2\mu}{s-1}\cdot\frac{1}{r^{s-1}}, \quad \text{.................(7)}$$

provided $s > 1$. If $s = 1$, or < 1, there is no critical velocity. Excluding these cases, we have

$$r^{s-1}\left(\frac{dr}{dt}\right)^2 = h^2\left(c^{s-3} - r^{s-3}\right), \quad \text{...............(8)}$$

where

$$c^{s-3} = \frac{2\mu}{(s-1)\,h^2}. \quad \text{.......................(9)}$$

Taking the square root, and dividing by (4), we find

$$\frac{r^{\frac{1}{2}(s-5)}\,dr}{\sqrt{(c^{s-3} - r^{s-3})}} = d\theta. \quad \text{....................(10)}$$

Hence

$$\cos^{-1}\left(\frac{r}{c}\right)^{\frac{1}{2}(s-3)} = \tfrac{1}{2}(s-3)\,\theta + \text{const.}, \quad \text{........(11)}$$

or

$$r^p = c^p \cos p(\theta - \alpha), \quad \text{.................(12)}$$

if

$$p = \tfrac{1}{2}(s-3). \quad \text{.......................(13)}$$

If $s=2$, $p=-\frac{1}{2}$ and the equation (12) is that of a parabola referred to the focus. If $s=5$, $p=1$ and the critical orbit is a circle through the centre of force. If $s=7$, $p=2$ and the orbit is a lemniscate, with the node at the centre of force.

The investigation requires modification in the case of $s=3$. We have then, from (7),

$$r\frac{dr}{dt}=\pm\sqrt{(\mu-h^2)}. \quad \text{.................(14)}$$

Dividing this by (4), we have

$$\frac{dr}{d\theta}=mr, \quad \text{..........................(15)}$$

where

$$m=\pm\sqrt{\left(\frac{\mu}{h^2}-1\right)}. \quad \text{....................(16)}$$

Hence

$$r=Ce^{m\theta}. \quad \text{........................(17)}$$

The critical orbit is therefore an equiangular spiral, unless $h=\sqrt{\mu}$, in which case it is a circle; cf. Art. 91.

It appears from (12) that a critical orbit will be finite if $s>3$, and will extend to infinity if $s<3$.

90. Differential Equation of Central Orbits.

The differential equation of central orbits assumes its simplest form if we take as our dependent variable the *reciprocal* of the radius vector.

Writing

$$u=\frac{1}{r}, \quad \text{..............................(1)}$$

we have

$$\frac{d\theta}{dt}=\frac{h}{r^2}=hu^2. \quad \text{......................(2)}$$

Hence

$$\frac{dr}{dt}=-\frac{1}{u^2}\frac{du}{dt}=-\frac{1}{u^2}\frac{du}{d\theta}\frac{d\theta}{dt}=-h\frac{du}{d\theta}, \quad \text{...........(3)}$$

$$\frac{d^2r}{dt^2}=-h\frac{d}{dt}\left(\frac{du}{d\theta}\right)=-h\frac{d^2u}{d\theta^2}\frac{d\theta}{dt}=-h^2u^2\frac{d^2u}{d\theta^2}. \quad \text{......(4)}$$

Substituting from (2) and (4) in the equation

$$\frac{d^2r}{dt^2}-r\left(\frac{d\theta}{dt}\right)^2=-\phi(r), \quad \text{...............(5)}$$

we obtain

$$\frac{d^2u}{d\theta^2}+u=\frac{f(u)}{h^2u^2}, \quad \text{.......................(6)}$$

where
$$f(u) = \phi(r) = \phi\left(\frac{1}{u}\right). \quad \text{..................(7)}$$

This is the differential equation required.

If we write it in the form

$$f(u) = h^2u^2\left(\frac{d^2u}{d\theta^2} + u\right), \quad \text{....................(8)}$$

it determines the law of force under which a given orbit can be described.

A first integral of (6) can be obtained at once; thus, multiplying by $2\,du/d\theta$, we have

$$2\frac{du}{d\theta}\frac{d^2u}{d\theta^2} + 2u\frac{du}{d\theta} = \frac{2f(u)}{h^2u^2}\frac{du}{d\theta},$$

whence
$$h^2\left\{\left(\frac{du}{d\theta}\right)^2 + u^2\right\} = 2\int\frac{f(u)\,du}{u^2} + C. \quad \text{...........(9)}$$

This is really identical with (3) of Art. 84, since

$$\left(\frac{du}{d\theta}\right)^2 + u^2 = \frac{1}{p^2}, \quad \text{......................(10)}$$

by a known formula of the Calculus, and

$$\int\frac{f(u)}{u^2}\,du = -\int\phi(r)\,dr. \quad \text{.................(11)}$$

The further integration of (9) is in general difficult. In the case, however, of

$$f(u) = \mu u^2, \quad \text{......................(12)}$$

which is the law of the inverse square, the differential equation takes the simple form

$$\frac{d^2u}{d\theta^2} + u = \frac{\mu}{h^2}. \quad \text{........................(13)}$$

Except for the notation, this is identical with (2) of Art. 12, and the solution is accordingly

$$u = \frac{\mu}{h^2} + A\cos(\theta - \alpha), \quad \text{.................(14)}$$

where the constants A and α are arbitrary. If we put

$$l = h^2/\mu, \quad \text{..........................(15)}$$

this is equivalent to

$$lu = 1 + e\cos(\theta - \alpha), \quad \text{.................(16)}$$

which is the polar equation of a conic referred to the focus. Cf. Art. 76.

Ex. 1. To apply (6) to the case of slightly disturbed circular motion, let us write for shortness

$$f(u)=u^2F(u), \qquad \text{.............................(17)}$$

so that the equation becomes

$$\frac{d^2u}{d\theta^2}+u=\frac{F(u)}{h^2}. \qquad \text{.............................(18)}$$

This is satisfied by $u=c$, provided

$$h^2c=F(c), \qquad \text{.............................(19)}$$

this being the condition for a circular orbit of radius c^{-1}.

If we now put

$$u=c+\xi, \qquad \text{.............................(20)}$$

where ξ is supposed small, the equation (18) becomes

$$\frac{d^2\xi}{d\theta^2}+\left(1-\frac{F'(c)}{h^2}\right)\xi=0,$$

or

$$\frac{d^2\xi}{d\theta^2}+\left\{1-\frac{cF'(c)}{F(c)}\right\}\xi=0, \qquad \text{.............................(21)}$$

provided c be supposed adjusted so as to satisfy (19). If the coefficient of ξ be positive, we may put

$$1-\frac{cF'(c)}{F(c)}=m^2, \qquad \text{.............................(22)}$$

and obtain

$$\xi=C\cos m(\theta-\beta), \qquad \text{.............................(23)}$$

where the constants C, β are arbitrary. The apsidal angle is therefore π/m.

If

$$f(u)=\mu u^s, \quad F(u)=\mu u^{s-2}, \qquad \text{.............................(24)}$$

we have

$$m^2=3-s, \qquad \text{.............................(25)}$$

if $s<3$, in agreement with Art. 88.

Ex. 2. Having given that the orbit

$$u=\chi(\theta) \qquad \text{.............................(26)}$$

can be described under the central force $f(u)$, to find the law under which the orbit

$$u'=\chi(\theta'), \qquad \text{.............................(27)}$$

can be described, where $u'=u$, $\theta'=\lambda\theta$. The values of the radius vector at corresponding points of the two orbits are equal, whilst the vectorial angles are in a constant ratio. If two particles describing these orbits keep step with one another, being always in corresponding positions, we may say that the second particle describes an orbit which is the same as that of the first, but revolves relatively to it at a rate $(\lambda-1)\dot{\theta}$, proportional always to the angular velocity of the former particle in its orbit.

If h, h' be the constants of angular momentum in the two cases, we have

$$\dot{\theta} = hu^2, \quad \dot{\theta}' = h'u'^2, \quad \text{.............................(28)}$$

and therefore $$h' = \lambda h. \quad \text{.................................(29)}$$

The required law is, by (8),

$$f(u') = h'^2 u'^2 \left(\frac{d^2u'}{d\theta'^2} + u'\right) = h^2u^2\left(\frac{d^2u}{d\theta^2} + \lambda^2 u\right)$$
$$= f(u) + (\lambda^2 - 1)\, h^2u^3. \quad \text{.......................(30)}$$

The original central force must therefore be modified by the addition of a term varying inversely as the cube of the distance.

This is Newton's theorem relating to 'revolving orbits,' obtained by him by a geometrical method *.

91. Law of the Inverse Cube.

It has been pointed out, in more than one connection, that the law of the inverse cube occupies a somewhat special position. For this reason, and from the fact that it is one of the few cases in which the forms of the various possible orbits can be ascertained without difficulty, considerable mathematical interest has attached to it. It was studied in particular by Cotes†, and the orbits are accordingly known as 'Cotes's Spirals.'

The physical peculiarities of the law are that the angular momentum is the same for all circular orbits, and that the velocity in a circular orbit is equal to the 'critical' velocity corresponding to the distance from the centre. For if

$$\phi(r) = \frac{\mu}{r^3}, \quad \text{...........................(1)}$$

the velocity v in a circular orbit of radius a is given by

$$\frac{v^2}{a} = \frac{\mu}{a^3}, \quad \text{..............................(2)}$$

whence $$h^2 = v^2a^2 = \mu. \quad \text{...........................(3)}$$

Again, the critical velocity for the distance a is given by

$$v^2 = 2\int_a^\infty \phi(r)\, dr = \frac{\mu}{a^2}, \quad \text{.....................(4)}$$

* *Principia*, lib. i., prop. xliii.

† Roger Cotes (1682–1716), first Plumian Professor of Astronomy at Cambridge 1706–16, editor of the second edition of the *Principia*.

and is therefore equal to the velocity in a circular orbit of radius a, as determined by (2).

To determine the orbits generally, we put

$$f(u) = \mu u^3 \quad \text{.........................(5)}$$

in the differential equation (6) of Art. 90, and obtain

$$\frac{d^2u}{d\theta^2} + \left(1 - \frac{\mu}{h^2}\right) u = 0. \quad \text{....................(6)}$$

Let us first suppose that the angular momentum is greater than that proper to a circular orbit, so that

$$h^2 > \mu. \quad \text{.............................(7)}$$

The velocity at any point is therefore greater than the critical velocity, for since $p < r$ we have $h/p > \sqrt{\mu}/r$. If we put

$$1 - \frac{\mu}{h^2} = m^2, \quad \text{.........................(8)}$$

the solution of (6) is

$$u = A \cos m\theta + B \sin m\theta, \quad \text{................(9)}$$

or, if the origin of θ be suitably chosen,

$$au = \cos m\theta. \quad \text{........................(10)}$$

There is therefore a minimum value of r, and the line $\theta = 0$ corresponding to this is an apse-line. There are two asymptotes whose directions are given by $m\theta = \pm \frac{1}{2}\pi$, so that the direction of motion in the orbit turns altogether through an angle $\pi/m - \pi$. The distances of these asymptotes from the centre of force are equal to a/m.

The diagram on the next page may assist the reader to trace the varying forms of the orbit as h diminishes from ∞ to $\sqrt{\mu}$. When $h = \infty$, $m = 1$ and the path is a straight line; this is indicated by the line marked I in the figure. The curves II, III, IV, V correspond to diminishing values of m, viz. $\frac{3}{5}$, $\frac{1}{2}$, $\frac{3}{8}$, $\frac{1}{4}$, respectively, the angles turned through being therefore 120°, 180°, 300°, 540°. The smaller the value of m, the greater is the number of convolutions round the origin.

When the angular momentum has exactly the 'circular' value, so that

$$h^2 = \mu, \quad \text{.........................(11)}$$

we have $$\frac{d^2u}{d\theta^2}=0,\quad u=A\theta+B. \qquad (12)$$

If $A=0$, we have a circle, which may be regarded as a limit to the preceding forms; whilst if $A\neq 0$, we have a 'reciprocal spiral*.' These may be regarded as the starting points of the two series of orbits next to be referred to.

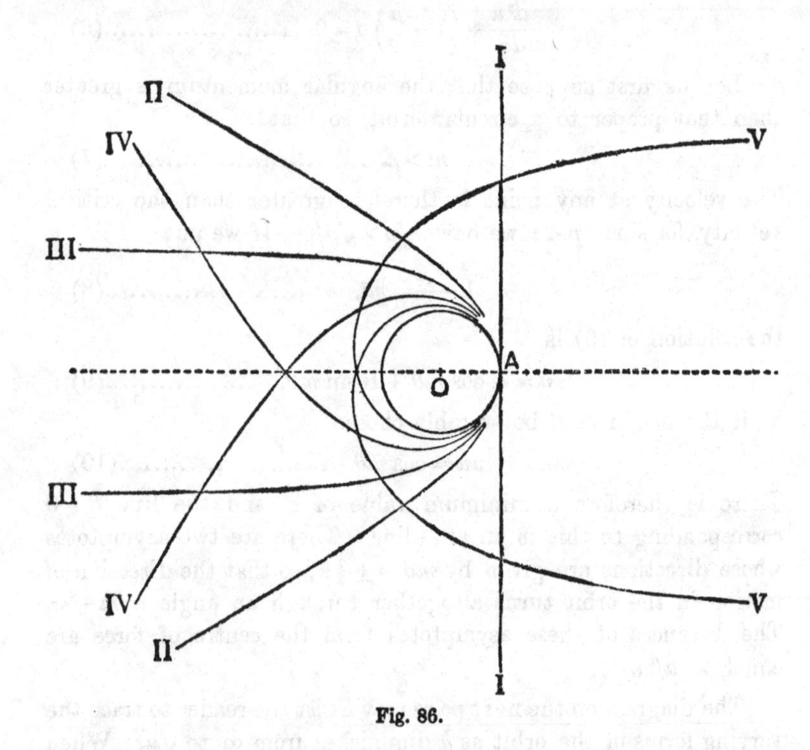

Fig. 86.

Finally we have the cases where the angular momentum has less than the 'circular' value, so that

$$h^2<\mu. \qquad (13)$$

If we now put $$\frac{\mu}{h^2}-1=n^2, \qquad (14)$$

we have $$u=A\,e^{n\theta}+Be^{-n\theta}. \qquad (15)$$

* *Inf. Calc.*, Art. 140.

At the point $\theta = 0$ we have

$$u = A + B, \quad \frac{du}{d\theta} = n(A - B), \qquad \text{.............(16)}$$

and therefore

$$v^2 - \mu u^2 = \frac{h^2}{p^2} - \mu u^2 = h^2 \left\{ u^2 + \left(\frac{du}{d\theta} \right)^2 \right\} - \mu u^2$$

$$= h^2 \left\{ \left(\frac{du}{d\theta} \right)^2 - n^2 u^2 \right\} = -4n^2 h^2 AB, \qquad \text{......(17)}$$

by (14). Hence A and B will have opposite signs, or the same sign, according as the velocity is greater or less than the critical velocity.

In the former case, we can, by adjustment of the origin of θ, reduce the equation (15) to the form

$$au = \sinh n\theta. \qquad \text{........................(18)}$$

Since $u = 0$, or $r = \infty$, for $\theta = 0$, we have now an asymptote

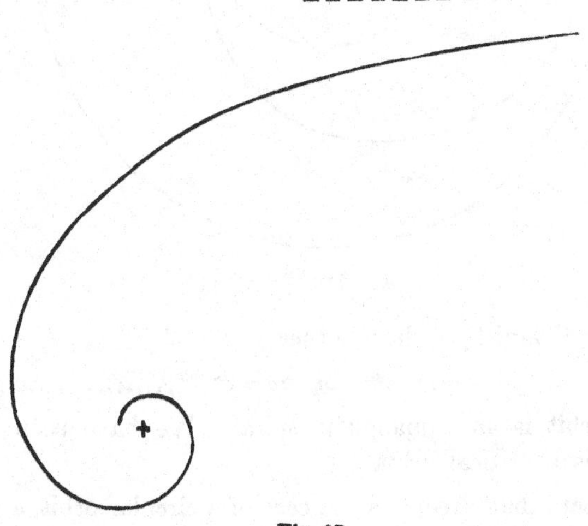

Fig. 87.

parallel to the initial line. Since $u = \infty$ for $\theta = \infty$, the path approaches the origin asymptotically, winding round it in an ever-narrowing spiral. Fig. 87 shews an orbit of this type.

If A and B have the same sign, the equation (15) can be brought to the form

$$au = \cosh n\theta. \qquad\qquad (19)$$

There is now a maximum value (a) of r, and the corresponding radius ($\theta = 0$) is an apse-line. Since $u = \infty$, or $r = 0$, for $\theta = \pm\infty$, the particle finally revolves round the origin in diminishing convolutions, as in the preceding case*. See Fig. 88.

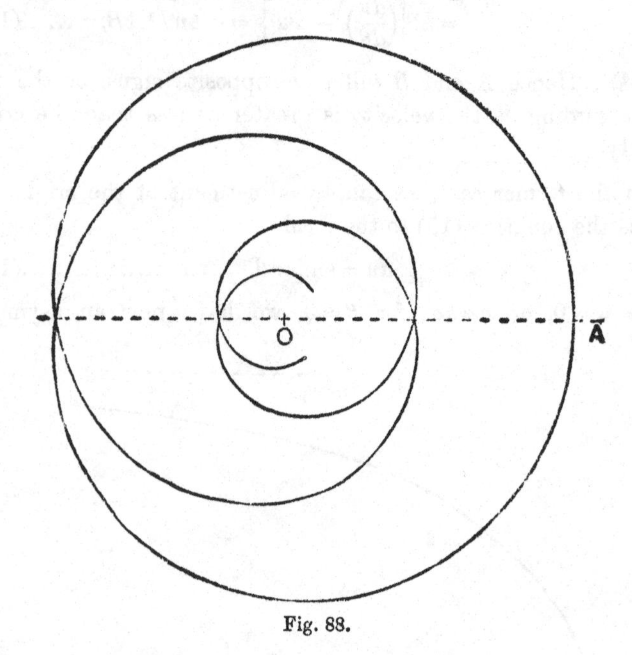

Fig. 88.

If A or B vanish, we have either

$$au = e^{m\theta}, \quad \text{or} \quad au = e^{-m\theta}; \qquad\qquad (20)$$

and the orbit is an equiangular spiral. We have seen already that this is a critical orbit.

It appears that, except in the case of a circular orbit, a particle moving under the law of the inverse cube will ultimately either

* Figs. 86, 88 together indicate the various types of orbit which result when the particle is projected from an apse (A) with different degrees of velocity. Fig. 86 shews cases where the velocity of projection is greater than the 'circular' value, whilst Fig. 88 shews (on a larger scale) a case where it is less.

escape to infinity, or will fall asymptotically to the centre. A circular orbit is therefore to be counted as unstable, as already proved (Art. 87).

Ex. The case of

$$f(u) = \mu u^2 + \nu u^3, \quad \text{.................................(21)}$$

where the force varies partly as the inverse square and partly as the inverse cube, is also readily integrable. We have

$$\frac{d^2u}{d\theta^2} + \left(1 - \frac{\nu}{h^2}\right)u = \frac{\mu}{h^2}. \quad \text{..............................(22)}$$

Assuming that $\nu < h^2$, and putting

$$1 - \frac{\nu}{h^2} = m^2, \quad \text{..(23)}$$

we have

$$u = \frac{\mu}{h^2 - \nu} + C\cos(m\theta - \alpha). \quad \text{.........................(24)}$$

This is of the form

$$lu = 1 + e\cos\{\theta - (1 - m)\theta - \alpha\}, \quad \text{......................(25)}$$

and may therefore be regarded as the equation of a conic which revolves about the focus with an angular velocity proportional to that of the radius vector, the coordinate of the apse-line being

$$(1 - m)\theta + \alpha.$$

This description is more appropriate, the more nearly m is equal to unity, i.e. the smaller the value of ν. Cf. Art. 90, Ex. 2.

A law of force of the form (21) was at one time proposed by Clairaut* as a modification of the law of gravitation, in order to account for the difference between the progressive motion of the apse of the lunar orbit (about 3° in one revolution) as actually observed, and that which was given by the calculation of the perturbations so far as this had at the time been carried. It was subsequently recognized by Clairaut himself that the calculations were defective, and that a more careful computation on the basis of the Newtonian law gives a result in agreement with the observations.

EXAMPLES. XX.

1. If a particle is subject to a constant force making a constant angle with the direction of motion, the path is an equiangular spiral.

2. A particle describing an ellipse about the centre receives a small impulse δv in the direction of motion. Prove that the consequent changes in the lengths of the principal semi-axes are given by

$$\frac{\delta a}{a} = \frac{\delta v}{v}\sin^2\phi, \qquad \frac{\delta b}{b} = \frac{\delta v}{v}\cos^2\phi,$$

where ϕ is the excentric angle.

* A. C. Clairaut (1713–65). His *Théorie de la lune* was published in 1752.

3. Prove that if in the hodograph of a central orbit equal areas are described about the pole in equal times the force must vary as the distance.

4. Prove that the hodograph of a central orbit is similar to the 'reciprocal polar' of the orbit with respect to the centre of force.

5. Prove that in the case of a circular orbit described about a centre of force at any given point the hodograph is an ellipse, parabola, or hyperbola, according as the point is inside, on, or outside the circumference of the circle.

6. Prove that the law of force for the orbit

$$a^{n-1}p=r^n$$

is $\phi(r)=\mu/r^{2n+1}$. Is this ever the general orbit for this law?

7. Prove that if the law of force in a central orbit be

$$\frac{\mu}{r^2}e^{-kr},$$

where k is small, the apse-line in a nearly circular orbit will advance in each revolution through the angle πka, if a be the radius of the orbit.

8. If in the preceding question the potential energy be

$$-\frac{\mu}{r}e^{-kr},$$

the angle of advance will be πk^2a^2, nearly.

9. Prove that under the conditions of Art. 88 the two apsidal distances cannot be equal unless the orbit be circular.

10. A particle subject to a central acceleration μ/r^3+f is projected from an apse at a distance a with the velocity $\sqrt{\mu/a}$. Prove that at any subsequent time t

$$r=a-\tfrac{1}{2}ft^2.$$

11. Prove that the law of the inverse cube is the only one which makes the critical velocity at any distance equal to that in a circle at the same distance.

12. Prove that in a central orbit

$$\frac{d^2(r^2)}{dt^2}=C-\frac{2}{r}\frac{d}{dr}\{r^2\int\phi(r)\,dr\}.$$

Examine the cases of $\phi(r)=\mu r$ and $\phi(r)=\mu/r^3$.

13. A ring can slide on a spoke of a wheel which revolves about its centre with constant angular velocity ω. Prove that if it start from rest at a distance a from the centre, its distance at any subsequent time t will be $a \cosh \omega t$.

14. Two particles m_1, m_2, connected by a string, are moving with equal velocities v at right angles to it. If the string strike a smooth peg which divides the length into segments a_1, a_2, find the initial tension.

$$\left[\frac{m_1 m_2 (a_1 + a_2) v^2}{(m_1 + m_2) a_1 a_2}.\right]$$

15. Work out the solution of the polar differential equations of central orbits (Art. 90 (6)) in the case of $\phi(r) = \mu r = \mu/u$.

16. A particle describing a circular orbit of radius a under a central acceleration μ/r^2 is disturbed by a small constant force f at right angles to the radius vector. Prove that the consequent changes in the radius vector and the angular velocity are given by the equations

$$r = a + \frac{2ft}{n}, \quad \dot{\theta} = n - \frac{3ft}{a}.$$

17. Find the law of force under which a cardioid $r = a(1 + \cos\theta)$ can be described under a centre of force at the pole; and find the relation between the angular momentum, the absolute acceleration, and the length a.

18. If a central orbit has the two apsidal distances a, b ($a < b$), prove that the velocity at a distance r is given by the equation

$$v^2 = \frac{2a^2}{b^2 - a^2}\int_a^r \phi(r)\, dr + \frac{2b^2}{b^2 - a^2}\int_r^b \phi(r)\, dr.$$

19. Prove that if a particle can describe the same free path under two fields of force (X_1, Y_1) and (X_2, Y_2) it can describe the same path under a field $(X_1 + X_2, Y_1 + Y_2)$, provided it be started from a given point A with the velocity $\sqrt{(v_1^2 + v_2^2)}$, where v_1, v_2 are the velocities at A in the two former cases.

Prove that a parabola can be described under the joint influence of an acceleration g parallel to the axis, and an acceleration ga^2/r from the focus, and that the velocity vanishes at the vertex. (The latus rectum is $4a$.)

CHAPTER XII

DISSIPATIVE FORCES

92. Resistance varying as the Velocity.

Except for a few problems in which sliding friction has been taken into account, it has been assumed up to this point that the conservation of mechanical energy holds in the various questions which have been considered. 'Dissipative' forces, as they are called, which convert energy into other than obviously mechanical forms, have been ignored.

The theoretical discussion of what precisely happens when a body traverses a resisting medium is difficult; and experiment has failed to elicit any simple law which shall hold for all velocities. We can only attempt here to examine the consequences of one or two simple empirical assumptions, each of which is probably fairly accurate under certain conditions.

We take first the case of rectilinear motion under no force except the resistance of the medium. The equation of motion is then of the type

$$\frac{du}{dt} = -\phi(u), \quad \text{or} \quad u\frac{du}{dx} = -\phi(u), \quad \text{..........(1)}$$

where $\phi(u)$ is some function of the velocity u. It is sometimes convenient to employ both forms of the equation, and to obtain a first integral of each. The first form gives a relation between u and t involving an arbitrary constant; thus

$$\frac{1}{\phi(u)}\frac{du}{dt} = -1,$$

$$\int\frac{du}{\phi(u)} = -t + A. \quad \text{......................(2)}$$

The second form leads to a relation between u and x, with a second arbitrary constant; thus

$$\frac{u}{\phi(u)}\frac{du}{dx} = -1,$$

$$\int\frac{udu}{\phi(u)} = -x + B. \quad\quad (3)$$

The elimination of u between (2) and (3) gives the relation between x and t, involving two arbitrary constants which may be used to satisfy the initial conditions.

As to the nature of the function $\phi(u)$, it is of course positive, and must vanish with u. A plausible assumption from the mathematical side is that for a certain range of u it can be expanded in a series of powers of u, thus

$$\phi(u) = Au + Bu^2 + Cu^3 + \ldots, \quad\quad (4)$$

and that consequently, when u is sufficiently small, the first term predominates.

We are thus led to the equations

$$\frac{du}{dt} = -ku, \quad \frac{du}{dx} = -k, \quad\quad (5)$$

whence $$\log u = -kt + A, \quad u = -kx + B. \quad\quad (6)$$

If we assume that $x = 0$, $u = u_0$, for $t = 0$, we have $A = \log u_0$, $B = u_0$, and therefore

$$u = u_0 e^{-kt}, \quad u = u_0 - kx. \quad\quad (7)$$

Eliminating u, we have

$$x = \frac{u_0}{k}(1 - e^{-kt}). \quad\quad (8)$$

If we put $u = 0$ in (7), or $t = \infty$, in (8), we find $x = u_0/k$. With this law of resistance there is therefore a limit to the space described.

The quantity k is the reciprocal of a time, viz. the time in which the velocity is diminished in the ratio $1/e$. Denoting this time by τ, we have

$$x = u_0\tau(1 - e^{-t/\tau}). \quad\quad (9)$$

93. Constant Propelling Force, with Resistance.

When the particle considered is subject to a constant propelling force, as well as to the resistance of the medium, the equation takes the form

$$\frac{du}{dt} = f - \phi(u). \qquad\qquad (1)$$

Since $\phi(u)$ increases indefinitely* with u, there is a certain speed V for which

$$\phi(V) = f, \qquad\qquad (2)$$

i.e. the resistance just balances the propelling force. This is called the 'terminal velocity,' or the 'full speed' of the body. If the body be started with a velocity less than this, the propelling force is greater than the resistance, and the velocity increases. If it be started with a velocity greater than V, the resistance exceeds the propelling force, and the velocity diminishes. In either case the velocity tends asymptotically to the value V.

In the case of resistance varying as the velocity, where $\phi(u) = ku$, we have $V = f/k$, and the equation (1) may be written

$$\frac{du}{dt} = \frac{f}{V}(V - u). \qquad\qquad (3)$$

Hence
$$\frac{1}{V-u}\frac{du}{dt} = \frac{f}{V}, \quad \log(V-u) = -\frac{ft}{V} + A. \qquad\qquad (4)$$

If, initially, $t = 0$, $u = u_0$, we have $A = \log(V - u_0)$, and

$$V - u = (V - u_0)\, e^{-ft/V}. \qquad\qquad (5)$$

Hence
$$\frac{dx}{dt} = u = V - (V - u_0)\, e^{-ft/V}, \qquad\qquad (6)$$

$$x = Vt + \frac{V(V-u_0)}{f} e^{-ft/V} + B. \qquad\qquad (7)$$

If $x = 0$ for $t = 0$, we have $B = -V(V - {}_0u)/f$, whence

$$x = Vt - \frac{V(V-u_0)}{f}(1 - e^{-ft/V}). \qquad\qquad (8)$$

* There are some exceptions to this statement in the case of the 'wave-making' resistance of boats in shallow water, but these need not be considered here.

We verify from (5) that $u = V$ for $t = \infty$. The last equation shews also that for large values of t

$$x = Vt - \frac{V(V-u_0)}{f}. \quad \text{.....................(9)}$$

If we put $f = g$, the above calculation may be taken to illustrate the descent of a *very small* globule of water through the air.

Ex. If a particle be projected vertically upwards under this law of resistance we find, if the positive direction of x be upwards,

$$u = (V + u_0)\,e^{-gt/V} - V, \quad \text{.................................(10)}$$

$$x = -Vt + \frac{V(V+u_0)}{g}(1 - e^{-gt/V}), \quad \text{.....................(11)}$$

provided $x = 0$, $u = u_0$, for $t = 0$. The particle will come to rest when

$$t = \frac{V}{g}\log\left(1 + \frac{u_0}{V}\right), \quad \text{.................................(12)}$$

at a height

$$h = \frac{V^2}{g}\left\{\frac{u_0}{V} - \log\left(1 + \frac{u_0}{V}\right)\right\}. \quad \text{.........................(13)}$$

94. Theory of Damped Oscillations.

The problem of the rectilinear motion of a particle under a central force varying as the distance, and a resistance varying as the velocity, is important not only in itself, but for the sake of its wide analogies. The differential equation on which it depends is of exactly the same type as that which applies to the small oscillations of a pendulum, or the torsional vibrations of a suspended bar, as affected by the resistance of the air, or to the vibrations of a galvanometer needle as affected by the currents induced in adjacent masses of metal, and so on.

The equation in question is

$$\frac{d^2x}{dt^2} = -\mu x - k\frac{dx}{dt}, \quad \text{.....................(1)}$$

where the first term on the right hand represents the acceleration due to the restoring force, whilst the second term is due to the resistance. It is usually studied under the form

$$\frac{d^2x}{dt^2} + k\frac{dx}{dt} + \mu x = 0. \quad \text{.....................(2)}$$

This may be transformed as follows. If we put

$$x = ye^{\lambda t}, \quad \text{.........................(3)}$$

we have
$$\frac{dx}{dt} = \left(\frac{dy}{dt} + \lambda y\right) e^{\lambda t}, \quad \text{....................(4)}$$

$$\frac{d^2x}{dt^2} = \left(\frac{d^2y}{dt^2} + 2\lambda \frac{dy}{dt} + \lambda^2 y\right) e^{\lambda t}. \quad \text{.............(5)}$$

Substituting in (2), we have

$$\frac{d^2y}{dt^2} + (2\lambda + k)\frac{dy}{dt} + (\lambda^2 + k\lambda + \mu) y = 0. \quad \text{........(6)}$$

The quantity λ is at our disposal. If we put $\lambda = -\frac{1}{2}k$, the second term in (6) disappears, and the equation reduces to the form which we have already met with in Arts. 5, 11, 57, &c. The equation (2) is therefore satisfied by

$$x = e^{-\frac{1}{2}kt} y, \quad \text{.........................(7)}$$

provided
$$\frac{d^2y}{dt^2} + (\mu - \tfrac{1}{4}k^2) y = 0. \quad \text{....................(8)}$$

Three cases now arise. The first, and most important, is where

$$\mu > \tfrac{1}{4}k^2, \quad \text{.........................(9)}$$

this condition being satisfied for all values of the frictional coefficient below a certain limit. Putting, then,

$$\mu - \tfrac{1}{4}k^2 = n^2, \quad \text{.........................(10)}$$

we have
$$y = A \cos nt + B \sin nt, \quad \text{....................(11)}$$

or
$$y = a \cos (nt + \epsilon), \quad \text{....................(12)}$$

where the constants A, B, or a, ϵ, are arbitrary. Hence

$$x = ae^{-\frac{1}{2}kt} \cos (nt + \epsilon). \quad \text{..................(13)}$$

The type of motion represented by this formula may be described as a simple-harmonic vibration whose amplitude $ae^{-\frac{1}{2}kt}$ diminishes exponentially to zero, as t increases. Since the trigonometrical factor oscillates between the values ± 1, the space-time curve is included between the curves $x = \pm ae^{-\frac{1}{2}kt}$. See Fig. 89.

It is evident, on differentiation of (2), that the velocity u ($=dx/dt$) satisfies an equation of exactly the same type. Hence we may write

$$u = ae^{-\frac{1}{2}kt} \cos(nt + \epsilon'), \quad \text{..................(14)}$$

as may be found otherwise from (13). The stationary values of x occur when $\cos(nt + \epsilon') = 0$, and therefore at equal intervals (π/n) of time. The interval between two successive *maxima* of x is $2\pi/n$, and this is accordingly reckoned as the 'period.' The ratio of one excursion to the next (on the opposite side) is $e^{k\pi/2n}$, and the logarithm of this ratio, viz.

$$\frac{k\pi}{2n} \log_{10} e = \cdot 2171 \frac{k\pi}{n}, \quad \text{..................(15)}$$

is called the 'logarithmic decrement' of the oscillations.

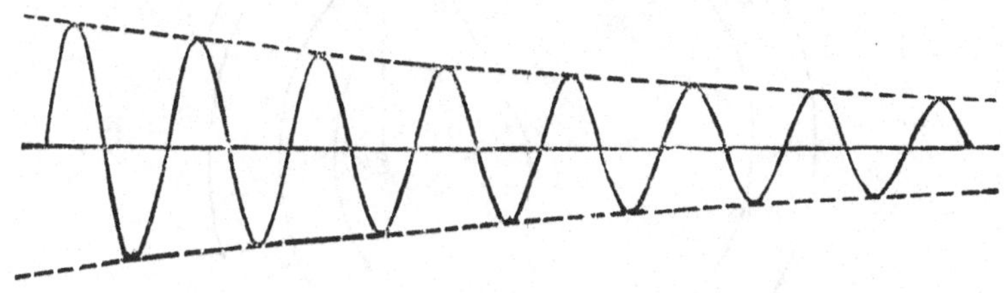

Fig. 89.

It appears from (10) that in consequence of the resistance the period is increased from

$$\frac{2\pi}{\sqrt{\mu}} \text{ to } \frac{2\pi}{\sqrt{(\mu - \frac{1}{4}k^2)}}.$$

If, as in many applications (e.g. in the case of the pendulum), the frictional coefficient is small compared with $\sqrt{\mu}$, the two periods differ only by a small quantity of the second order. A small degree of friction therefore hardly affects the *period* of oscillation, its influence being mainly on the *amplitude*.

It was shewn in Art. 10 that the rectilinear motion of a particle subject only to a central force varying as the distance is identical with that of the orthogonal projection of a point describing a circle with constant angular velocity. This representation may be modified so as to meet the altered circumstances if we replace the circle by an equiangular spiral*. To see this we note that the equations

$$x = ae^{-\frac{1}{2}kt}\cos(nt+\epsilon), \quad y = ae^{-\frac{1}{2}kt}\sin(nt+\epsilon), \quad \text{...(16)}$$

in which x, y are regarded as rectangular coordinates of a moving point, determine such a spiral, being equivalent to

$$x = r\cos\theta, \quad y = r\sin\theta, \quad \text{..................(17)}$$

provided $$r = ae^{-\frac{1}{2}kt}, \quad \theta = nt+\epsilon. \quad \text{..................(18)}$$

For, eliminating t, we have

$$r = a'e^{\theta\cot\alpha}, \quad \text{..........................(19)}$$

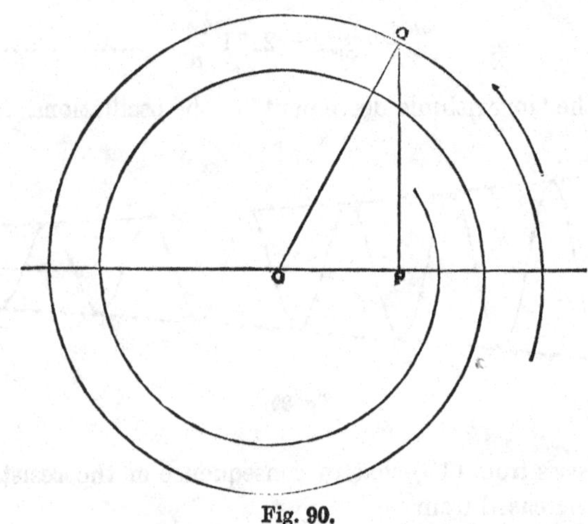

Fig. 90.

where $$a' = ae^{\frac{1}{2}k\epsilon/n}, \quad \cot\alpha = -\tfrac{1}{2}k/n. \quad \text{...............(20)}$$

The angle of the spiral is therefore given by

$$\alpha = \tfrac{1}{2}\pi + \tan^{-1}\frac{k}{2n}. \quad \text{......................(21)}$$

When k/n is small, this only slightly exceeds a right angle.

When the frictional coefficient k is greater than $2\sqrt{\mu}$, the equation (8) takes the form

$$\frac{d^2y}{dt^2} - n^2y = 0, \quad \text{..........................(22)}$$

* This remark is due to P. G. Tait (1867).

where $$n^2 = \tfrac{1}{4}k^2 - \mu. \quad\text{.......................(23)}$$

The solution is $$y = Ae^{nt} + Be^{-nt}, \quad\text{.....................(24)}$$

whence, by (7), $$x = Ae^{-\lambda_1 t} + Be^{-\lambda_2 t}, \quad\text{....................(25)}$$

provided

$$\lambda_1 = \tfrac{1}{2}k - \sqrt{(\tfrac{1}{4}k^2 - \mu)}, \quad \lambda_2 = \tfrac{1}{2}k + \sqrt{(\tfrac{1}{4}k^2 - \mu)}. \quad\text{......(26)}$$

This solution may, however, be obtained more directly from the original equation (2). If we assume

$$x = Ae^{-\lambda t}, \quad\text{...........................(27)}$$

we find that (2) is satisfied provided

$$\lambda^2 - k\lambda + \mu = 0. \quad\text{.........................(28)}$$

This determines the two admissible values of λ, which are in fact the two quantities denoted above by λ_1, λ_2. Superposing the two solutions of the type (27) thus obtained, we have the result (25).

Since λ_1, λ_2 are both positive, the value of x tends ultimately to zero. Moreover, in order that x may vanish, we must have

$$e^{(\lambda_2 - \lambda_1)t} = -B/A. \quad\text{.....................(29)}$$

If A and B have opposite signs, this equation is satisfied by one, and only one, real value of t. If A and B have the same sign there is no real solution. Hence the particle, however started, cannot pass more than once through its equilibrium position, to which it finally creeps asymptotically. The present type of motion is therefore called 'aperiodic.' It is realized in the case of a pendulum immersed in a very viscous medium, in 'dead beat' galvanometers, where the needle is closely surrounded by metal of high conductivity in which currents are induced opposing the motion, and in modern forms of seismograph.

In the transition case, where

$$k^2 = 4\mu, \quad\text{...........................(30)}$$

exactly, the equation (8) becomes

$$\frac{d^2y}{dt^2} = 0, \quad\text{...........................(31)}$$

whence $$y = At + B, \quad\text{.........................(32)}$$

and $$x = e^{-\frac{1}{2}kt}(At + B). \quad\text{.....................(33)}$$

The diminution of the exponential factor ultimately prevails over the increase of t in the second factor; moreover x can only vanish for one finite value of t. This type of motion is therefore also classed as 'aperiodic.' The annexed figure shews the space-time curve for this case. For the sake of comparison, the first half-period of an unresisted particle started from the origin with the same velocity is also represented (by the dotted curve).

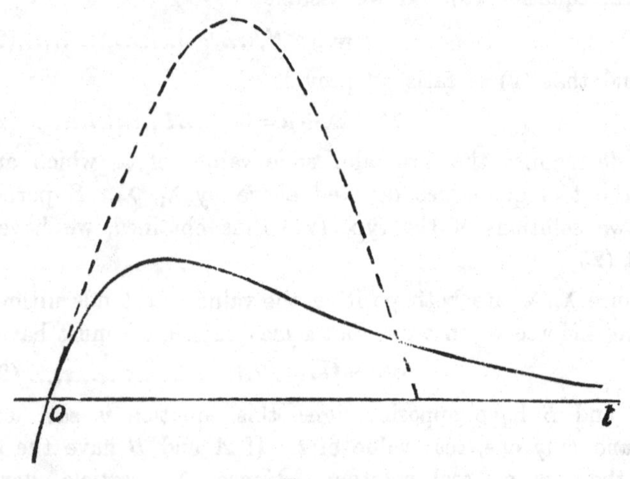

Fig. 91.

95. Forced Oscillations.

We have next to consider how *forced* oscillations, such as have been discussed in Art. 13, are modified by the resistance. Adding a term X on the right-hand side of Art. 94 (1), to represent the accelerative effect of the disturbing force, we have

$$\frac{d^2x}{dt^2} + k\frac{dx}{dt} + \mu x = X. \quad \text{.....................(1)}$$

This equation gives at once the force which is necessary to maintain a motion of any prescribed type. Thus in order that a simple-harmonic vibration

$$x = a\cos pt \quad \text{..........................(2)}$$

may be maintained, we must have

$$X = a\,\{(\mu - p^2)\cos pt - kp\sin pt\}. \quad \text{...........(3)}$$

If we put

$$\mu - p^2 = R\cos\alpha, \quad kp = R\sin\alpha, \text{ ....(4)}$$

as is always possible by a proper choice of R and α, this becomes

$$X = R a \cos(pt + \alpha). \text{ ....(5)}$$

Hence, putting $Ra = f$, and writing $pt - \alpha$ for pt, which is merely equivalent to changing the origin of t, we find that to the disturbing force

$$X = f\cos pt \text{ ....(6)}$$

will correspond the forced vibration

$$x = \frac{f}{R}\cos(pt - \alpha). \text{ ....(7)}$$

This does not of course constitute the *complete* solution of (1), with the value (6) of X. We are at liberty to add any value of x which makes the expression on the left-hand side of (1) vanish. Thus in the case where $k^2 < 4\mu$ we have the form

$$x = e^{-\frac{1}{2}kt}(A\cos nt + B\sin nt) + \frac{f}{R}\cos(pt - \alpha), \text{ ......(8)}$$

where n is given by (10) of Art. 94. The arbitrary constants A, B will depend on the initial conditions.

The first part of the solution (8) may be interpreted as representing a free vibration, such as the particle could execute in the absence of disturbing force, which is superposed on the forced oscillation. Owing to the indefinite decrease of the exponential factor, the free vibrations, and therewith the effects of the initial conditions, gradually die out, until after a time the forced vibration represented by the last term is alone sensible. The same conclusion holds also in the cases of $k^2 > 4\mu$, and $k^2 = 4\mu$.

With regard to the forced vibration, there are several important points to be noticed.

First, with respect to the amplitude f/R, we have from (4)

$$R^2 = (\mu - p^2)^2 + k^2p^2 = \{p^2 - (\mu - \tfrac{1}{2}k^2)\}^2 + k^2(\mu - \tfrac{1}{4}k^2). \text{ ...(9)}$$

If $k^2 < 2\mu$ this is a minimum, as regards variation of p, for

$$p^2 = \mu - \tfrac{1}{2}k^2, \text{ ....(10)}$$

and the maximum amplitude is

$$\frac{f}{kn}. \quad \text{..............................(11)}$$

If k is small compared with $\sqrt{\mu}$, the amplitude is greatest when $p = \sqrt{\mu}$, nearly, i.e. when the period of the disturbing force is equal to that of unresisted free vibration. Owing to the factor k in the denominator of (11) the maximum amplitude is relatively large. In the case of a pendulum swinging in air it may easily become so great as to impair the legitimacy of the approximations on which the fundamental equation (1) is based.

The relations of phase are also important. It appears from (7) that the phase of the forced vibration lags behind that of the disturbing force by an amount α, which is determined by (4). Since $\sin \alpha$ is positive, whilst $\cos \alpha$ is positive or negative according as $p^2 \lessgtr \mu$, the angle α may be taken to be in the first or second quadrant according as the period of the disturbing force is longer or shorter than that of free vibration in the absence of friction. Moreover, since

$$\tan \alpha = \frac{kp}{\mu - p^2}, \quad \text{....................(12)}$$

α will *in general*, if the damping be small, be nearly equal to 0 or π in the respective cases. This is as we should anticipate from Art. 13, where we found exact agreement or opposition of phase in the two cases, on the hypothesis of no friction. But when there is almost exact coincidence between the periods, or, more precisely, when the difference between $p/\sqrt{\mu}$ and unity is small compared with k/p, the angle α approaches $\frac{1}{2}\pi$ from one side or the other, and in the case of maximum resonance when $p = \sqrt{\mu}$, we have $\alpha = \frac{1}{2}\pi$; i.e. the maximum displacement follows the maximum force at an interval of a quarter-period. The force now synchronizes with the *velocity*.

The question of phase is closely related to that of the supply of energy. When $\alpha = 0$ the work done in one half-period is cancelled by the work restored in the other half, and there is on the whole no absorption of energy. In the general case, the rate at which work is being done in maintaining the vibration against resistance is, from (6) and (7),

$$X\frac{dx}{dt} = -\frac{pf^2}{R}\sin(pt-\alpha)\cos pt$$

$$= \frac{pf^2}{2R}\{\sin\alpha - \sin(2pt-\alpha)\}. \quad \text{.........(13)}$$

The average value of the second term in the bracket is zero, and the mean rate of absorption of energy is therefore

$$\frac{pf^2\sin\alpha}{2R} = \frac{kp^2f^2}{2R^2} = \frac{\frac{1}{2}kf^2}{k^2+(p-\mu/p)^2}. \quad \text{.........(14)}$$

The absorption is therefore a maximum when $p=\sqrt{\mu}$, i.e. when the imposed period coincides with the free period in the absence of friction, exactly. The maximum absorption is

$$\frac{f^2}{2k}, \quad \text{.........(15)}$$

and is therefore greater, the smaller the value of k.

To illustrate the effect of a slight deviation from exact coincidence, in producing a falling off from the maximum, we may put

$$\frac{p}{\sqrt{\mu}} = 1+z, \quad \text{.........(16)}$$

where z is supposed small. The expression (14) for the absorption may then be put in the form

$$\frac{f^2}{4\sqrt{\mu}}\cdot\frac{\beta}{\beta^2+z^2}, \quad \text{.........(17)}$$

approximately, where

$$\beta = \frac{k}{2\sqrt{\mu}}. \quad \text{.........(18)}$$

The absorption therefore sinks to one-half its maximum when $z=\beta$, or

$$p=\sqrt{\mu}+\tfrac{1}{2}k, \quad \text{.........(19)}$$

The graph of the function

$$\frac{\beta}{\beta^2+z^2} \quad \text{.........(20)}$$

is shewn in Fig. 92 for three different values of β. The area included between the curve and the zero line is independent of β, the higher values of the maximum ordinate for small values of β being compensated by a more rapid falling off on either side. This principle that, the greater the intensity of the resonance in the case of exact agreement between the periods, the narrower the range over which it is approximately equal to the maximum, has many striking illustrations in Acoustics. Thus a system such as a tuning fork, or a piano wire, whose free vibrations are only slowly extinguished by dissipation, is not easily set into vigorous sympathetic vibration unless there be very close concordance

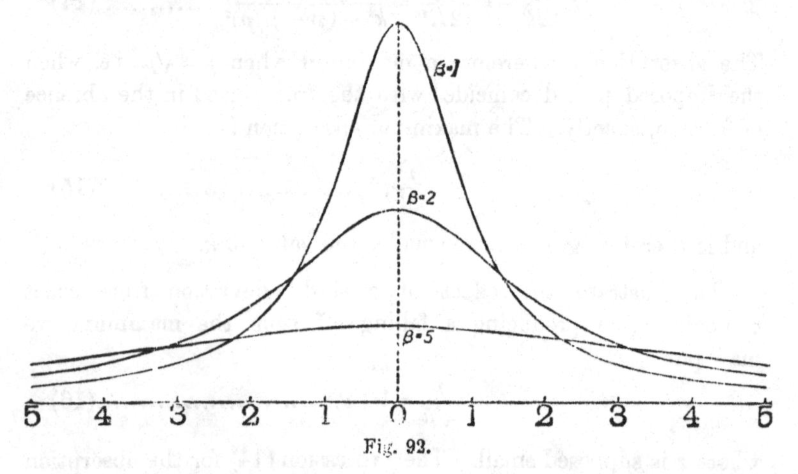

Fig. 92.

between the imposed and the free periods; whilst the air column of an organ pipe will respond to a comparatively wide range of frequencies.

Similar conclusions hold with regard to the square of the amplitude, or to the total energy of the vibration.

In modern seismographs damping contrivances are introduced with a view of diminishing the influence of free vibrations. In the Galitzin type, for example, the pendulum carries a metal plate which swings in its own plane in a magnetic field, so that the reaction of the magnets on the electric currents thus induced in the plate produces a resistance varying as the velocity.

Taking, for instance, the case of the horizontal pendulum (Arts. 55, 67), the equation of motion, when corrected for damping, takes the shape

$$\frac{d^2\theta}{dt^2} + k\frac{d\theta}{dt} + \mu\theta = -\frac{1}{l}\frac{d^2\xi}{dt^2}, \quad \text{.......................(21)}$$

where ξ is the displacement of the axis at right angles to the equilibrium position of the boom. The forced oscillation due to a motion

$$\xi = a\cos pt \quad \text{.............................(22)}$$

of the frame of the instrument is therefore

$$\theta = \frac{a}{l}\cdot\frac{p^2}{R}\cos(pt-\alpha), \quad \text{..........................(23)}$$

in the preceding notation. The ratio of the amplitude of vibration of a point on the boom, relative to the frame, to that of the frame itself depends on the factor

$$\frac{p^2}{R} = \frac{p^2}{\sqrt{\{(\mu-p^2)^2+k^2p^2\}}}. \quad \text{...........................(24)}$$

In the Galitzin instruments, the damping is adjusted so that the free vibration is on the border line of aperiodicity, or $k^2 = 4\mu$. The factor (24) then takes the simpler form

$$\frac{p^2}{p^2+\mu}. \quad \text{...................................(25)}$$

This is sensibly independent of p, so long as p is large compared with $\sqrt{\mu}$, i.e. so long as the period of the tremors is short compared with that of an undamped free vibration.

In the instruments referred to, the record is made, however, by the galvanometric method referred to in Art. 67, in which the indications depend not on the angular displacement θ but on the angular *velocity* $d\theta/dt$. A further factor is thus introduced by the consideration of the law of oscillation of the galvanometer mirror, which is also damped so as to be on the border line of aperiodicity. The scale on which the velocity of the ground is recorded is accordingly determined by a factor of the form

$$\frac{p^2}{(p^2+\mu)(p^2+\mu')}. \quad \text{...........................(26)}$$

This varies very slowly for values of p^2 in the neighbourhood of $\sqrt{(\mu\mu')}$, for which it is a maximum. In practice the undamped periods of the galvanometer and seismograph are adjusted as nearly as possible to equality, so that $\mu = \mu'$.

96. The Spherical Pendulum.

The effect of friction on the small oscillations of the spherical pendulum (Art. 29) is easily ascertained. The equations of motion have the forms

$$\frac{d^2x}{dt^2} + k\frac{dx}{dt} + \mu x = 0, \quad \frac{d^2y}{dt^2} + k\frac{dy}{dt} + \mu y = 0, \quad \text{........(1)}$$

whatever the directions of the axes; and if $k^2 < 4\mu$ the solution is

$$x = e^{-\frac{1}{2}kt}(A \cos nt + B \sin nt), \quad y = e^{-\frac{1}{2}kt}(C \cos nt + D \sin nt). \quad \text{........(2)}$$

If we choose the origin of t at an instant when the particle crosses the axis of x, at the point $(a, 0)$, say, we have $A = a$, $C = 0$; and if we make the axis of y pass through the position which the particle occupies after an interval $\frac{1}{2}\pi/n$, we have $B = 0$. The solution thus takes the form

$$x = ae^{-\frac{1}{2}kt} \cos nt, \quad y = be^{-\frac{1}{2}kt} \sin nt. \quad \text{............(3)}$$

The particle may therefore be said to move in an ellipse which continually contracts according to the exponential law, remaining always of the same shape, with the same orientation.

If we draw two perpendicular radii, it is evident that in some positions the time of passing from one of these to the other will be less, and in others greater, than $\frac{1}{2}\pi/n$, the period of a quarter-revolution. There is therefore one pair of perpendicular radii such that the interval is exactly $\frac{1}{2}\pi/n$, and accordingly one set of *rectangular* axes to which the formulæ (3) may be supposed to refer.

The actual path may be regarded as a kind of elliptic spiral. If $a = b$, we get the equiangular spiral of Art. 94.

The problem of the spherical pendulum is identical with that of the motion of a particle in a spherical bowl. An interesting variation is obtained if we suppose the bowl to rotate with constant angular velocity ω about a vertical axis through its centre. If we take fixed horizontal axes of x, y as before, with the origin at the lowest point, the component velocities of a point (x, y) of the bowl will be $-\omega y$, ωx, respectively. The velocities of a particle relative to the bowl will therefore be given by the expressions

$$\dot{x} + \omega y, \quad \dot{y} - \omega x. \quad \text{..................................(4)}$$

Hence, assuming that the friction varies as the *relative* velocity, we have

$$\ddot{x} = -n^2 x - k(\dot{x} + \omega y), \quad \ddot{y} = -n^2 y - k(\dot{y} - \omega x), \quad \text{..............(5)}$$

where

$$n^2 = g/a, \quad \text{....................................(6)}$$

if a be the radius of the bowl.

The solution of simultaneous linear equations of the type (5) is usually best conducted in terms of imaginary quantities, but for the present purpose we may obtain a sufficient solution as follows. We assume

$$x = r \cos(\lambda t + \epsilon), \quad y = r \sin(\lambda t + \epsilon), \quad \text{......................(7)}$$

where r is a function of t to be determined. If we substitute in (5) we find that both equations are satisfied provided

$$\ddot{r}+k\dot{r}+(n^2-\lambda^2)\,r=0, \qquad \text{.......(8)}$$

$$2\lambda\dot{r}+k\,(\lambda-\omega)\,r=0. \qquad \text{.......(9)}$$

Eliminating r, we find

$$\lambda^2(\lambda^2-n^2)+\tfrac{1}{4}k^2(\lambda^2-\omega^2)=0. \qquad \text{.......(10)}$$

The real roots of this are

$$\lambda=\pm n, \qquad \text{.......(11)}$$

approximately, if we neglect the square of k*. Also, from (9) we have

$$r=Ce^{-\frac{1}{2}k\,(1-\omega/\lambda)\,t}. \qquad \text{.......(12)}$$

We thus obtain the solutions

$$x=Ce^{-\nu t}\cos(nt+\epsilon), \quad y=Ce^{-\nu t}\sin(nt+\epsilon), \qquad \text{.......(13)}$$

and

$$x=C'e^{-\nu' t}\cos(nt+\epsilon'), \quad y=-C'e^{-\nu' t}\sin(nt+\epsilon'), \qquad \text{.......(14)}$$

where

$$\nu=\tfrac{1}{2}k\left(1-\frac{\omega}{n}\right), \qquad \nu'=\tfrac{1}{2}k\left(1+\frac{\omega}{n}\right). \qquad \text{.......(15)}$$

Since the differential equations are linear, these results can be superposed.

The formulæ (13) and (14) represent circular vibrations of gradually varying amplitude, the directions of revolution being opposite in the two cases. If $\omega<n$, the amplitude in each case diminishes exponentially. But if $\omega>n$, the sign of ν is changed, and the amplitude of the circular vibration whose sense is the same as that of the angular velocity ω continually increases. The particle then works its way outwards in an ever-widening spiral path, approximating to the state in which it revolves with the bowl in relative equilibrium, like the bob of a conical pendulum (Art. 35, Ex. 1), the angular distance from the vertical being

$$\cos^{-1}\frac{g}{\omega^2 a}.$$

The relative equilibrium of the particle when at the lowest point of the bowl is accordingly now unstable.

It has already been indicated (Art. 33) that the criterion of stability has to be modified when *relative* equilibrium is in question. The argument given on p. 87, applied to the equation of relative energy (Art. 33 (10)), shews that for practical stability in the present case the expression

$$V-\tfrac{1}{2}m\omega^2(x^2+y^2) \qquad \text{.......(16)}$$

* If the imaginary solutions were examined it would be found that they lead to no new results. The fact that the formulæ (13) and (14) involve *four* arbitrary constants C, C', ϵ, ϵ' shews that the solution which we obtain is complete.

must be a minimum. If z denote altitude above the tangent plane at the lowest point we have $x^2+y^2=2az$, approximately, and therefore

$$V-\tfrac{1}{2}m\omega^2(x^2+y^2)=m(g-\omega^2 a)z. \quad\text{.....................(17)}$$

This will be positive, and the lowest position accordingly stable, only if $\omega^2<g/a$.

If θ be the angular distance of the particle from the lowest point, we have

$$V-\tfrac{1}{2}m\omega^2(x^2+y^2)=ma\{g(1-\cos\theta)-\tfrac{1}{2}\omega^2 a\sin^2\theta\}, \quad\text{.........(18)}$$

and it is easily shewn that if $\omega^2>g/a$ this expression is a minimum when $\cos\theta=g/\omega^2 a$. The inclined position of relative equilibrium, when it exists, is therefore stable.

97. Quadratic Law of Resistance.

The linear law of resistance hitherto assumed has practical validity only for very small velocities. For a certain range of velocities, not very small and not very great, the hypothesis of a resistance varying as the *square* of the velocity is found to give better results.

This law is not without some theoretical support. In the case of a body moving through the air, for example, a mass of air proportional to $u\delta t$, where u is the velocity, is (so to speak) overtaken in the time δt, and an average velocity proportional to u is communicated to it. The *momentum* given to the air, and accordingly lost to the body, is therefore proportional to u^2, per unit time*. This argument is defective in that it assumes the flow of air relative to the body to have the same geometrical distribution for all velocities, but it avails to shew that under certain conditions a considerable part of the resistance may follow the law above stated.

Assuming the law, we have, in the case of a particle subject to no force except the resistance,

$$\frac{du}{dt}=-ku^2, \quad \frac{du}{dx}=-ku. \quad\text{.................(1)}$$

Hence

$$-\frac{1}{u^2}\frac{du}{dt}=k, \quad\text{or}\quad \frac{1}{u}=kt+A. \quad\text{...............(2)}$$

If $u=u_0$ for $t=0$, we have $A=1/u_0$, whence

$$u=\frac{u_0}{1+ku_0t}. \quad\text{.........................(3)}$$

* Newton, *Principia*, lib. ii, prop. iv (scholium).

Again, from the second form of the equation,

$$\frac{1}{u}\frac{du}{dx} = -k, \quad \log u = -kx + B. \quad \text{..............(4)}$$

If $x = 0$ for $u = u_0$, we have $B = \log u_0$, and therefore

$$u = u_0 e^{-kx}. \quad \text{..........................(5)}$$

Eliminating u, or by integration of (3), we find

$$x = \frac{1}{k}\log(1 + ku_0 t). \quad \text{....................(6)}$$

It appears from (5) that u does not vanish for any finite value of x, and that consequently there is on the present hypothesis no limit to the space described. The law ceases however to have a practical value for small velocities, the real resistance diminishing ultimately as u, rather than as u^2.

The coefficient k is the reciprocal of a length, viz. the distance in which the velocity is diminished in the ratio $1/e$. Denoting this length by a, we have

$$x = a\log\left(1 + \frac{u_0 t}{a}\right). \quad \text{....................(7)}$$

98. Case of a Constant Propelling Force.

In the case of a body subject to a constant propelling force, as well as to a resistance varying as the square of the velocity, we have

$$\frac{du}{dt} = f - ku^2, \quad \text{or} \quad u\frac{du}{dx} = f - ku^2. \quad \text{...........(1)}$$

The terminal velocity, when the acceleration vanishes, is given by

$$kV^2 = f, \quad \text{........................(2)}$$

whence $$\frac{du}{dt} = \frac{f}{V^2}(V^2 - u^2), \quad u\frac{du}{dx} = \frac{f}{V^2}(V^2 - u^2). \quad \text{.........(3)}$$

Hence $$\frac{1}{V^2 - u^2}\frac{du}{dt} = \frac{f}{V^2}, \quad \tanh^{-1}\frac{u}{V} = \frac{ft}{V} + A. \quad \text{.........(4)}$$

If the initial velocity be zero, we have $A = 0$, and

$$u = V\tanh\frac{ft}{V}, \quad \text{.....................(5)}$$

tending to the limit V for $t \to \infty$. Putting $u = dx/dt$, and integrating with respect to t, we have

$$x = \frac{V^2}{f} \log \cosh \frac{ft}{V}, \quad \text{.....................(6)}$$

no additive constant being necessary if $x = 0$ for $t = 0$.

The second of equations (3) leads to

$$x = \frac{V^2}{2f} \log \frac{V^2}{V^2 - u^2}, \quad \text{.....................(7)}$$

which may of course be derived from (5) and (6).

If we put $f = g$, the investigation may be taken to illustrate the fall of a body from rest, under gravity. It applies also to the starting of a steamer. For large values of t we have

$$\cosh(ft/V) = \tfrac{1}{2} e^{ft/V},$$

nearly, and therefore

$$x = Vt - \frac{V^2}{f} \log 2, \quad \text{.....................(8)}$$

where the last term represents the distance lost in getting up speed. The time lost is

$$V/f \,.\, \log 2.$$

In the case of a constant force *opposing* the motion we have

$$\frac{du}{dt} = -f - ku^2, \text{ or } u\frac{du}{dx} = -f - ku^2, \quad \text{.........(9)}$$

these equations being valid only so long as u is positive. If V be the terminal velocity under the given force, as determined by (2), we have

$$\frac{du}{dt} = -\frac{f}{V^2}(V^2 + u^2), \quad u\frac{du}{dx} = -\frac{f}{V^2}(V^2 + u^2). \quad \text{...(10)}$$

Hence

$$\frac{1}{V^2 + u^2}\frac{du}{dt} = -\frac{f}{V^2}, \quad \tan^{-1}\frac{u}{V} = -\frac{ft}{V} + A\,; \quad \text{......(11)}$$

or, if $u = u_0$ for $t = 0$,

$$\tan^{-1}\frac{u}{V} = \tan^{-1}\frac{u_0}{V} - \frac{ft}{V}. \quad \text{..............(12)}$$

Again, from the second of equations (10),

$$\log (V^2 + u^2) = -\frac{2fx}{V^2} + B. \quad \text{..............(13)}$$

If $x = 0$ for $u = u_0$, we have $B = \log (V^2 + u_0^2)$, and

$$x = \frac{V^2}{2f} \log \frac{V^2 + u_0^2}{V^2 + u^2}. \quad \text{..................(14)}$$

When the body comes to rest we have $u = 0$, and therefore

$$t = \frac{V}{f} \tan^{-1} \frac{u_0}{V}, \quad x = \frac{V^2}{2f} \log \left(1 + \frac{u_0^2}{V^2}\right). \quad \text{.........(15)}$$

In the case of a steamer whose engines are reversed when going at full speed, we have $u_0 = V$. The time required to stop the steamer is therefore

$$t = \frac{1}{4} \frac{\pi V}{f}, \quad \text{........................(16)}$$

and the distance travelled in this time is

$$x = \frac{V^2}{2f} \log 2. \quad \text{........................(17)}$$

Ex. In the case of a particle projected vertically upwards with velocity u_0, the height h attained is found by putting $u = 0$ in (14); thus

$$h = \frac{V^2}{2g} \log \left(1 + \frac{u_0^2}{V^2}\right), \quad \text{............................(18)}$$

and the time of ascent is, by (15),

$$t = \frac{V}{g} \tan^{-1} \frac{u_0}{V}. \quad \text{..............................(19)}$$

The time (t_1) occupied in the subsequent descent is given by (6); thus

$$t_1 = \frac{V}{g} \cosh^{-1} e^{gh/V^2} = \frac{V}{g} \cosh^{-1} \left(1 + \frac{u_0^2}{V^2}\right)^{\frac{1}{2}}$$

$$= \frac{V}{g} \log \left\{\frac{u_0}{V} + \left(1 + \frac{u_0^2}{V^2}\right)^{\frac{1}{2}}\right\}. \quad \text{..............................(20)}$$

The velocity (u_1) acquired in the descent is to be found from (7); thus

$$\frac{u_1^2}{V^2} = 1 - e^{-2gh/V^2} = \frac{u_0^2}{V^2 + u_0^2}, \quad \text{........................(21)}$$

by (18).

When the ratio u_0/V is small, we find

$$h = \frac{u_0^2}{2g}\left(1 - \frac{u_0^2}{2V^2}\right), \qquad t = \frac{u_0}{g}\left(1 - \frac{u_0^2}{3V^2}\right), \quad \text{...............(22)}$$

and

$$t_1 = \frac{u_0}{g}\left(1 - \frac{u_0^2}{6V^2}\right), \qquad u_1 = u_0\left(1 - \frac{u_0^2}{2V^2}\right), \quad \text{...............(23)}$$

approximately.

99. Effect of Resistance on Projectiles.

The *general* effect of resistance on the motion of a projectile is easily understood if we consider that of a succession of instantaneous tangential impulses, each of which diminishes the velocity. Since the normal acceleration v^2/ρ, being equal to the component of gravity normal to the path, undergoes no instantaneous change, the diminution of v implies an increased curvature. The path is therefore continually deflected inwards from the parabolic course which it would proceed to describe if there were no further resistance, and the motion tends ultimately to become vertical, with the terminal velocity.

If l be the semi-latus-rectum of the 'instantaneous parabola' we have, by Art. 27,

$$l = \frac{v^2 \cos^2 \psi}{g}, \quad \text{........................(1)}$$

where ψ is the inclination of the path to the horizontal. Hence, in the case of a small tangential impulse,

$$\frac{\delta l}{l} = \frac{2\delta v}{v}. \quad \text{........................(2)}$$

Putting $\delta v = -f\delta t$, where f is the retardation due to the medium, we have

$$\frac{1}{l}\frac{dl}{dt} = -\frac{2f}{v}. \quad \text{........................(3)}$$

For example, if

$$f = kv^2, \quad \text{........................(4)}$$

we have

$$\frac{1}{l}\frac{dl}{dt} = -2kv = -2k\frac{ds}{dt}, \quad \text{.................(5)}$$

whence

$$l = Ce^{-2ks}, \quad \text{........................(6)}$$

shewing the gradual diminution in the parameter of the parabola.

The particular case of a resistance varying as the velocity can be solved completely, by means of the Cartesian equations, which are in this case independent of one another. With the usual axes (horizontal and vertical), we have

$$\frac{d^2x}{dt^2} = -k\frac{ds}{dt}\cos\psi = -k\frac{dx}{dt}, \quad \text{.....................(7)}$$

and

$$\frac{d^2y}{dt^2} = -g - k\frac{ds}{dt}\sin\psi = -g - k\frac{dy}{dt}. \quad \text{.........(8)}$$

The first of these gives

$$\log\frac{dx}{dt} = -kt + A, \quad \frac{dx}{dt} = Ce^{-kt}, \quad x = D - \frac{C}{k}e^{-kt}. \quad \text{...(9)}$$

If $x=0$, $dx/dt = u_0$, for $t=0$, we have $C = u_0$, $D = u_0/k$, whence

$$x = \frac{u_0}{k}(1 - e^{-kt}), \quad \text{.....................(10)}$$

as in Art. 92 (8).

Again, writing (8) in the form

$$\frac{d^2y}{dt^2} + k\frac{dy}{dt} = -g, \quad \text{.....................(11)}$$

we see that a particular first integral is $dy/dt = -g/k$; hence, completing the integral, we have

$$\frac{dy}{dt} = -\frac{g}{k} + C'e^{-kt}. \quad \text{.....................(12)}$$

Integrating again, we have

$$y = -\frac{gt}{k} - \frac{C'}{k}e^{-kt} + D'. \quad \text{..................(13)}$$

If $y=0$, $dy/dt = v_0$, for $t=0$, we find $C' = g/k + v_0$, $D' = C'/k$. Hence

$$y = -\frac{gt}{k} + \left(\frac{v_0}{k} - \frac{g}{k^2}\right)(1 - e^{-kt}). \quad \text{...........(14)}$$

It appears from (10) that the straight line

$$x = \frac{u_0}{k} = \frac{u_0 V}{g}, \quad \text{.......................(15)}$$

where V is the terminal velocity, is an asymptote to the path.

It should be remarked, however, that this investigation has merely an illustrative value, since the linear law of resistance is far from being valid in the case of a projectile.

The hypothesis of a resistance varying as the square of the velocity is from a practical point of view to be preferred, but does not lend itself so easily to mathematical treatment. The equations of motion now are

$$\frac{d^2x}{dt^2} = -kv^2 \cos\psi, \quad \frac{d^2y}{dt^2} = -g - kv^2 \sin\psi. \quad \text{......(16)}$$

Since $dx/dt = v\cos\psi$, the former of these may be written

$$\frac{\ddot{x}}{\dot{x}} = -kv = -k\dot{s}, \quad \text{......................(17)}$$

whence $$\log \dot{x} = -ks + \text{const.},$$

or $$\dot{x} = u_0 e^{-ks}, \quad \text{........................(18)}$$

if u_0 be the horizontal velocity corresponding to $s = 0$. It is easily seen that this is equivalent to (6).

To carry the matter further it is convenient to resolve along the normal, instead of employing the second of equations (16). We have, with the usual convention as to the sign of ρ,

$$\frac{v^2}{\rho} = -g\cos\psi. \quad \text{........................(19)}$$

Putting $$v = \dot{x}\sec\psi = u_0 e^{-ks}\sec\psi, \quad \text{..............(20)}$$

$$\frac{1}{\rho} = \frac{d^2y}{dx^2}\cos^3\psi, \quad \text{........................(21)}$$

we find $$\frac{d^2y}{dx^2} = -\frac{g}{u_0^2} e^{2ks}. \quad \text{........................(22)}$$

If we confine our attention to a part of the trajectory that is nearly horizontal, we may replace s by x, the difference being of the second order of small quantities. Thus

$$\frac{d^2y}{dx^2} = -\frac{g}{u_0^2} e^{2kx}. \quad \text{........................(23)}$$

Integrating, we find

$$\frac{dy}{dx} = -\frac{g}{2ku_0^2} e^{2kx} + A, \quad \text{..................(24)}$$

$$y = -\frac{g}{4k^2u_0^2} e^{2kx} + Ax + B. \quad \text{..............(25)}$$

If we assume that $y=0$, $dy/dx=\tan\alpha$, where α is of course assumed to be small, for $x=0$, we have

$$A=\frac{g}{2ku_0^2}+\tan\alpha,\quad B=\frac{g}{4k^2u_0^2},$$

and the equation to the (nearly horizontal) path is

$$y=x\left(\tan\alpha+\frac{g}{2ku_0^2}\right)+\frac{g}{4k^2u_0^2}(1-e^{2kx}).\ \text{.........}(26)$$

It has been already noticed (Art. 97) that $k=1/a$, where a is the distance within which a particle subject only to the resistance would have its velocity reduced in the ratio $1/e$. If x be small compared with this distance, we find, expanding the exponential in (26),

$$y=x\tan\alpha-\frac{g}{2u_0^2}x^2-\frac{gk}{3u_0^2}x^3-\dots\ \text{............}(27)$$

The first two terms of this series give the parabolic path which would be described in the absence of resistance.

100. Effect of Resistance on Planetary Orbits.

The general effect of a resisting medium on the motion of a planet can be inferred from the formulæ of Art. 82. It was there shewn that the change in the mean distance due to a tangential impulse δv is given by

$$\frac{\delta a}{a}=\frac{2v\,\delta v}{n^2a^2},\ \text{..........................}(1)$$

where n is the mean angular velocity. Putting $\delta v=-f\delta t$, where f is the retardation, we have

$$\frac{da}{dt}=-\frac{2vf}{n^2a}.\ \text{..........................}(2)$$

Also, from Art. 82 (4),

$$\frac{db}{dt}=-\left(1+\frac{v^2}{n^2a^2}\right)\frac{bf}{v}.\ \text{.....................}(3)$$

The orbit therefore continually contracts.

The change in the mean motion is given by

$$\frac{dn}{dt}=\frac{3vf}{na^2};\ \text{..........................}(4)$$

the angular velocity therefore continually increases.

The mean kinetic energy (K) also increases, the formula being

$$\frac{dK}{dt}=vf. \qquad\qquad (5)$$

The total energy, of course, continually diminishes.

EXAMPLES. XXI.

1. A particle is projected vertically upwards. Prove that if the resistance of the air were constant, and equal to $1/n$th of the weight of the particle, the times of ascent and descent would be as $\sqrt{(n-1)}:\sqrt{(n+1)}$.

2. If x_1, x_2, x_3 be the scale readings of three consecutive points of rest of a damped galvanometer needle, prove that the equilibrium reading is

$$\frac{x_1 x_3 - x_2^2}{x_1+x_3-2x_2}.$$

If the damping be slight, prove that this is equal to

$$\tfrac{1}{4}(x_1+2x_2+x_3),$$

approximately.

3. If a particle be subject to gravity and to a resistance varying as the velocity, the hodograph is a straight line.

4. If chords be drawn from either extremity of the vertical diameter of a circle, the time of descent down each, in a medium whose resistance varies as the velocity, is the same.

5. A particle moves on a smooth cycloid whose axis is vertical and vertex downwards. Prove that if in addition to gravity it be subject to a resistance varying as the velocity, the oscillations will still be isochronous.

6. Shew that in the problem of Art. 95 the mean energy of the forced vibration is a maximum when

$$p^2=\mu-\tfrac{1}{4}k^2,$$

approximately, if $k^2/4\mu$ be small.

7. Prove that the solution of the equation

$$\frac{d^2x}{dt^2}+k\frac{dx}{dt}+\mu x=f(t)$$

is $\quad x=-\dfrac{1}{n}e^{-\frac{1}{2}kt}\cos nt\displaystyle\int e^{\frac{1}{2}kt}f(t)\sin nt\,dt+\dfrac{1}{n}e^{-\frac{1}{2}kt}\sin nt\displaystyle\int e^{\frac{1}{2}kt}f(t)\cos nt\,dt,$

where $\qquad n=\sqrt{(\mu-\frac{1}{4}k^2)}.$

8. Prove that if $k^2=4\mu$ the solution of the preceding equation is

$$x=e^{-\frac{1}{2}kt}\iint e^{\frac{1}{2}kt}f(t)\,dt\,dt.$$

9. Prove that in the case of rectilinear motion subject only to a resistance varying as the square of the velocity the mean retardation in any interval of time is equal to the geometric mean of the retardations at the beginning and end of the interval.

10. A body moves in a medium whose resistance varies as the square of the velocity. Prove that if the speed vary according to the law

$$u = u_0 (1 + a \cos pt),$$

the mean value (per unit time) of the propulsive force required is greater than if the speed had been constant, and equal to u_0, in the ratio $1 + \frac{1}{2}a^2$.

Prove also that the mean rate of expenditure of energy is increased in the ratio $1 + \frac{3}{2}a^2$.

11. If the speed vary according to the law

$$u = u_0 (1 + \beta \cos mx),$$

where β is small, the mean propulsive force (per unit time) is less than if the speed had been u_0 in the ratio $1 - \frac{1}{2}\beta^2$, and that the mean time of accomplishing a given distance is greater in the ratio $1 + \frac{1}{2}\beta^2$.

Prove that the mean rate of expenditure of energy is unaltered.

12. A particle is projected vertically upwards in a medium whose resistance varies as the square of the velocity, and the coefficient of resistance is such that the terminal velocity would be 300 ft./sec. If the velocity of projection be such as would in the absence of resistance carry the particle to a height of 100 ft., what will be the height actually attained? [96·4 ft.]

13. A ship of W tons is steaming at full speed (V knots), under a horse-power H, against a resistance varying as the square of the velocity; prove that the thrust of the screw is

$$\cdot 145 H/V \text{ tons}.$$

Prove that if the engines be reversed, the ship will be brought to rest in

$$\cdot 283\, WV^2/H \text{ secs.},$$

after travelling a distance of

$$\cdot 211\, WV^3/H \text{ ft.}$$

Work out these results for the case of $W = 2000$, $H = 5000$, $V = 20$.

(Assume that a sea-mile = 6080 ft.) [45 sec.; 675 ft.]

14. A particle oscillates in a straight line about a centre of force varying as the distance, and is subject to a retardation $k \times (\text{vel.})^2$. If a, b be two successive elongations, on opposite sides, prove that

$$(1 + 2ka)\, e^{-2ka} = (1 - 2kb)\, e^{2kb}.$$

What form does the result take if a is infinite?

15. If a particle, started with the velocity u_0, be subject only to a resistance

$$ku + k'u^2,$$

prove that the space described before it comes to rest is

$$\frac{1}{k'} \log\left(1 + \frac{k'}{k} u_0\right).$$

16. If the retardation be equal to $k\dot{x}^3$, prove that

$$t = Ax^2 + Bx + C,$$

where A, B, C are constants; and determine the constants, having given that $x = x_0$, $\dot{x} = u_0$, for $t = 0$.

17. A bullet moving horizontally pierces in succession three screens placed at distances a apart. Assuming that the resistance varies as the cube of the velocity, prove that the velocity at the middle screen is equal to the mean velocity in the interval between passing the first and third screens.

Also that if t_1, t_2 be the times of passing from the first screen to the second, and from the second to the third, respectively, the initial and final velocities are

$$\frac{2a}{3t_1 - t_2} \text{ and } \frac{2a}{3t_2 - t_1},$$

respectively.

18. Prove that in the preceding question the bullet must have been started at a distance from the middle screen less than

$$\frac{t_1 + t_2}{t_2 - t_1} \cdot \frac{a}{2},$$

and that the time that has elapsed up to this point must be less than

$$\frac{1}{8} \frac{(t_1 + t_2)^2}{t_2 - t_1}.$$

19. A point subject to a retardation varying as the cube of the velocity starts with velocity u_0, and the initial retardation is f. Prove that the space described in time t is

$$\frac{u_0^2}{f}\left\{\sqrt{\left(1 + \frac{2ft}{u_0}\right)} - 1\right\}.$$

20. A particle is subject to a retardation $f(t)$ which brings it to rest in a time τ; prove that the distance travelled is

$$\int_0^\tau t f(t)\, dt.$$

21. Prove that, in the motion of a projectile,

$$\frac{d^2y}{dx^2} = -\frac{g}{u^2},$$

where u is the horizontal velocity, whatever the law of resistance, the axes of x, y being horizontal and vertical respectively.

CHAPTER XIII

SYSTEMS OF TWO DEGREES OF FREEDOM

101. Motion of a Particle on a Smooth Surface.

A mechanical system is said to have n 'degrees of freedom' when n independent variables are necessary and sufficient to specify the positions of its various parts. These variables are called, in a generalized sense, the 'coordinates' of the system. Thus the position of a particle moveable on a spherical surface may be specified by its latitude and longitude; the configuration of the double pendulum in Fig. 64 (Art. 68) is specified by the angles θ, ϕ; the position of a rigid body moveable in two dimensions may be defined as in Art. 63 by the two coordinates of its mass-centre and the angle through which it has been turned from some standard position; and so on.

The theory of a conservative system having *one* degree of freedom has been treated in a general manner in Art. 65. Since there is only one coordinate, the equation of energy, together with the initial conditions, completely determines the motion.

When we proceed to systems of greater freedom, the equation of energy is no longer sufficient, and a further appeal to dynamical principles becomes necessary. In the case of systems having *two* degrees of freedom, especially when the motion is in two dimensions, the principle of angular momentum can sometimes be made to furnish the additional information required, in a form free from unknown reactions. We have had an instance of this procedure in the theory of Central Forces (Arts. 76, 84).

As a further example we may take the case of a particle moving on a smooth surface of revolution under no forces except the reaction of the surface. Since this reaction does no work, the velocity v is constant. Again, since the direction of the reaction intersects the axis of symmetry, the angular momentum of the particle about this axis is constant. To express the result analytically, let r denote the distance of the particle from the axis, and ϕ the angle which the direction of motion makes with the circle of latitude. The velocity may be resolved into two components $v \cos \phi$, $v \sin \phi$, along this circle, and along the meridian, respectively, of which the former alone has a moment about the axis. Hence, since v is constant,

$$r \cos \phi = \text{const.} \quad \text{.........................(1)}$$

This is, virtually, a differential equation of the first order, to determine the path.

Since $\cos \phi$ cannot exceed unity, there is in any given case a lower limit to the value of r. For instance, if the surface resemble an ellipsoid of revolution, and if the particle cross the equator (whose radius is a, say) at an angle α, we have

$$r \cos \phi = a \cos \alpha, \quad \text{.........................(2)}$$

and the path therefore lies between two parallel circles of radius $a \cos \alpha$ which it alternately touches.

Since there is no force on the particle except in the direction of the normal, this must be the direction of the resultant acceleration. It has been remarked in Art. 34 that this resultant lies always in the osculating plane of the path. It follows that the particle follows the 'straightest' path possible to it on the surface, i.e. in the language of Solid Geometry it describes a 'geodesic.'

Ex. A particle is slightly disturbed from motion along the equator of a convex surface of revolution.

If z denote distance from the plane of the equatorial circle, a the radius of this circle, and ρ the radius of curvature of the meridian at the equator, we have

$$a - r = \frac{z^2}{2\rho}, \quad \text{....................................(3)}$$

if z be small. It follows that $\dot{r}$ may be neglected in comparison with $\dot{z}$. If θ be the longitude, so that r, θ are the polar coordinates of the projection of the particle on the equatorial plane, we have

$$r^2 \dot{\theta} = h, \quad \text{...(4)}$$

a constant, by the principle of angular momentum. Also

$$v^2 = \dot{r}^2 + r^2\dot{\theta}^2 + \dot{z}^2 = \dot{z}^2 + \frac{h^2}{r^2}, \qquad \text{.........................(5)}$$

if $\dot{r}^2$ be neglected. Substituting from (3), we have

$$v^2 = \dot{z}^2 + \frac{h^2}{a^2}\left(1 - \frac{z^2}{2\rho a}\right)^{-2} = \dot{z}^2 + \frac{h^2}{a^2}\left(1 + \frac{z^2}{\rho a}\right), \qquad \text{...........(6)}$$

approximately. Hence, putting

$$h = a^2\omega, \qquad \text{....................................(7)}$$

we have

$$\dot{z}^2 + \frac{\omega^2 a}{\rho} z^2 = \text{const.}, \qquad \text{.............................(8)}$$

whence

$$\ddot{z} + \frac{\omega^2 a}{\rho} z = 0. \qquad \text{..................................(9)}$$

The variations of z are therefore simple-harmonic, of period

$$\frac{2\pi}{\omega}\sqrt{\frac{\rho}{a}}. \qquad \text{..................................(10)}$$

In the half-period the particle describes an angle $\pi\sqrt{(\rho/a)}$, approximately, about the axis.

In the case of an ellipsoid, if the length of the axis of symmetry be $2b$, we have $\rho = b^2/a$, and the angle is $\pi b/a$. That is, the distance between consecutive intersections of the path with the equator is πb.

102. Motion on a Spherical Surface.

There are a number of questions relating to motion on a spherical surface which can be treated in a similar manner.

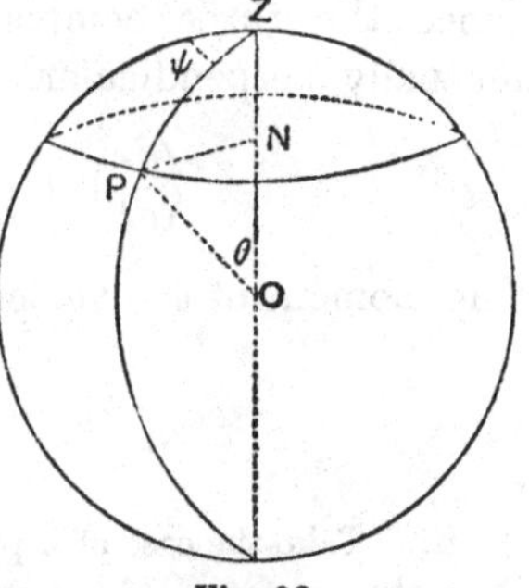

Fig. 93.

The position of a point P on such a surface may be specified by two angular coordinates analogous to polar distance and longitude on the earth, or to zenith distance and azimuth on the celestial sphere of any station.

If O be the centre, we denote by θ the angular distance ZOP from a fixed point Z on the sphere, measured along a great circle, and by ψ the angle which the plane of this great circle makes with a fixed plane through OZ. If θ alone be slightly varied, the point P describes an arc $a\delta\theta$, where a is the radius, along the great circle ZP;

whilst if ψ alone be varied, P moves along a small circle of radius $a \sin\theta$, and therefore through an arc $a \sin\theta\, \delta\psi$. The component velocities of P along and at right angles to the great circle ZP are therefore

$$a\frac{d\theta}{dt} \quad \text{and} \quad a \sin\theta \frac{d\psi}{dt}, \quad \text{.................(1)}$$

respectively. The square of the velocity of P is therefore

$$a^2\left\{\left(\frac{d\theta}{dt}\right)^2 + \sin^2\theta\left(\frac{d\psi}{dt}\right)^2\right\}. \quad \text{.................(2)}$$

Again, the second of the components (1) has alone a moment about OZ, and since the distance of P from this axis is $a \sin\theta$, the moment in question is

$$a^2 \sin^2\theta \frac{d\psi}{dt}. \quad \text{..........................(3)}$$

It may be added that if we regard the radius of the sphere as variable, and denote it accordingly by r instead of a, we are able to specify the position of any point P whatever in space by the three 'spherical polar' coordinates r, θ, ψ. If r alone be varied, the displacement of P is denoted by δr, and the radial velocity is therefore

$$\frac{dr}{dt}. \quad \text{..............................(4)}$$

Since the three components which we have investigated are mutually perpendicular, the square of the velocity is

$$\left(\frac{dr}{dt}\right)^2 + r^2\left(\frac{d\theta}{dt}\right)^2 + r^2 \sin^2\theta\left(\frac{d\psi}{dt}\right)^2. \quad \text{...........(5)}$$

The moment of the velocity about OZ is

$$r^2 \sin^2\theta \frac{d\psi}{dt}. \quad \text{..........................(6)}$$

Ex. Take the case of a particle constrained to move along a meridian of the earth's surface. If the axis OZ coincide with the polar axis, the angular momentum about this axis is $ma^2\omega \sin^2\theta$, where ω is the earth's angular velocity of rotation. Hence if S be the constraining force towards the east, we have

$$\frac{d}{dt}(ma^2\omega \sin^2\theta) = S \cdot a \sin\theta, \quad \text{.......................(7)}$$

whence $$S = 2ma\omega \cos\theta \frac{d\theta}{dt}, \quad \text{.........................(8)}$$

or $$S = 2mv\omega \cos\theta, \quad \text{.........................(9)}$$

if v be the (relative) velocity along the meridian, from the pole. For instance, a train travelling southwards in the northern hemisphere exerts a pressure $2mv\omega \cos\theta$ westwards on the rails, owing to its tendency to retain its angular momentum about the polar axis unchanged*. The ratio of this lateral pressure to the weight of the train is

$$\frac{S}{mg} = \frac{2\omega^2 a}{g} \cdot \frac{v}{\omega a} \cdot \cos\theta = \frac{1}{144} \frac{v}{\omega a} \cdot \cos\theta, \quad \text{..................(10)}$$

where the fraction $v/\omega a$ is the ratio of the velocity of the train to the rotational velocity of the earth's surface at the equator. This latter velocity is of course 1000 sea-miles per (sidereal) hour.

103. The Spherical Pendulum.

An important application of the preceding formulae is to the theory of the spherical pendulum, or of the motion of a particle (under gravity) on a smooth spherical surface.

If a be the length of the string (or the radius of the sphere), and if the axis OZ of the spherical coordinates be directed vertically downwards, the equation of energy is

$$\tfrac{1}{2} ma^2 \left\{ \left(\frac{d\theta}{dt}\right)^2 + \sin^2\theta \left(\frac{d\psi}{dt}\right)^2 \right\} = mga \cos\theta + \text{const.}, \quad \text{...(1)}$$

whilst the principle of angular momentum gives

$$\sin^2\theta \frac{d\psi}{dt} = \text{const.} = h, \text{ say.} \quad \text{..................(2)}$$

Eliminating $d\psi/dt$, we have

$$\left(\frac{d\theta}{dt}\right)^2 + \frac{h^2}{\sin^2\theta} = \frac{2g}{a} \cos\theta + \text{const.} \quad \text{............(3)}$$

Differentiating, and dividing by $d\theta/dt$, we obtain

$$\frac{d^2\theta}{dt^2} - \frac{h^2 \cos\theta}{\sin^3\theta} = -\frac{g}{a} \sin\theta. \quad \text{..................(4)}$$

With a view to a future reference, we note that this is equivalent to

$$\frac{d^2\theta}{dt^2} - \sin\theta\cos\theta \left(\frac{d\psi}{dt}\right)^2 = -\frac{g}{a} \sin\theta. \quad \text{............(5)}$$

* The same tendency is appealed to in the familiar explanation of the trade winds.

It is evident that if the path goes through the lowest or highest point of the sphere it must be a vertical circle. If we exclude this case, there will be an upper and a lower limit to the value of θ; moreover it is evident by the same kind of reasoning as in the theory of apse-lines of central orbits (Art. 88), that the path is symmetrical with respect to the meridian plane through any point where the direction of motion is horizontal. The path therefore lies between two horizontal circles which it alternately touches at equal intervals of azimuth.

If the two limiting circles coincide, we have the case of the conical pendulum. Since θ is then constant, we have from (5)

$$\left(\frac{d\psi}{dt}\right)^2 = \frac{g}{a\cos\theta}, \quad \ldots\ldots\ldots\ldots\ldots\ldots\ldots(6)$$

in agreement with Art. 35 (5).

If this state of steady motion in a horizontal circle be slightly disturbed, the limiting circles will slightly separate, and the path will undulate between them. If we write

$$\theta = \alpha + \xi, \quad \ldots\ldots\ldots\ldots\ldots\ldots\ldots\ldots(7)$$

where ξ is supposed small, we have, on substituting in (4) and retaining only the first power of ξ,

$$\frac{d^2\xi}{dt^2} + \left(\frac{h^2}{\sin^2\alpha} + \frac{3h^2\cos^2\alpha}{\sin^4\alpha}\right)\xi = -\frac{g}{a}\cos\alpha\,.\,\xi, \quad \ldots\ldots\ldots(8)$$

provided

$$\frac{h^2\cos\alpha}{\sin^3\alpha} = \frac{g}{a}\sin\alpha. \quad \ldots\ldots\ldots\ldots\ldots\ldots(9)$$

This is, by (4), the condition of steady motion in a horizontal circle of angular radius α. By hypothesis, it is very nearly fulfilled in the disturbed state, and by a slight adjustment of the value of α it may be fulfilled exactly (cf. Art. 87). If we put

$$\omega^2 = \frac{g}{a\cos\alpha}, \quad \ldots\ldots\ldots\ldots\ldots\ldots\ldots(10)$$

we have

$$h^2 = \omega^2\sin^4\alpha, \quad \ldots\ldots\ldots\ldots\ldots\ldots\ldots(11)$$

and the equation (8) reduces to

$$\frac{d^2\xi}{dt^2} + (1 + 3\cos^2\alpha)\,\omega^2\xi = 0. \quad \ldots\ldots\ldots\ldots(12)$$

The period of a small oscillation in the value of θ is therefore

$$\frac{2\pi}{\omega}\cdot\frac{1}{\sqrt{(1+3\cos^2\alpha)}}. \quad\text{..................(13)}$$

Since, from (6) and (10),

$$\frac{d\psi}{dt}=\omega, \quad\text{..........................(14)}$$

approximately, the interval of azimuth between a maximum of θ and the consecutive minimum is

$$\frac{\pi}{\sqrt{(1+3\cos^2\alpha)}}. \quad\text{.........................(15)}$$

This may be called the apsidal angle; if the inclination α be small it is $\frac{1}{2}\pi$, nearly, as already known from Art. 29.

Ex. As a variation on this question we may consider the motion of a particle on a smooth paraboloid of revolution whose axis is vertical.

We denote by r, θ the polar coordinates of the projection of the particle on the horizontal plane, and by z the altitude above the level of the vertex. We have then

$$r^2=4az,\quad \dot{z}=r\dot{r}/2a, \quad\text{..............................(16)}$$

if $4a$ be the latus-rectum of the generating parabola. The kinetic energy is therefore

$$\tfrac{1}{2}m(\dot{z}^2+\dot{r}^2+r^2\dot{\theta}^2)=\tfrac{1}{2}m\left\{\left(1+\frac{r^2}{4a^2}\right)\dot{r}^2+r^2\dot{\theta}^2\right\}, \quad\text{...........(17)}$$

and the potential energy is

$$mgz=\frac{mgr^2}{4a}. \quad\text{................................(18)}$$

We have, then,

$$\left(1+\frac{r^2}{4a^2}\right)\dot{r}^2+r^2\dot{\theta}^2+\frac{gr^2}{2a}=\text{const.}, \quad\text{..................(19)}$$

and

$$r^2\dot{\theta}=h, \quad\text{..................................(20)}$$

the latter being the equation of angular momentum.

Eliminating $\dot{\theta}$, we have

$$\left(1+\frac{r^2}{4a^2}\right)\dot{r}^2+\frac{h^2}{r^2}+\frac{gr^2}{2a}=\text{const.}, \quad\text{..................(21)}$$

whence, by differentiation,

$$\left(1+\frac{r^2}{4a^2}\right)\ddot{r}+\frac{r\dot{r}^2}{4a^2}-\frac{h^2}{r^3}+\frac{gr}{2a}=0. \quad\text{.....................(22)}$$

In order that the particle may describe a horizontal circle this equation must be satisfied by $r=$ const., whence

$$h^2=gr^4/2a. \quad\text{..............................(23)}$$

If ω be the angular velocity in this circle, we have $h=r^2\omega$, and therefore

$$\omega^2=g/2a. \qquad (24)$$

This is independent of the radius of the circle.

To investigate the effect of a slight disturbance from motion in a horizontal circle, we put

$$r=r_0+x, \qquad (25)$$

in (22), and retain only terms of the first order in x, which is assumed to be small. We get

$$\left(1+\frac{r_0^2}{4a^2}\right)\ddot{x}+\frac{3h^2}{r_0^4}x+\frac{g}{2a}x=0 \qquad (26)$$

or, putting $h^2=gr_0^4/2a$, as suggested by (23),

$$\left(1+\frac{r_0^2}{4a^2}\right)\ddot{x}+\frac{2g}{a}x=0. \qquad (27)$$

If we write

$$r_0^2=4az_0, \qquad (28)$$

this becomes

$$\ddot{x}+\frac{2g}{z_0+a}x=0. \qquad (29)$$

The period of a small oscillation about the horizontal circle is therefore that of a simple pendulum whose length is $\frac{1}{2}(z_0+a)$, or one-half the distance of the circumference of the circle from the focus of the parabola.

104. General Motion of a Particle. Lagrange's Equations.

In the preceding examples the elimination of unknown reactions was easy because advantage could be taken of special features in the problems considered. In other cases, and especially when the number of degrees of freedom is considerable, the direct appeal to fundamental principles may lead to equations of some complexity*, and the elimination may be troublesome.

The problem of effecting this elimination once for all, for conservative systems of any constitution, was first attacked, and solved, by Lagrange†. He shewed that the dynamical properties of a conservative system are completely defined by the expressions for the kinetic and potential energies in terms of the n generalized coordinates of the system, and their differential coefficients with respect to the time, and that when these expressions are known the n equations of motion, free from unknown reactions, can be written down at once without any further reference to the particular constitution of the system.

* This is illustrated to a slight extent by the investigation of Art. 68.

† *Méchanique analitique*, Paris, 1788.

In obtaining Lagrange's equations we shall limit ourselves to the case of *two* degrees of freedom. This suffices for the treatment of a number of interesting questions, and avoids the necessity for an elaborate system of notation. At the same time the manner in which the results can be extended to the general case will be easily perceived.

Further, we consider in the first instance the case of a single particle. We denote by θ, ϕ the independent variables, or coordinates, which specify its position. These may be Cartesian coordinates (rectangular or oblique) in a plane, or the spherical coordinates of Art. 103, or *any* two quantities which are convenient for specifying the position in a particular problem. The differential coefficients $\dot{\theta}$, $\dot{\phi}$ of these coordinates with respect to the time may be called the (generalized) 'velocities' of the particle.

If, in any position, the coordinate θ be slightly varied, whilst ϕ is constant, the particle will undergo a displacement

$$\delta s_1 = \alpha \delta\theta \qquad (1)$$

in some definite direction, where the coefficient α will in general depend on the position from which the displacement is made, and therefore on the values of θ, ϕ. Similarly, if ϕ alone be varied the displacement will be

$$\delta s_2 = \beta \delta\phi, \qquad (2)$$

in another definite direction. If both variations coexist, the resulting displacement δs will be given as to magnitude by

$$\begin{aligned}\delta s^2 &= \delta s_1^2 + 2\delta s_1 \delta s_2 \cos\epsilon + \delta s_2^2 \\ &= \alpha^2\delta\theta^2 + 2\alpha\beta\cos\epsilon\,\delta\theta\,\delta\phi + \beta^2\delta\phi^2, \qquad (3)\end{aligned}$$

where ϵ is the angle between the directions of δs_1 and δs_2. The square of the velocity v is therefore given by

$$v^2 = \alpha^2\dot{\theta}^2 + 2\alpha\beta\cos\epsilon\,\dot{\theta}\dot{\phi} + \beta^2\dot{\phi}^2. \qquad (4)$$

Hence if, to conform to an established usage, we denote the kinetic energy by T, we have, m being the mass,

$$T = \tfrac{1}{2}mv^2 = \tfrac{1}{2}(A\dot{\theta}^2 + 2H\dot{\theta}\dot{\phi} + B\dot{\phi}^2), \qquad (5)$$

where $$A = m\alpha^2, \quad H = m\alpha\beta\cos\epsilon, \quad B = m\beta^2. \qquad (6)$$

The coefficients A, H, B are in general functions of θ, ϕ. They may be called the 'coefficients of inertia' of the particle.

The Cartesian coordinates x, y, z of m, referred to any system of fixed rectangular axes, are by hypothesis functions of θ and ϕ. We have, then,

$$\dot{x} = \frac{\partial x}{\partial \theta}\dot{\theta} + \frac{\partial x}{\partial \phi}\dot{\phi}, \quad \dot{y} = \frac{\partial y}{\partial \theta}\dot{\theta} + \frac{\partial y}{\partial \phi}\dot{\phi}, \quad \dot{z} = \frac{\partial z}{\partial \theta}\dot{\theta} + \frac{\partial z}{\partial \phi}\dot{\phi}. \;\ldots(7)$$

Hence

$$T = \tfrac{1}{2}m(\dot{x}^2 + \dot{y}^2 + \dot{z}^2)$$
$$= \tfrac{1}{2}(A\dot{\theta}^2 + 2H\dot{\theta}\dot{\phi} + B\dot{\phi}^2), \;\ldots\ldots\ldots\ldots(8)$$

where

$$\left.\begin{aligned} A &= m\left\{\left(\frac{\partial x}{\partial \theta}\right)^2 + \left(\frac{\partial y}{\partial \theta}\right)^2 + \left(\frac{\partial z}{\partial \theta}\right)^2\right\}, \\ H &= m\left\{\frac{\partial x}{\partial \theta}\frac{\partial x}{\partial \phi} + \frac{\partial y}{\partial \theta}\frac{\partial y}{\partial \phi} + \frac{\partial z}{\partial \theta}\frac{\partial z}{\partial \phi}\right\}, \\ B &= m\left\{\left(\frac{\partial x}{\partial \phi}\right)^2 + \left(\frac{\partial y}{\partial \phi}\right)^2 + \left(\frac{\partial z}{\partial \phi}\right)^2\right\}. \end{aligned}\right\} \;\ldots\ldots\ldots(9)$$

It is easily seen that these forms of the coefficients are equivalent to those given by (6).

If (X, Y, Z) be the force acting on the particle, the Cartesian equations of motion are

$$m\ddot{x} = X, \quad m\ddot{y} = Y, \quad m\ddot{z} = Z. \;\ldots\ldots\ldots\ldots(10)$$

Multiplying these by the partial differential coefficients $\partial x/\partial\theta$, $\partial y/\partial\theta$, $\partial z/\partial\theta$, respectively, and adding, we obtain

$$m\left(\ddot{x}\frac{\partial x}{\partial \theta} + \ddot{y}\frac{\partial y}{\partial \theta} + \ddot{z}\frac{\partial z}{\partial \theta}\right) = X\frac{\partial x}{\partial \theta} + Y\frac{\partial y}{\partial \theta} + Z\frac{\partial z}{\partial \theta}. \;\ldots(11)$$

It will be noticed that this procedure is equivalent to resolving in the direction of the displacement δs_1.

To transform the equation (11), we note that

$$\ddot{x}\frac{\partial x}{\partial \theta} = \frac{d}{dt}\left(\dot{x}\frac{\partial x}{\partial \theta}\right) - \dot{x}\frac{d}{dt}\frac{\partial x}{\partial \theta}, \;\ldots\ldots\ldots\ldots(12)$$

where the differentiations with respect to t are *total*. Now from (7) we have

$$\frac{\partial \dot{x}}{\partial \dot{\theta}} = \frac{\partial x}{\partial \theta}, \;\ldots\ldots\ldots\ldots\ldots\ldots(13)$$

and

$$\frac{d}{dt}\frac{\partial x}{\partial \theta} = \frac{\partial^2 x}{\partial \theta^2}\dot{\theta} + \frac{\partial^2 x}{\partial \theta \partial \phi}\dot{\phi} = \frac{\partial \dot{x}}{\partial \theta}. \;\ldots\ldots\ldots\ldots(14)$$

Hence (12) may be written

$$\ddot{x}\frac{\partial x}{\partial \theta}=\frac{d}{dt}\left(\dot{x}\frac{\partial \dot{x}}{\partial \dot{\theta}}\right)-\dot{x}\frac{\partial \dot{x}}{\partial \theta}. \qquad\qquad (15)$$

Hence, and by similar formulæ, we have

$$\begin{aligned} m&\left(\ddot{x}\frac{\partial x}{\partial \theta}+\ddot{y}\frac{\partial y}{\partial \theta}+\ddot{z}\frac{\partial z}{\partial \theta}\right) \\ &=\frac{d}{dt}m\left(\dot{x}\frac{\partial \dot{x}}{\partial \dot{\theta}}+\dot{y}\frac{\partial \dot{y}}{\partial \dot{\theta}}+\dot{z}\frac{\partial \dot{z}}{\partial \dot{\theta}}\right)-m\left(\dot{x}\frac{\partial \dot{x}}{\partial \theta}+\dot{y}\frac{\partial \dot{y}}{\partial \theta}+\dot{z}\frac{\partial \dot{z}}{\partial \theta}\right) \\ &=\frac{d}{dt}\frac{\partial T}{\partial \dot{\theta}}-\frac{\partial T}{\partial \theta}, \qquad\qquad (16)\end{aligned}$$

by (8).

If the work done by the forces in a small displacement be expressed in the form

$$\Theta\delta\theta+\Phi\delta\phi, \qquad\qquad (17)$$

the quantities Θ, Φ are called the 'generalized components of force.' It is evident that in computing them we may ignore all forces (such as the reactions of smooth fixed surfaces, &c.) which do no work. In our previous notation we have

$$\begin{aligned} X\delta x+Y\delta y+Z\delta z&=\left(X\frac{\partial x}{\partial \theta}+Y\frac{\partial y}{\partial \theta}+Z\frac{\partial z}{\partial \theta}\right)\delta\theta \\ &\quad+\left(X\frac{\partial x}{\partial \phi}+Y\frac{\partial y}{\partial \phi}+Z\frac{\partial z}{\partial \phi}\right)\delta\phi, \qquad (18)\end{aligned}$$

so that

$$\left.\begin{aligned} \Theta&=X\frac{\partial x}{\partial \theta}+Y\frac{\partial y}{\partial \theta}+Z\frac{\partial z}{\partial \theta}, \\ \Phi&=X\frac{\partial x}{\partial \phi}+Y\frac{\partial y}{\partial \phi}+Z\frac{\partial z}{\partial \phi}. \end{aligned}\right\} \qquad\qquad (19)$$

The equations of motion of the system are thus reduced to the Lagrangian forms*

$$\left.\begin{aligned} \frac{d}{dt}\frac{\partial T}{\partial \dot{\theta}}-\frac{\partial T}{\partial \theta}&=\Theta, \\ \frac{d}{dt}\frac{\partial T}{\partial \dot{\phi}}-\frac{\partial T}{\partial \phi}&=\Phi. \end{aligned}\right\} \qquad\qquad (20)$$

* The above method of proof by direct transformation is due to Sir W. R. Hamilton (1835). The terminology of generalized 'coordinates,' 'velocities,' 'forces,' 'momenta,' was introduced by Thomson and Tait, *Natural Philosophy*, 1st ed., Oxford, 1867.

So far the forces X, Y, Z, and consequently Θ, Φ, are subject to no restrictions. They may be functions of the position, of the velocity, and explicitly also of the time. But in the case of a particle moving in a conservative field, and subject to no extraneous forces, the work done on it in a small displacement is equal to the decrement of the potential energy, whence

$$\Theta\delta\theta + \Phi\delta\phi = -\delta V = -\frac{\partial V}{\partial\theta}\delta\theta - \frac{\partial V}{\partial\phi}\delta\phi. \quad\text{.........(21)}$$

Since the variations $\delta\theta, \delta\phi$ are independent, this implies

$$\Theta = -\frac{\partial V}{\partial\theta}, \quad \Phi = -\frac{\partial V}{\partial\phi}. \quad\text{..............(22)}$$

The equations thus take the classical forms

$$\left.\begin{aligned} \frac{d}{dt}\frac{\partial T}{\partial\dot\theta} - \frac{\partial T}{\partial\theta} &= -\frac{\partial V}{\partial\theta}, \\ \frac{d}{dt}\frac{\partial T}{\partial\dot\phi} - \frac{\partial T}{\partial\phi} &= -\frac{\partial V}{\partial\phi}. \end{aligned}\right\} \quad\text{.................(23)}$$

An interpretation of the expressions

$$\frac{\partial T}{\partial\dot\theta}, \quad \frac{\partial T}{\partial\dot\phi},$$

which occur in Lagrange's equations may be obtained as follows. The impulse (X', Y', Z') which would be necessary to start the actual motion of the particle instantaneously from rest is given by

$$m\dot x = X', \quad m\dot y = Y', \quad m\dot z = Z'. \quad\text{............(24)}$$

If we multiply these equations by $\partial x/\partial\theta, \partial y/\partial\theta, \partial z/\partial\theta$, respectively, and add, we obtain in virtue of (13),

$$\left.\begin{aligned} \frac{\partial T}{\partial\dot\theta} &= X'\frac{\partial x}{\partial\theta} + Y'\frac{\partial y}{\partial\theta} + Z'\frac{\partial z}{\partial\theta}, \\ \frac{\partial T}{\partial\dot\phi} &= X'\frac{\partial x}{\partial\phi} + Y'\frac{\partial y}{\partial\phi} + Z'\frac{\partial z}{\partial\phi}. \end{aligned}\right\} \quad\text{...........(25)}$$

The expressions on the right hand may be called the generalized 'components of impulse'; and the equivalent expressions on the left may by an obvious dynamical analogy be called the 'components of momentum.' Denoting them by λ, μ, so that

$$\lambda = \frac{\partial T}{\partial\dot\theta}, \quad \mu = \frac{\partial T}{\partial\dot\phi}, \quad\text{.....................(26)}$$

the equations of motion may be written

$$\frac{d\lambda}{dt} = \Theta + \frac{\partial T}{\partial \theta}, \quad \frac{d\mu}{dt} = \Phi + \frac{\partial T}{\partial \phi}, \quad \text{............(27)}$$

or, in the case of a conservative field,

$$\frac{d\lambda}{dt} = -\frac{\partial}{\partial \theta}(V - T), \quad \frac{d\mu}{dt} = -\frac{\partial}{\partial \phi}(V - T). \quad \text{......(28)}$$

105. Applications.

The significance of the terminology which has been introduced will be best understood from the study of a few examples.

In the motion of a particle referred to plane polar coordinates we have

$$T = \tfrac{1}{2} m (\dot{r}^2 + r^2\dot{\theta}^2), \quad \text{.....................(1)}$$

by Art. 86 (1), (2). The generalized components of momentum are therefore

$$\frac{\partial T}{\partial \dot{r}} = m\dot{r}, \quad \frac{\partial T}{\partial \dot{\theta}} = mr^2\dot{\theta}. \quad \text{.....................(2)}$$

The former of these is recognized as the linear momentum of the particle in the direction of the radius vector, and the latter as the moment of momentum (in the ordinary sense) about the origin.

If R denote the radial, and S the transverse component of force, the work done in a small displacement is

$$R\delta r + Sr\delta\theta, \quad \text{.........................(3)}$$

and the generalized components of force are accordingly

$$R, \quad Sr,$$

the latter being the moment of the force about the origin.

The equations of motion are therefore

$$\frac{d}{dt}\left(\frac{\partial T}{\partial \dot{r}}\right) - \frac{\partial T}{\partial r} = R, \quad \frac{d}{dt}\left(\frac{\partial T}{\partial \dot{\theta}}\right) - \frac{\partial T}{\partial \theta} = rS. \quad \text{........(4)}$$

Since

$$\frac{\partial T}{\partial r} = mr\dot{\theta}^2, \quad \frac{\partial T}{\partial \theta} = 0, \quad \text{...................(5)}$$

these reduce to

$$m(\ddot{r} - r\dot{\theta}^2) = R, \quad \frac{m}{r}\frac{d}{dt}(r^2\dot{\theta}) = S, \quad \text{...........(6)}$$

in agreement with Art. 86 (7), (8).

Again in the case of motion on a spherical surface we have

$$T=\tfrac{1}{2}ma^2(\dot\theta^2+\sin^2\theta\,\dot\psi^2). \quad\text{.................(7)}$$

Hence

$$\frac{\partial T}{\partial\dot\theta}=ma^2\dot\theta,\quad \frac{\partial T}{\partial\dot\psi}=ma^2\sin^2\theta\,\dot\psi. \quad\text{...........(8)}$$

The former of these expressions is the angular momentum about an axis through the centre O perpendicular to the plane of the meridian; whilst the latter is the angular momentum about the axis OZ (Fig. 93).

If R, S be the components of force along the meridian and the parallel of latitude, tending to increase θ and ψ, respectively, the work done in a small displacement is

$$Ra\,\delta\theta+Sa\sin\theta\,\delta\psi, \quad\text{....................(9)}$$

so that the generalized components of force are

$$Ra \text{ and } Sa\sin\theta, \quad\text{....................(10)}$$

respectively.

The equations of motion are therefore

$$\frac{d}{dt}\frac{\partial T}{\partial\dot\theta}-\frac{\partial T}{\partial\theta}=Ra,\quad \frac{d}{dt}\frac{\partial T}{\partial\dot\psi}-\frac{\partial T}{\partial\psi}=Sa\sin\theta, \quad\text{......(11)}$$

or

$$ma(\ddot\theta-\sin\theta\cos\theta\,\dot\psi^2)=R,\quad \frac{1}{\sin\theta}\frac{d}{dt}(ma\sin^2\theta\,\dot\psi)=S. \quad\text{...(12)}$$

In the case of the spherical pendulum we have, if OZ coincide with the downward vertical, $R=-mg\sin\theta$, $S=0$, and the equations reduce to

$$\ddot\theta-\sin\theta\cos\theta\,\dot\psi^2=-\frac{g}{a}\sin\theta,\quad \sin^2\theta\,\dot\psi=h, \quad\text{......(13)}$$

as in Art. 103.

Ex. A particle is contained in a smooth circular tube of radius a which is constrained to rotate with constant angular velocity ω about a vertical diameter.

Putting $R=-mg\sin\theta$, $\dot\psi=\omega$, in (12), we have

$$\ddot\theta-\omega^2\sin\theta\cos\theta=-\frac{g}{a}\sin\theta. \quad\text{.....................(14)}$$

This shews that the particle can be in relative equilibrium in the position $\theta=\alpha$, provided

$$\sin\alpha\left(\omega^2\cos\alpha-\frac{g}{a}\right)=0. \quad\text{...........................(15)}$$

This has the obvious solution

$$\sin \alpha = 0, \qquad (16)$$

whilst if $\omega^2 a > g$ there is a second solution

$$\cos \alpha = g/\omega^2 a. \qquad (17)$$

To examine the stability of the relative equilibrium we put

$$\theta = \alpha + \xi, \qquad (18)$$

where α is a solution of (15), and ξ is supposed to be small. Substituting in (14) we find

$$\ddot{\xi} + \left(\frac{g}{a}\cos \alpha - \omega^2 \cos 2\alpha\right)\xi = 0. \qquad (19)$$

The position $\alpha = 0$ is therefore stable only if $\omega^2 a < g$, whilst the position $\alpha = \pi$ is always unstable, as we should expect. When $\omega^2 a > g$, we have, in the case of the intermediate position given by (17),

$$\ddot{\xi} + \omega^2 \sin^2 \alpha \, . \, \xi = 0. \qquad (20)$$

This position, when it exists, is therefore stable, and the period of a small oscillation about it is

$$\frac{2\pi}{\omega \sin \alpha}. \qquad (21)$$

The lateral pressure S exerted by the tube on the particle is given by the second of equations (12).

106. Mechanical Systems of Double Freedom. Lagrange's Equations.

We proceed to the most general case of a mechanical system of two degrees of freedom.

By hypothesis, the rectangular coordinates x, y, z of any one particle m of the system are definite functions of two independent variables, which we denote as before by the letters θ, ϕ, but the forms of these functions will in general vary from particle to particle. Denoting by (X, Y, Z) the force acting on the particle m, we form, as in Art. 104, the equation

$$m\left(\ddot{x}\frac{\partial x}{\partial \theta} + \ddot{y}\frac{\partial y}{\partial \theta} + \ddot{z}\frac{\partial z}{\partial \theta}\right) = X\frac{\partial x}{\partial \theta} + Y\frac{\partial y}{\partial \theta} + Z\frac{\partial z}{\partial \theta}. \qquad (1)$$

We have an equation of this type for each particle, and by addition we deduce

$$\Sigma m\left(\ddot{x}\frac{\partial x}{\partial \theta} + \ddot{y}\frac{\partial y}{\partial \theta} + \ddot{z}\frac{\partial z}{\partial \theta}\right) = \Sigma\left(X\frac{\partial x}{\partial \theta} + Y\frac{\partial y}{\partial \theta} + Z\frac{\partial z}{\partial \theta}\right), \qquad (2)$$

where the summation Σ embraces all the particles of the system.

Since the kinetic energy of the system is the sum of the kinetic energies of its various particles, the investigation of Art. 104 shews that the left-hand member is equivalent to

$$\frac{d}{dt}\frac{\partial T}{\partial \dot{\theta}} - \frac{\partial T}{\partial \theta}, \qquad (3)$$

where T now denotes the total kinetic energy of the system, viz.

$$\begin{aligned} T &= \tfrac{1}{2}\Sigma m(\dot{x}^2 + \dot{y}^2 + \dot{z}^2) \\ &= \tfrac{1}{2}(A\dot{\theta}^2 + 2H\dot{\theta}\dot{\phi} + B\dot{\phi}^2), \end{aligned} \qquad (4)$$

where

$$\left.\begin{aligned} A &= \Sigma m\left\{\left(\frac{\partial x}{\partial \theta}\right)^2 + \left(\frac{\partial y}{\partial \theta}\right)^2 + \left(\frac{\partial z}{\partial \theta}\right)^2\right\}, \\ H &= \Sigma m\left\{\frac{\partial x}{\partial \theta}\frac{\partial x}{\partial \phi} + \frac{\partial y}{\partial \theta}\frac{\partial y}{\partial \phi} + \frac{\partial z}{\partial \theta}\frac{\partial z}{\partial \phi}\right\}, \\ B &= \Sigma m\left\{\left(\frac{\partial x}{\partial \phi}\right)^2 + \left(\frac{\partial y}{\partial \phi}\right)^2 + \left(\frac{\partial z}{\partial \phi}\right)^2\right\}. \end{aligned}\right\} \qquad (5)$$

Hence T is a homogeneous quadratic function of the 'generalized velocities' $\dot{\theta}$, $\dot{\phi}$, with coefficients which are, in any given case, known functions of θ, ϕ.

As regards the right-hand member of (2), if we put

$$\left.\begin{aligned} \Theta &= \Sigma\left(X\frac{\partial x}{\partial \theta} + Y\frac{\partial y}{\partial \theta} + Z\frac{\partial z}{\partial \theta}\right), \\ \Phi &= \Sigma\left(X\frac{\partial x}{\partial \phi} + Y\frac{\partial y}{\partial \phi} + Z\frac{\partial z}{\partial \phi}\right), \end{aligned}\right\} \qquad (6)$$

the expression $\qquad \Theta\delta\theta + \Phi\delta\phi \qquad (7)$

will represent the work done by all the forces acting on the system in an infinitesimal displacement. These quantities Θ, Φ are the generalized components of force; in computing them we may neglect all forces, such as the tensions of inextensible strings or rods, or the mutual reactions of the various parts of a rigid body, which on the whole do no work.

The equations of motion now take the same forms

$$\left.\begin{aligned} \frac{d}{dt}\frac{\partial T}{\partial \dot{\theta}} - \frac{\partial T}{\partial \theta} &= \Theta, \\ \frac{d}{dt}\frac{\partial T}{\partial \dot{\phi}} - \frac{\partial T}{\partial \phi} &= \Phi, \end{aligned}\right\} \qquad (8)$$

as in Art. 104.

If the system be conservative, and if there are no extraneous forces, we have

$$\Theta\delta\theta + \Phi\delta\phi = -\delta V, \qquad (9)$$

where V is the potential energy. Since this is a definite function of the configuration of the system, i.e. of the variables θ, ϕ, we have

$$\Theta = -\frac{\partial V}{\partial\theta}, \quad \Phi = -\frac{\partial V}{\partial\phi}; \qquad (10)$$

and the equations (8) become

$$\left.\begin{aligned} \frac{d}{dt}\frac{\partial T}{\partial\dot\theta} - \frac{\partial T}{\partial\theta} &= -\frac{\partial V}{\partial\theta}, \\ \frac{d}{dt}\frac{\partial T}{\partial\dot\phi} - \frac{\partial T}{\partial\phi} &= -\frac{\partial V}{\partial\phi}. \end{aligned}\right\} \qquad (11)$$

Ex. 1. To form the accurate equations of motion of the double pendulum of Fig. 39 (p. 129).

A mass m was supposed to hang from a fixed point O, whilst a second particle m' hangs from m by a string of length l'. For greater generality we may imagine the strings to be replaced by light rods, capable of exerting thrust as well as tension.

Let θ, ϕ be the angles which the rods respectively make with the vertical. The velocity of m is $l\dot\theta$. That of m' is compounded of its velocity relative to m and the velocity of m itself. These two components are at right angles to l', l, respectively; their amounts are $l'\dot\phi$, $l\dot\theta$; and their directions are inclined at an angle $\phi-\theta$. Hence, for the kinetic energy of the system,

$$2T = ml^2\dot\theta^2 + m'\{l^2\dot\theta^2 + 2ll'\dot\theta\dot\phi\cos(\phi-\theta) + l'^2\dot\phi^2\}. \qquad (12)$$

The potential energy is

$$V = -mgl\cos\theta - m'g(l\cos\theta + l'\cos\phi). \qquad (13)$$

The equations of motion are therefore

$$\frac{d}{dt}\{(m+m')\,l^2\dot\theta + m'll'\dot\phi\cos(\phi-\theta)\} - m'll'\dot\theta\dot\phi\sin(\phi-\theta)$$
$$= -(m+m')\,gl\sin\theta, \qquad (14)$$

$$\frac{d}{dt}\{m'll'\dot\theta\cos(\phi-\theta) + m'l'^2\dot\phi\} + m'll'\dot\theta\dot\phi\sin(\phi-\theta)$$
$$= -m'gl'\sin\phi. \qquad (15)$$

If we add these equations together we obtain a result which expresses that the total angular momentum about O is increasing at a rate equal to the moment of gravity about O.

Ex. 2. The case of the double pendulum of Fig. 64, p. 196, is hardly more complicated. With the notation explained in Art. 68 the kinetic energy of the upper body is $\frac{1}{2}Mk^2\dot{\theta}^2$; that of the lower body relative to its mass-centre is $\frac{1}{2}m\kappa^2\dot{\phi}^2$, whilst the kinetic energy of a particle m at G' would be

$$\tfrac{1}{2}m\{a^2\dot{\theta}^2+2ab\dot{\theta}\dot{\phi}\cos(\phi-\theta)+b^2\dot{\phi}^2\}.$$

Hence for the whole system

$$2T=(Mk^2+ma^2)\dot{\theta}^2+2mab\dot{\theta}\dot{\phi}\cos(\phi-\theta)+m(b^2+\kappa^2)\dot{\phi}^2, \quad \text{...(16)}$$

whilst

$$V=-Mgh\cos\theta-mg(a\cos\theta+b\cos\phi). \quad \text{...............(17)}$$

The equations of motion are substantially of the same type as in (14), (15), the differences being merely in the forms of the constant coefficients.

Ex. 3. If the two axes represented by O, O' in Fig. 64 be vertical instead of horizontal, gravity will have no effect on the motion. An interesting and comparatively simple problem of steady motion here suggests itself, but it is convenient, with a view to this, to make a slight change in the variables, writing

$$\phi=\theta+\chi. \quad \text{...............(18)}$$

This makes

$$2T=\{Mk^2+m(a^2+2ab\cos\chi+b^2+\kappa^2)\}\dot{\theta}^2$$
$$+2m(ab\cos\chi+b^2+\kappa^2)\dot{\theta}\dot{\chi}+m(b^2+\kappa^2)\dot{\chi}^2. \quad \text{...............(19)}$$

The equations are therefore

$$\frac{d}{dt}[\{Mk^2+m(a^2+2ab\cos\chi+b^2+\kappa^2)\}\dot{\theta}$$
$$+m(ab\cos\chi+b^2+\kappa^2)\dot{\chi}]=0, \quad \text{...............(20)}$$

$$\frac{d}{dt}\{(ab\cos\chi+b^2+\kappa^2)\dot{\theta}+(b^2+\kappa^2)\dot{\chi}\}$$
$$+ab\dot{\theta}(\dot{\theta}+\dot{\chi})\sin\chi=0. \quad \text{...............(21)}$$

The equation (20) expresses the constancy of angular momentum about the fixed vertical axis O.

The question of steady motion is, under what condition is the angle χ, viz. the inclination of OO', $O'G$ in Fig. 64, constant, and $\dot{\chi}$ accordingly zero? It is easily seen from (20) and (21), that this will not be the case unless

$$\sin\chi=0, \quad \text{...............(22)}$$

i.e. the angle must be either 0 or π; the mass-centre of the whole system must therefore be at its greatest or least distance from the fixed axis O. It is evident also that $\dot{\theta}$ must be constant.

To investigate the stability of the steady motions we will suppose, in the first place, that χ is small. Writing

$$\dot{\theta}=\omega+\dot{\xi}, \quad \text{...............(23)}$$

where $\dot{\xi}$ is also supposed to be small, we have, neglecting terms of the second order in χ, $\dot{\chi}$, $\ddot{\chi}$, $\dot{\xi}$, $\ddot{\xi}$,

$$\{Mk^2+m(a^2+2ab+b^2+\kappa^2)\}\ddot{\xi}+m(ab+b^2+\kappa^2)\ddot{\chi}=0, \quad \text{.........(24)}$$

$$(ab+b^2+\kappa^2)\ddot{\xi}+(b^2+\kappa^2)\ddot{\chi}+ab\omega^2\chi=0. \quad \text{...............(25)}$$

If we eliminate $\ddot{\xi}$ we find, after a little reduction,

$$\{Mk^2(\kappa^2+b^2)+m\kappa^2a^2\}\ddot{\chi}+[Mk^2+m\{(a+b)^2+\kappa^2\}]\,\omega^2ab\chi=0. \quad \text{...(26)}$$

That type of steady motion in which the mass-centre is furthest from the axis O is accordingly stable, and the period of a small oscillation about it is

$$\left\{\frac{Mk^2(b^2+\kappa^2)+m\kappa^2a^2}{Mk^2ab+mab\{(a+b)^2+\kappa^2\}}\right\}^{\frac{1}{2}}\cdot\frac{2\pi}{\omega}. \quad \text{..................(27)}$$

The particular case where the system reduces to two particles is derived by putting $k=a$, $\kappa=0$; the period is then

$$\left\{\frac{Mab}{Ma^2+m(a+b)^2}\right\}^{\frac{1}{2}}\cdot\frac{2\pi}{\omega}. \quad \text{..........................(28)}$$

To investigate the character of the steady motion when the mass-centre is nearest the axis we should write $\chi=\pi+\chi'$, and assume χ' to be small. The sign of the last term in the equation corresponding to (26) would be reversed, shewing that the steady motion is now unstable.

Ex. 4. If, in the problem of Art. 63, Ex. 3, the outer cylinder in Fig. 54 be free to rotate, and if ϕ be its angular coordinate, the angular velocity of the inner cylinder *relative to the outer one* will be given by

$$\omega a=(b-a)(\dot{\theta}-\dot{\phi}); \quad \text{..........................(29)}$$

and its true angular velocity will therefore be

$$\omega-\dot{\phi}=\frac{b-a}{a}\dot{\theta}-\frac{b}{a}\dot{\phi}. \quad \text{..........................(30)}$$

Hence if I be the moment of inertia of the outer cylinder about its axis, we have

$$2T=I\dot{\phi}^2+M(b-a)^2\dot{\theta}^2+M\kappa^2\left(\frac{b-a}{a}\dot{\theta}-\frac{b}{a}\dot{\phi}\right)^2. \quad \text{............(31)}$$

If the mass-centre of the outer cylinder be on its axis, we have

$$V=-Mg(b-a)\cos\theta, \quad \text{..........................(32)}$$

as before.

Lagrange's equations give, on reduction,

$$\left(I+\frac{M\kappa^2b^2}{a^2}\right)\ddot{\phi}-\frac{M\kappa^2b(b-a)}{a^2}\ddot{\theta}=0, \quad \text{..................(33)}$$

$$(b-a)\left(1+\frac{\kappa^2}{a^2}\right)\ddot{\theta}-\frac{\kappa^2b}{a^2}\ddot{\phi}=-g\sin\theta, \quad \text{..................(34)}$$

whence, eliminating $\ddot{\phi}$, we find

$$(b-a)\left(1+\frac{I\kappa^2}{Ia^2+M\kappa^2b^2}\right)\ddot{\theta}+g\sin\theta=0. \quad \text{...............(35)}$$

The period of oscillation is therefore that of a simple pendulum of length

$$l = (b-a)\left(1 + \frac{I\kappa^2}{Ia^2 + M\kappa^2 b^2}\right), \quad \text{.....................(36)}$$

and is accordingly prolonged by the inertia of the outer cylinder. If we put $I = \infty$, we fall back on the result of Art. 63 (15).

107. Energy. Momentum. Impulse.

The equation of energy may be deduced from Lagrange's equations as follows. We have

$$\left(\frac{d}{dt}\frac{\partial T}{\partial \dot\theta} - \frac{\partial T}{\partial \theta}\right)\dot\theta + \left(\frac{d}{dt}\frac{\partial T}{\partial \dot\phi} - \frac{\partial T}{\partial \phi}\right)\dot\phi$$

$$= \frac{d}{dt}\left(\dot\theta\frac{\partial T}{\partial \dot\theta} + \dot\phi\frac{\partial T}{\partial \dot\phi}\right) - \left(\frac{\partial T}{\partial \dot\theta}\ddot\theta + \frac{\partial T}{\partial \dot\phi}\ddot\phi + \frac{\partial T}{\partial \theta}\dot\theta + \frac{\partial T}{\partial \phi}\dot\phi\right). \quad \text{...(1)}$$

Since T is a homogeneous quadratic function of $\dot\theta$, $\dot\phi$, we have

$$\dot\theta\frac{\partial T}{\partial \dot\theta} + \dot\phi\frac{\partial T}{\partial \dot\phi} = 2T, \quad \text{....................(2)}$$

and the right-hand member of (1) therefore reduces to

$$2\frac{dT}{dt} - \frac{dT}{dt} \text{ or } \frac{dT}{dt}.$$

The equations (8) of Art. 106 therefore lead to

$$\frac{dT}{dt} = \Theta\dot\theta + \Phi\dot\phi, \quad \text{........................(3)}$$

shewing that the kinetic energy is at any instant increasing at a rate equal to that at which work is being done on the various parts of the system.

If the system be conservative, and if there are no extraneous forces, the right-hand member of (3) is equal to $-dV/dt$, by Art. 106 (10). The equation has then the integral

$$T + V = \text{const.} \quad \text{........................(4)}$$

If we write

$$\lambda = \frac{\partial T}{\partial \dot\theta}, \quad \mu = \frac{\partial T}{\partial \dot\phi}, \quad \text{........................(5)}$$

the quantities λ, μ are called the generalized 'components of momentum.' The reason for this name may be seen as in Art. 104, or we may justify it by an application of the equations

(8) of Art. 106. If we imagine the actual motion at any instant to be generated from rest in an infinitely short time τ, we have, by integration of the equations referred to,

$$\frac{\partial T}{\partial \dot{\theta}} = \int_0^\tau \Theta\, dt, \quad \frac{\partial T}{\partial \dot{\phi}} = \int_0^\tau \Phi\, dt. \quad \text{..............(6)}$$

For the terms $\partial T/\partial \theta$, $\partial T/\partial \phi$, since they depend on the velocities and the coordinates only, are essentially finite, and therefore disappear when integrated over the evanescent interval τ; whilst the *initial* values of $\partial T/\partial \dot{\theta}$, $\partial T/\partial \dot{\phi}$, which are linear functions of the velocities $\dot{\theta}$, $\dot{\phi}$, are zero in the imagined process. The formulæ (6) shew that the generalized momenta, as above defined, are equal to the generalized components of the impulse which would be required to start the system, in its actual state, from rest.

Ex. 1. The double pendulum of Art. 68 is struck, when at rest in its stable position, by a horizontal impulse ξ acting on the lower body in a line at a distance x below the axis O'.

Since the work done in a small displacement by a finite force X in this line would be

$$X(a\delta\theta + x\delta\phi),$$

the generalized components of impulse are ξa, ξx, respectively. Hence the values of $\dot{\theta}$, $\dot{\phi}$ immediately after the impulse are given by

$$\frac{\partial T}{\partial \dot{\theta}} = \xi a, \quad \frac{\partial T}{\partial \dot{\phi}} = \xi x, \quad \text{..............................(7)}$$

where the value of T is given by Art. 106 (16) with $\theta = 0$, $\phi = 0$. Thus

$$\left.\begin{aligned} (Mk^2 + ma^2)\,\dot{\theta} + mab\dot{\phi} &= \xi a, \\ mab\dot{\theta} + m\,(b^2 + \kappa^2)\,\dot{\phi} &= \xi x. \end{aligned}\right\} \quad \text{..........................(8)}$$

Hence $\dot{\theta} = 0$, i.e. the upper body will remain (initially) at rest, if

$$x = b + \kappa^2/b, \quad \text{..................................(9)}$$

as is known from Art. 71.

If the two bodies are two equal uniform bars of length $2b$, we have

$$M = m, \quad a = 2b, \quad k^2 = \tfrac{4}{3}b^2, \quad \kappa^2 = \tfrac{1}{3}b^2,$$

and we find

$$\left.\begin{aligned} a\dot{\theta} &= \frac{9\xi}{7mb}\left(\tfrac{4}{3}b - x\right), \\ b\dot{\phi} &= \frac{12\xi}{7mb}\left(x - \tfrac{3}{4}b\right). \end{aligned}\right\} \quad \text{..........................(10)}$$

The velocity of the point struck is

$$a\dot{\theta}+x\dot{\phi}=\frac{6\xi}{7mb^2}(2b^2-3bx+2x^2). \qquad \text{.....................(11)}$$

Ex. 2. A rhombus $ABCD$, formed of four equal jointed rods, is struck by an impulse ξ at A, in the direction AC.

Let x be the coordinate of the centre G of the rhombus, measured from some fixed point in the line AC. We will suppose that the mass-centre of each rod is at its middle point; let $2a$ be the length of each, κ its radius of gyration about its centre, and 2θ the angle ABC. The formula for the kinetic energy is found by an easy calculation to be

$$2T=M\dot{x}^2+M(\kappa^2+a^2)\dot{\theta}^2, \qquad \text{...........................(12)}$$

where M is the total mass. Since the coordinate of A is $x-2a\sin\theta$, the work done by a finite force X at A in a small displacement would be

$$X(\delta x-2a\cos\theta\,\delta\theta).$$

The generalized components of impulse are therefore ξ and $-2a\xi\cos\theta$. Hence

$$M\dot{x}=\xi, \quad M(\kappa^2+a^2)\dot{\theta}=-2a\xi\cos\theta. \qquad \text{.................(13)}$$

The initial velocity of A is therefore

$$\dot{x}-2a\cos\theta\,\dot{\theta}=\left(1+\frac{4a^2\cos^2\theta}{\kappa^2+a^2}\right)\frac{\xi}{M}; \qquad \text{.................(14)}$$

and the energy generated is

$$\frac{1}{2}\left(1+\frac{4a^2\cos^2\theta}{\kappa^2+a^2}\right)\frac{\xi^2}{M}. \qquad \text{...........................(15)}$$

This is greater than if the frame had been rigid, in the ratio given by the second factor (cf. Art. 108).

108. General Theorems.

Writing $\qquad T=\frac{1}{2}(A\dot{\theta}^2+2H\dot{\theta}\dot{\phi}+B\dot{\phi}^2), \qquad$ ....(1)

we have $\qquad \lambda=A\dot{\theta}+H\dot{\phi}, \quad \mu=H\dot{\theta}+B\dot{\phi}. \qquad$ ....(2)

If in another state of motion through the same configuration the velocities be denoted by $\dot{\theta}'$, $\dot{\phi}'$, and the momenta by λ', μ', we have

$$\lambda\dot{\theta}'+\mu\dot{\phi}'=\lambda'\dot{\theta}+\mu'\dot{\phi}, \qquad \text{.....................(3)}$$

each of these expressions being equal to

$$A\dot{\theta}\dot{\theta}'+H(\dot{\theta}'\dot{\phi}+\dot{\theta}\dot{\phi}')+B\dot{\phi}\dot{\phi}'. \qquad \text{...............(4)}$$

This leads to a remarkable theorem of reciprocity. If we put

$$\lambda'=0, \quad \mu=0,$$

the formula (3) gives $\qquad \dfrac{\dot{\phi}}{\lambda}=\dfrac{\dot{\theta}'}{\mu'}. \qquad$ ....(5)

The interpretation of this is simplest when the two coordinates θ, ϕ are of the same geometrical character, e.g. both linear magnitudes, or both angles. Thus if both be of the nature of lines, the two momenta will be of the nature of ordinary impulsive forces, and we may put $\lambda = \mu'$, in which case $\dot{\phi} = \dot{\theta}'$. Hence the velocity of one type due to an impulse of the second type is equal to the velocity of the second type due to an impulse of the first type.

Thus, suppose we have two bars AB, BC, freely jointed at B, of which the former is free to turn about A as a fixed point. For simplicity we will suppose the bars to be in a straight line. The two coordinates may be taken to be the displacements, at right angles to the lengths, of any two points P, Q of the system. The theorem then asserts that the velocity of Q due to an impulse at P is equal to the velocity at P due to an equal impulse at Q. Again, if we use angular coordinates, the theorem shews that the angular velocity of BC due to an impulsive couple applied to AB is equal to the angular velocity of AB due to an equal couple applied to BC. Finally, as an instance where the coordinates are of different kinds, we infer that if an impulse ξ applied at any point P of AB generates an angular velocity ω in BC, an impulsive couple ξa applied to BC would produce the velocity ωa at P.

Another important theorem relates to the effect of constraints on the amount of energy generated when a mechanical system is started in different ways.

If in (1) we substitute the value of $\dot{\phi}$ in terms of $\dot{\theta}$ and μ, from (2), we find

$$2T = \frac{AB - H^2}{B}\dot{\theta}^2 + \frac{\mu^2}{B}, \quad \text{.....................(6)}$$

where the coefficient of $\dot{\theta}^2$ is necessarily positive.

The energy due to an impulse μ is therefore greater than if $\dot{\theta} = 0$, i.e. it is *greater* than if a constraint had been applied to prevent the coordinate θ from varying (cf. Art. 73, Ex. 2 and Art. 107, Ex. 2).

Again if the system be started with a prescribed *velocity* $\dot{\theta}$ by an impulse of the type θ alone, the energy is *less* than if a constraining impulse of the other type had been operative.

By a proper choice of the coordinates the conclusions of this Article can be extended to systems of any number of degrees of freedom; but we must here content ourselves with a statement of the two theorems just proved, in their generalized form:

(1) (Delaunay's Theorem*.) A system started from rest by given *impulses* acquires *greater* energy than if it had been constrained in any way.

(2) (Thomson's Theorem†.) A system started with given *velocities* has *less* energy than if it had been constrained.

In a system with one degree of freedom, the kinetic energy may be expressed in either of the forms

$$T = \tfrac{1}{2}\alpha\dot{\theta}^2 = \tfrac{1}{2}\lambda^2/\alpha, \quad\text{.......................(7)}$$

where α is the coefficient of inertia. The preceding theorems therefore come to this, that the effect of a constraint is to increase the apparent inertia of the system‡.

109. Oscillations about Equilibrium.

In order that the equations (11) of Art. 106 may be satisfied by $\dot{\theta}=0$, $\dot{\phi}=0$ we must have

$$\frac{\partial V}{\partial \theta} = 0, \quad \frac{\partial V}{\partial \phi} = 0, \quad\text{.........................(1)}$$

as is otherwise known. These may be regarded as two equations to determine the values of θ, ϕ corresponding to the various possible configurations of equilibrium. They express that the potential energy is stationary for small displacements [*S.* 57].

To ascertain the nature of the equilibrium, we may suppose the coordinates modified, by the addition of constants, so that they shall vanish in the configuration considered. Hence, in investigating the effect of a small disturbance, the values of θ, ϕ, as well as of $\dot{\theta}$, $\dot{\phi}$, are treated as small quantities.

* This theorem is associated with the names of Euler, Lagrange, Delaunay (1840), and Bertrand (1853). The maximum property was definitely established by Delaunay.

† Sir W. Thomson, afterwards Lord Kelvin (1863).

‡ Lord Rayleigh (1886).

The terms $\partial T/\partial\theta$, $\partial T/\partial\phi$ in the dynamical equations, being of the second order in $\dot\theta$, $\dot\phi$, are omitted, and the equations reduce to

$$\frac{d}{dt}\frac{\partial T}{\partial\dot\theta}+\frac{\partial V}{\partial\theta}=0,\quad \frac{d}{dt}\frac{\partial T}{\partial\dot\phi}+\frac{\partial V}{\partial\phi}=0. \quad\text{.............(2)}$$

Moreover, the coefficients A, H, B in the expression for T, viz.

$$T=\tfrac{1}{2}(A\dot\theta^2+2H\dot\theta\dot\phi+B\dot\phi^2) \quad\text{..................(3)}$$

may be treated as constants, and equal to the values which they have in the equilibrium configuration, since the resulting errors in (2) are of the second order.

Again, the value of V may be supposed expanded, for small values of θ, ϕ, in the form

$$V=V_0+\tfrac{1}{2}(a\theta^2+2h\theta\phi+b\phi^2)+\dots, \quad\text{...........(4)}$$

the terms of the first order being absent, since (1) must be satisfied by $\theta=0$, $\phi=0$. The coefficients a, h, b may be called the 'coefficients of stability' of the system.

Hence, if we keep only terms of the first order, the equations (2) become

$$\left.\begin{array}{l} A\ddot\theta+H\ddot\phi+a\theta+h\phi=0, \\ H\ddot\theta+B\ddot\phi+h\theta+b\phi=0. \end{array}\right\} \quad\text{...............(5)}$$

To solve these, we assume

$$\theta=C\cos(nt+\epsilon),\quad \phi=kC\cos(nt+\epsilon), \quad\text{........(6)}$$

where C, k, ϵ are constants. The equations are satisfied provided

$$\left.\begin{array}{l} (n^2A-a)+(n^2H-h)k=0, \\ (n^2H-h)+(n^2B-b)k=0. \end{array}\right\} \quad\text{...............(7)}$$

Eliminating k, we have

$$\begin{vmatrix} n^2A-a, & n^2H-h \\ n^2H-h, & n^2B-b \end{vmatrix}=0, \quad\text{..................(8)}$$

which is a quadratic in n^2. The formulæ (6) therefore constitute a solution of the equations (5) provided n satisfies (8), and provided the coefficient k be determined in accordance with either of the equations (7).

Since the expression (3) for T is essentially positive, whatever the values of $\dot{\theta}$, $\dot{\phi}$, the coefficients of inertia are subject to the relations

$$A > 0, \quad B > 0, \quad AB - H^2 > 0. \quad\text{..............(9)}$$

Hence, if we put $n^2 = \pm\infty$ in the above determinant, its sign is $+$; if we put $n^2 = a/A$ or b/B, its sign is $-$; whilst if we put $n^2 = 0$, its sign is that of $ab - h^2$. The two roots of (8), considered as an equation in n^2, are therefore always real.

Moreover, it is evident that the roots are unequal unless $a/A = b/B$. It is further necessary for equality that n^2 should be equal to the common value of these fractions, and it follows from (8) that n^2 must then also be equal to h/H. The conditions for equal roots are therefore

$$\frac{a}{A} = \frac{h}{H} = \frac{b}{B}; \quad\text{........................(10)}$$

and it appears from (7) that the values of k are then indeterminate. We leave this case aside for the present.

Again, if the expression for $V - V_0$ in (4) is essentially positive, i.e. if V be a minimum in the configuration of equilibrium, we have

$$a > 0, \quad b > 0, \quad ab - h^2 > 0. \quad\text{...............(11)}$$

The two values of n^2 are therefore both positive, one being greater than the greater, and the other less than the lesser of the two quantities a/A, b/B.

If, on the other hand, V be a maximum in the equilibrium configuration we have

$$a < 0, \quad b < 0, \quad ab - h^2 > 0, \quad\text{...............(12)}$$

and both values of n^2 are negative. The form (6) is then imaginary, and is to be replaced by one involving real exponentials (cf. Art. 32).

If
$$ab - h^2 < 0$$
one value of n^2 is negative and the other positive.

Confining ourselves to the case of stable equilibrium, and superposing the solutions corresponding to the two values of n^2, we have

$$\left.\begin{aligned} \theta &= C_1 \cos(n_1 t + \epsilon_1) + C_2 \cos(n_2 t + \epsilon_2), \\ \phi &= k_1 C_1 \cos(n_1 t + \epsilon_1) + k_2 C_2 \cos(n_2 t + \epsilon_2), \end{aligned}\right\} \quad\text{......(13)}$$

where the values of C_1, C_2, ϵ_1, ϵ_2 are arbitrary, whilst k_1, k_2 are given by either of the relations (7), with the respective values of n^2 inserted. Since there are four arbitrary constants, this solution is general.

The same fundamental equation (8) presents itself in a slightly different connection. Suppose that, by the introduction of a frictionless constraint, the coordinate ϕ is made to bear a given constant ratio to θ*, say

$$\phi = k\theta. \quad \text{.........................(14)}$$

The system is now reduced to one degree of freedom, the expressions for the kinetic and potential energies being

$$T = \tfrac{1}{2}(A + 2Hk + Bk^2)\,\dot{\theta}^2, \quad \text{...........(15)}$$

$$V - V_0 = \tfrac{1}{2}(a + 2hk + bk^2)\,\theta^2. \quad \text{..............(16)}$$

The period $(2\pi/n)$ of a small oscillation in this constrained mode is therefore given by

$$n^2 = \frac{a + 2hk + bk^2}{A + 2Hk + Bk^2}, \quad \text{..................(17)}$$

by Art. 65. A negative value of the fraction would indicate of course that the solution involves real exponentials, and that the constrained mode is accordingly unstable.

Since the value of T in (15) is essentially positive, the denominator in (17) cannot vanish. The fraction is therefore finite for all values of k, and must have a maximum and a minimum value. To find these, we write the equation in the form

$$n^2(A + 2Hk + Bk^2) = a + 2hk + bk^2, \quad \text{........(18)}$$

and differentiate on the supposition that $d(n^2)/dk = 0$. We obtain

$$n^2(H + Bk) = h + bk; \quad \text{..................(19)}$$

and combining this with (18) we have, also,

$$n^2(A + Hk) = a + hk. \quad \text{..................(20)}$$

These equations are identical with (7); and if we eliminate k, we find that the stationary values of n^2 are the roots of the quadratic (8), which are therefore by the present argument real. Moreover, if $V - V_0$ is essentially positive the values of n^2 are both

* For example, in the case of Blackburn's pendulum (Art. 29), the particle may be restricted to move in a particular vertical plane through the equilibrium position.

positive, by (17); whilst if it is essentially negative they are both negative. If $V - V_0$ is capable of both signs, the maximum value of n^2 is of course positive, and the minimum negative.

110. Normal Modes of Vibration.

The process followed in the first part of Art. 109 consisted in ascertaining whether a mode of motion of the system is possible in which each of the independent coordinates θ, ϕ is a simple-harmonic function of the time, with the same period and phase; and we found that (in the case of stability) there are two such modes. Each of these is called a 'normal mode' of vibration of the system; its period is determined solely by the constitution of the system; and its type is also determinate, since the *relative* amplitudes of θ, ϕ are fixed, although the absolute amplitude and phase are arbitrary.

The formulæ relating to the two modes are

$$\theta = C_1 \cos(n_1 t + \epsilon_1), \quad \phi = k_1 C_1 \cos(n_1 t + \epsilon_1), \quad \text{......(1)}$$

and

$$\theta = C_2 \cos(n_2 t + \epsilon_2), \quad \phi = k_2 C_2 \cos(n_2 t + \epsilon_2). \quad \text{......(2)}$$

We have seen that by superposition of these, with arbitrary amplitudes C_1, C_2, and phases ϵ_1, ϵ_2, the most general motion of the system, consequent on a small disturbance, can be represented.

When the system oscillates in one of these modes alone, every particle executes a simple-harmonic vibration in a straight line, and the various particles keep step with one another, passing simultaneously through their respective equilibrium positions. The relative amplitudes are also determinate. To verify these statements we need only write down the expressions for the component displacements of a particle whose mean position is (x, y, z). On account of the assumed smallness of θ, ϕ, we have, in the first mode,

$$\left.\begin{aligned} \delta x &= \frac{\partial x}{\partial \theta}\theta + \frac{\partial x}{\partial \phi}\phi = \left(\frac{\partial x}{\partial \theta} + k_1 \frac{\partial x}{\partial \phi}\right) C_1 \cos(n_1 t + \epsilon_1), \\ \delta y &= \frac{\partial y}{\partial \theta}\theta + \frac{\partial y}{\partial \phi}\phi = \left(\frac{\partial y}{\partial \theta} + k_1 \frac{\partial y}{\partial \phi}\right) C_1 \cos(n_1 t + \epsilon_1), \\ \delta z &= \frac{\partial z}{\partial \theta}\theta + \frac{\partial z}{\partial \phi}\phi = \left(\frac{\partial z}{\partial \theta} + k_1 \frac{\partial z}{\partial \phi}\right) C_1 \cos(n_1 t + \epsilon_1). \end{aligned}\right\} \quad \text{...(3)}$$

The ratios $\delta x : \delta y : \delta z$ are seen to be fixed for each particle, being independent of the time and of the absolute amplitude C_1.

We have seen in Art. 109 that the normal modes are characterized by the property that the period is stationary for a small variation in the type of vibration; i.e. if the system be constrained to vibrate in a mode slightly different from a 'normal' one, the period is unaltered, to the first order. As an example, we may refer again to the case of motion in a smooth bowl (Art. 29). If the particle be constrained to oscillate in a vertical section through the lowest point the period is $2\pi\sqrt{(R/g)}$, where R is the radius of curvature of the section. This is a maximum or minimum when the section coincides with one of the principal planes of curvature at the lowest point.

The coordinates θ, ϕ which determine the configuration of the system can of course be chosen in an infinite variety of ways. There is one choice, however, which is specially interesting from a theoretical point of view, since each normal mode then involves the variation of one coordinate alone.

We have seen that

$$\left.\begin{aligned} n_1^2(A + Hk_1) &= a + hk_1, \\ n_1^2(H + Bk_1) &= h + bk_1, \end{aligned}\right\} \quad \text{.................(4)}$$

with similar equations involving n_2^2. If we multiply the second of these equations by k_2, and add to the first, we obtain

$$n_1^2\{A + H(k_1 + k_2) + Bk_1k_2\} = a + h(k_1 + k_2) + bk_1k_2. \quad \text{...(5)}$$

In a similar way we should find

$$n_2^2\{A + H(k_1 + k_2) + Bk_1k_2\} = a + h(k_1 + k_2) + bk_1k_2. \quad \text{...(6)}$$

We infer, by subtraction, that if $n_1^2 \neq n_2^2$, we must have

$$A + H(k_1 + k_2) + Bk_1k_2 = 0, \quad \text{..............(7)}$$

and

$$a + h(k_1 + k_2) + bk_1k_2 = 0. \quad \text{..............(8)}$$

Hence if we put

$$\theta = q_1 + q_2, \quad \phi = k_1q_1 + k_2q_2, \quad \text{..............(9)}$$

the expressions for the kinetic and potential energies take the simplified forms

$$T = \tfrac{1}{2}(A_1\dot{q}_1^2 + A_2\dot{q}_2^2), \quad \text{..............(10)}$$

$$V - V_0 = \tfrac{1}{2}(a_1q_1^2 + a_2q_2^2), \quad \text{..............(11)}$$

where $A_1 = A + 2Hk_1 + Bk_1^2, \quad A_2 = A + 2Hk_2 + Bk_2^2$, ...(12)

and $a_1 = a + 2hk_1 + bk_1^2, \quad a_2 = a + 2hk_2 + bk_2^2$. ......(13)

The terms involving the products $\dot{q}_1\dot{q}_2$, and q_1q_2, respectively, vanish in consequence of the relations (7), (8).

The new variables q_1, q_2 are called the 'normal coordinates' of the system*. In terms of them, the equations of small motion are

$$A_1\ddot{q}_1 + a_1q_1 = 0, \quad A_2\ddot{q}_2 + a_2q_2 = 0, \text{ ....(14)}$$

which are independent of one another. In each normal mode, one coordinate varies alone; and the two periods are determined by

$$n_1^2 = a_1/A_1, \quad n_2^2 = a_2/A_2. \text{ ....(15)}$$

We have seen that when the roots of the quadratic in n^2 coincide, the quantities k, and therefore the characters of the normal modes, are indeterminate. The solution of the differential equations (5) of Art. 109 is then

$$\theta = C\cos(nt + \epsilon), \quad \phi = C'\cos(nt + \epsilon'), \text{ ....(16)}$$

where C, C', ϵ, ϵ' are arbitrary, and

$$n^2 = \frac{a}{A} = \frac{h}{H} = \frac{b}{B}. \text{ ....(17)}$$

An example is furnished by the spherical pendulum (Art. 29).

Ex. 1. If we have two equal particles attached symmetrically to a tense string, as in Fig. 40, p. 132, the work required to stretch the three segments of the string, against the tension P, into the position shewn in the figure is

$$V = P[\sqrt{\{a^2+x^2\}} - a + \sqrt{\{4b^2+(y-x)^2\}} - 2b + \sqrt{\{a^2+y^2\}} - a]$$

$$= \frac{1}{2}P\left(\frac{x^2}{a} + \frac{(x-y)^2}{2b} + \frac{y^2}{a}\right), \text{ ....(18)}$$

* Analytically, the investigation of the normal modes and normal coordinates is identical with the problem of finding the pair of diameters which are conjugate with respect to each of the conics

$$Ax^2 + 2Hxy + By^2 = \text{const.},$$

$$ax^2 + 2hxy + by^2 = \text{const.},$$

and ascertaining the forms which the equations assume when referred to these diameters. Since the former of the two conics is an ellipse, the common conjugate diameters are real, as may be seen by projecting orthogonally so that the ellipse becomes a circle.

approximately. Since the kinetic energy is

$$T=\tfrac{1}{2}m(\dot{x}^2+\dot{y}^2), \qquad \text{.......(19)}$$

the equations (2) of Art. 109 give

$$\left.\begin{aligned} m\ddot{x}+P\left(\frac{x}{a}+\frac{x-y}{2b}\right)&=0,\\ m\ddot{y}+P\left(\frac{y-x}{2b}+\frac{y}{a}\right)&=0,\end{aligned}\right\} \qquad \text{.......(20)}$$

in agreement with Art. 44 (13).

The results of the Article referred to shew that in this case the normal coordinates are proportional to $x+y$ and $x-y$, respectively. Writing, in fact,

$$x=u+v, \quad y=u-v, \qquad \text{.......(21)}$$

we have

$$T=m(\dot{u}^2+\dot{v}^2), \qquad \text{.......(22)}$$

and

$$V=P\left\{\frac{u^2}{a}+\left(\frac{1}{a}+\frac{1}{b}\right)v^2\right\}. \qquad \text{.......(23)}$$

The Lagrangian equations in terms of these variables are

$$m\ddot{u}+\frac{P}{a}u=0, \quad m\ddot{v}+P\left(\frac{1}{a}+\frac{1}{b}\right)v=0. \qquad \text{.......(24)}$$

In one normal mode we have $u=0$, and in the other $v=0$; and the frequencies are as determined by Art. 44 (17).

Ex. 2. In the case of the double pendulum of Fig. 39 (p. 129), we have, with the approximations above explained,

$$T=\tfrac{1}{2}\{(m+m')l^2\dot{\theta}^2+2m'll'\dot{\theta}\dot{\phi}+m'l'^2\dot{\phi}^2\}, \qquad \text{.......(25)}$$

$$V-V_0=\tfrac{1}{2}\{(m+m')gl\theta^2+m'gl'\phi^2\}. \qquad \text{.......(26)}$$

The formulæ (2) of Art. 109 then give, after discarding unnecessary factors,

$$\left.\begin{aligned} (m+m')l\ddot{\theta}+m'l'\ddot{\phi}+(m+m')g\theta&=0,\\ l\ddot{\theta}+\ l'\ddot{\phi}+\qquad\qquad g\phi&=0,\end{aligned}\right\} \qquad \text{.......(27)}$$

and the solution can be completed as in Art. 44. The equations just written are seen, in fact, to be equivalent to (1) of that Article if we write

$$x=l\theta, \quad y=l\theta+l'\phi. \qquad \text{.......(28)}$$

The double pendulum of Fig. 64 can be treated in exactly the same manner, and the results compared with those obtained in Art. 68.

111. Forced Oscillations.

If, with a slight change from our previous notation, we denote by Θ, Φ the components of *extraneous* force acting on a conservative system, Lagrange's equations become

$$\left.\begin{aligned} \frac{d}{dt}\frac{\partial T}{\partial \dot{\theta}}-\frac{\partial T}{\partial \theta}&=-\frac{\partial V}{\partial \theta}+\Theta,\\ \frac{d}{dt}\frac{\partial T}{\partial \dot{\phi}}-\frac{\partial T}{\partial \phi}&=-\frac{\partial V}{\partial \phi}+\Phi.\end{aligned}\right\} \qquad \text{.......(1)}$$

In the case of small oscillations about equilibrium these reduce to

$$\left.\begin{aligned} A\ddot{\theta} + H\ddot{\phi} + a\theta + h\phi = \Theta, \\ H\ddot{\theta} + B\ddot{\phi} + h\theta + b\phi = \Phi. \end{aligned}\right\} \quad \text{.................(2)}$$

As in Arts. 13, 67, the most important case is where the disturbing forces Θ, Φ are periodic in character. Assuming, then, that Θ and Φ both vary as $\cos pt$, we find that the equations (2) are satisfied by the assumption that θ, ϕ also vary as $\cos pt$, provided

$$\left.\begin{aligned} (p^2A - a)\,\theta + (p^2H - h)\,\phi = -\Theta, \\ (p^2H - h)\,\theta + (p^2B - b)\,\phi = -\Phi. \end{aligned}\right\} \quad \text{...........(3)}$$

These determine θ, ϕ in terms of Θ, Φ; thus

$$\left.\begin{aligned} \theta &= \frac{(p^2H - h)\,\Phi - (p^2B - b)\,\Theta}{\Delta(p^2)}, \\ \phi &= \frac{(p^2H - h)\,\Theta - (p^2A - a)\,\Phi}{\Delta(p^2)}, \end{aligned}\right\} \quad \text{..............(4)}$$

where

$$\Delta(p^2) = \begin{vmatrix} p^2A - a, & p^2H - h \\ p^2H - h, & p^2B - b \end{vmatrix}. \quad \text{...........(5)}$$

Since, by hypothesis, θ and ϕ both vary as $\cos pt$, the motion of every particle of the system is periodic, with the period $2\pi/p$ of the disturbing forces; the various particles moreover keep step. The amplitudes, relative as well as absolute, of the different particles will however depend on the period. In particular, the amplitudes become very great when p^2 approximates to a root of

$$\Delta(p^2) = 0, \quad \text{.........................(6)}$$

i.e. when the period of the imposed vibration coincides nearly with one of the free periods as determined by Art. 109 (8). To obtain a more practical result in such cases, as well as in that of exact coincidence, we should have to take account of dissipative forces, as in Art. 95.

On the forced oscillation above found we may superpose the free oscillations of Art. 109, with arbitrary amplitudes and phases.

There is another important kind of forced vibration in which a prescribed variation is imposed on one of the coordinates by means of a suitable force of the same type, whilst extraneous force of the other type is absent. Thus we may have the equations

$$A\ddot{\theta} + H\ddot{\phi} + a\theta + h\phi = 0, \quad \text{.................(7)}$$

$$H\ddot{\theta} + B\ddot{\phi} + h\theta + b\phi = \Phi, \quad \text{.................(8)}$$

where ϕ is given as a function of t, and the second equation serves merely to determine the force Φ which is necessary to maintain the given variation of ϕ. An example is furnished by the forced oscillations of a compound pendulum, or of a seismograph (Art. 67) due to a prescribed motion of the axis.

Thus if
$$\phi \propto \cos pt, \quad \text{..........................(9)}$$

we have, on the assumption that θ also varies as $\cos pt$,

$$(p^2A - a)\,\theta + (p^2H - h)\,\phi = 0, \quad \text{.............(10)}$$

or
$$\theta = -\frac{p^2H - h}{p^2A - a}\,\phi, \quad \text{...................(11)}$$

which gives the forced oscillation in the coordinate θ. It will be noticed that the inertia-coefficient B does not enter into this determination.

Ex. Let ξ denote the lateral displacement of the axis of a seismograph (Art. 67) from its mean position. The velocity of the mass-centre is, with sufficient approximation, $\dot{\xi} + h\dot{\theta}$, where θ is the angular coordinate, and the expression for the kinetic energy, so far as we require it, is therefore

$$2T = M(h^2 + \kappa^2)\,\dot{\theta}^2 + 2Mh\dot{\theta}\dot{\xi} + M\dot{\xi}^2. \quad \text{.....................(12)}$$

The potential energy has the form

$$V = \tfrac{1}{2}K\theta^2, \quad \text{....................................(13)}$$

where K is some constant. The equation corresponding to (7) is therefore

$$M(h^2 + \kappa^2)\,\ddot{\theta} + Mh\ddot{\xi} + K\theta = 0, \quad \text{........................(14)}$$

or
$$\ddot{\theta} + n^2\theta = -\frac{\ddot{\xi}}{l}, \quad \text{..............................(15)}$$

if
$$n^2 = \frac{K}{M(h^2 + \kappa^2)}, \qquad l = \frac{\kappa^2}{h} + h. \quad \text{........................(16)}$$

In the case of the horizontal pendulum (Art. 67) we have

$$K = Mgh\sin\beta, \qquad n^2 = \frac{g\sin\beta}{l} \quad \text{.....................(17)}$$

EXAMPLES. XXII.

1. A smooth open tube of length $2a$ contains a rod which just fills it, and the masses of the tube and rod are M, m, respectively. If the tube be set in rotation about its centre (which is fixed), in a horizontal plane, with angular velocity ω_0, find the angular velocity when the rod has left the tube, it being supposed that the centres of the tube and rod are initially all but coincident. Also find the velocity of the centre of the rod.

$$\left[\frac{M+m}{M+13m}\omega_0; \quad \frac{\sqrt{\{8(M+m)(M+7m)\}}}{M+13m}\omega_0 a.\right]$$

2. Two masses are connected by a helical spring. When they vibrate freely in a straight line the period is $2\pi/p$. If they are set rotating about one another with the constant angular velocity ω, prove that the period of a small disturbance is

$$\frac{2\pi}{\sqrt{(p^2+3\omega^2)}}.$$

3. A particle moveable on a smooth spherical surface of radius a is projected along the horizontal great circle with a velocity v which is great compared with $\sqrt{(2ga)}$. Prove that its path lies between this great circle and a parallel circle whose plane is at a depth $2ga^2/v^2$ below the centre, approximately.

4. A particle moves on the surface of a smooth cone of semi-angle α whose axis is vertical and vertex downwards. Prove that it can describe a horizontal circle of radius a with angular velocity ω provided

$$\omega^2 a = g \cot \alpha.$$

If this motion be slightly disturbed, the period of a small oscillation is

$$\frac{2\pi}{\omega\sqrt{3}\,.\,\sin\alpha}.$$

5. Form by Lagrange's method the equations of motion of a particle on a paraboloid of revolution, in terms of the distance (r) from the axis, and the azimuth (θ). (Cf. Art. 103.)

6. A particle is contained in a smooth parabolic tube which is constrained to rotate with constant angular velocity ω about the axis, which is vertical, the concavity being upwards. Prove that if $\omega^2 = g/l$, where l is the semi-latus-rectum, the particle can be in relative equilibrium in any position, and that if it be disturbed from this state, the distance r from the axis will ultimately increase indefinitely.

7. Prove the following expressions for the component accelerations in spherical polar coordinates:

$$\ddot{r} - r\dot{\theta}^2 - r\sin^2\theta\,\dot{\psi}^2, \qquad \frac{1}{r}\frac{d}{dt}(r^2\dot{\theta}) - r\sin\theta\cos\theta\,\dot{\psi}^2, \qquad \frac{1}{r\sin\theta}\frac{d}{dt}(r^2\sin^2\theta\,\dot{\psi})$$

8. A particle m is contained in a smooth circular tube of radius a which is free to rotate about a vertical diameter. When the tube is rotating with angular velocity ω the particle is in relative equilibrium at an angular distance α from the lowest point. Prove that if this state be slightly disturbed the period of a small oscillation is $2\pi/n$, where n is given by

$$\frac{n^2}{\omega^2}=\frac{\{I+ma^2(1+3\cos^2\alpha)\}\sin^2\alpha}{I+ma^2\sin^2\alpha},$$

I denoting the moment of inertia of the tube about the diameter.

Examine the cases of $I=0$, $I=\infty$, respectively.

9. A particle moves on a smooth surface generated by the revolution of a circle of radius b about a vertical axis at a distance a ($>b$) from its centre. Prove that it can describe a horizontal circle of radius $a+b\sin\alpha$ with constant angular velocity ω provided

$$\omega^2\cos\alpha\,(a+b\sin\alpha)=g\sin\alpha.$$

Prove that if this motion be slightly disturbed the period of a small oscillation about it is $2\pi/n$, if

$$n^2=\omega^2\left(1+3\cos^2\alpha+\frac{a}{b\sin\alpha}\right).$$

10. A particle is free to move in a smooth circular tube of radius b which is constrained to rotate with constant angular velocity ω about a vertical axis in its own plane at a distance a ($>b$) from its centre.

Prove that the period of a small oscillation about a position of relative equilibrium is $2\pi/n$, if

$$n^2=\omega^2\left(\sin^2\alpha+\frac{a}{b\sin\alpha}\right).$$

11. Shew that the equations of motion of a system of two degrees of freedom may be written

$$A\ddot{\theta}+H\ddot{\phi}+\frac{1}{2}\frac{\partial A}{\partial\theta}\dot{\theta}^2+\frac{\partial A}{\partial\phi}\dot{\theta}\dot{\phi}+\left(\frac{\partial H}{\partial\phi}-\frac{1}{2}\frac{\partial B}{\partial\theta}\right)\dot{\phi}^2=\Theta,$$

$$H\ddot{\theta}+B\ddot{\phi}+\left(\frac{\partial H}{\partial\theta}-\frac{1}{2}\frac{\partial A}{\partial\phi}\right)\dot{\theta}^2+\frac{\partial B}{\partial\theta}\dot{\theta}\dot{\phi}+\frac{1}{2}\frac{\partial B}{\partial\phi}\dot{\phi}^2=\Phi.$$

12. Prove that in the notation of Art. 108 the kinetic energy of a system of two degrees of freedom is given, in terms of the component momenta, by

$$2T=\frac{B\lambda^2-2H\lambda\mu+A\mu^2}{AB-H^2}.$$

Prove that if T be expressed in this form

$$\dot{\theta}=\frac{\partial T}{\partial\lambda},\qquad \dot{\phi}=\frac{\partial T}{\partial\mu}$$

13. A uniform bar AB hangs by two strings OA, OB from a fixed point O, the triangle OAB being equilateral. Prove that if one string be cut the tension of the other is instantaneously reduced in the ratio $6:13$.

14. Four equal particles are connected by strings forming the sides of a rhombus $ABCD$. Prove that the system can rotate in its own plane about the centre O, without change of form, with any given angular velocity. Also prove that if this state of motion be disturbed the diagonals AC, BD will still rotate with constant angular velocity, and that the angles of the rhombus will change at a constant rate.

15. If, in the preceding Question, the masses (M) of the particles at A and C are different from those (m) of the particles at B and D, form Lagrange's equations of motion, in terms of the angle θ which OA makes with a fixed direction, and the angle ϕ which AB makes with AO.

Verify that the equations have the first integrals

$$(M\cos^2\phi + m\sin^2\phi)\,\dot{\theta} = \text{const.},$$

$$(M\cos^2\phi + m\sin^2\phi)\,\dot{\theta}^2 + (M\sin^2\phi + m\cos^2\phi)\,\dot{\phi}^2 = \text{const.}$$

16. Four equal bars, smoothly jointed together, form a rhombus $ABCD$, which is supported in a vertical plane at the middle point of AB. The mass-centre of each bar is at its middle point; the radius of gyration about this point is κ; and the length is $2a$. If θ be the angle which AB or DC makes with the horizontal, and ϕ that which AD or BC makes with the vertical, prove that the kinetic energy of the system when swinging in a vertical plane is

$$M(a^2+\kappa^2)\,\dot{\theta}^2 + M(\kappa^2+3a^2)\,\dot{\phi}^2,$$

where M is the mass of a bar.

Form the equations of motion, and interpret the results.

17. A horizontal bar AB hangs by two equal vertical strings of length l attached to its ends, and a similar bar CD hangs from AB by two equal strings AC, BD, of length l'. If the system makes small angular oscillations about the vertical through the centres of gravity (which are the middle points) prove that the periods $2\pi/n$ of the two normal modes are given by the equation

$$n^4 - \frac{2ga^2}{\kappa^2}\left(\frac{1}{l'}+\frac{1}{l}\right)n^2 + \frac{2g^2a^4}{\kappa^4 ll'} = 0,$$

where a is the half-length of a bar, and κ its radius of gyration about the centre.

18. A compound pendulum hangs from the axis of a hollow circular cylinder which is free to roll along a horizontal plane. If ϕ be the angular coordinate of the cylinder, θ the inclination of the pendulum to the vertical, prove that

$$2T = (I+Ma^2)\,\dot{\phi}^2 - 2Mah\,\dot{\theta}\,\dot{\phi}\cos\theta + Mk^2\dot{\theta}^2,$$

where a is the radius of the cylinder, I its moment of inertia about the line of contact with the plane, and the letters M, h, k have their usual meanings in relation to the compound pendulum.

Prove that in the case of small oscillations the length of the equivalent simple pendulum is less than if the cylinder were fixed, in the ratio

$$1 - \frac{Ma^2h^2}{(I + Ma^2)\,k^2}.$$

19. A particle is attached at Q to a string PQ of length l. The point P is made to describe a circle of radius a about a fixed point O with the constant angular velocity ω, and the whole motion is in one plane, gravity being neglected. If χ be the angle which PQ makes with OP produced, prove that

$$\frac{d^2\chi}{dt^2} + \frac{\omega^2 a}{l}\sin\chi = 0.$$

20. A bar of mass M hangs in a horizontal position by two equal vertical strings of length l. From it are suspended at different points two particles of equal mass m by two strings of length l'. Prove that the periods $2\pi/n$ of the three normal modes of oscillation in the vertical plane through the equilibrium position of the bar are given by

$$n^2 = g/l',$$

and

$$n^4 - g\left(\frac{1}{l} + \frac{1}{l'}\right)\left(1 + \frac{2m}{M}\right)n^2 + \frac{g^2}{ll'}\left(1 + \frac{2m}{M}\right) = 0.$$

21. The three knife-edges A, O, B of the beam of a balance are in one plane, and the line OG joining the central one to the mass-centre G is perpendicular to AB. Two equal masses m are suspended from A and B by similar helical springs of stiffness K. Having given that $OA = OB = a$, $OG = h$, that the mass of the beam is M, and its radius of gyration about O is k, prove that, in the case of small oscillations about equilibrium,

$$2T = (Mk^2 + 2ma^2)\,\dot\theta^2 + 2ma\,(\dot x - \dot y)\,\dot\theta + m\,(\dot x^2 + \dot y^2),$$

$$2V = Mgh\,\theta^2 + K\,(x^2 + y^2),$$

where θ is the inclination of AB to the horizontal, and x, y are the increments of length of the two springs from their equilibrium values.

22. Prove that, in the preceding Question, the periods $(2\pi/n)$ of the three normal modes of vibration are given by the equations

$$n^2 = K/m,$$

and

$$n^4 - \left\{\left(1 + \frac{2ma^2}{Mk^2}\right)\frac{K}{m} + \frac{gh}{k^2}\right\}n^2 + \frac{K}{m}\cdot\frac{gh}{k^2} = 0.$$

EXAMPLES XXIII.

(Miscellaneous.)

1. The engines of an airship moving on a straight horizontal course are shut off, and a series of observations of the velocity are taken at subsequent instants. When these are plotted with the reciprocal of the velocity as ordinate and the time as abscissa, the corresponding points are found to lie on a straight line. What is the law of resistance?

What would have been the form of the graph if the resistance had been proportional to the *cube* of the velocity?

2. An aeroplane whose speed is 80 miles an hour flies to an aerodrome 80 miles distant in a direction making an angle 60° E. of N., and returns. There is a N. wind of 20 miles an hour. Prove that the times out and home are 70·5 minutes and 54·5 minutes, respectively.

3. An aeroplane has a speed v miles per hour and a range of action (out and home) of R miles in calm weather. Prove that in a N. wind w its range in a direction whose true bearing is ϕ is

$$\frac{R(v^2 - w^2)}{v\sqrt{(v^2 - w^2\sin^2\phi)}}.$$

4. The centre of gravity of an airship is moving with velocity V in a direction making an angle θ with the fore and aft line, and the ship is turning with angular velocity ω. Prove that the local yaw (θ') at a distance x from the centre of gravity is given by

$$\tan\theta' = \tan\theta + \frac{\omega x}{V}\sec\theta.$$

5. A point describes an equiangular spiral with a constant angular velocity about the pole; prove that the hodograph is a similar spiral.

6. A point P describes a straight line with constant velocity u, and a second point Q moves so that its (constant) velocity v is always at right angles to QP. Prove that the path of Q relative to P is a conic of eccentricity u/v.

7. The motion of a point relative to axes rotating with angular velocity ω is elliptic harmonic about the origin O, the period being $2\pi/\omega'$. Prove that the path referred to fixed axes through O is an epicyclic.

8. A curve rolls on a straight line with angular velocity ω. Prove that the acceleration of that point of the curve which is in contact with the line is normal to the latter and equal to $\omega^2\rho$.

What is the corresponding result in the case of one curve rolling on another which is fixed?

9. A lamina moves in any manner in its own plane, its angular velocity at any instant being ω. Prove that the component accelerations of a point P of the lamina at a distance r from the point of contact A of the two pole-curves [S, 16] are

$$r\frac{d\omega}{dt}+\frac{\omega^2}{c+c'}\sin\theta, \text{ and } \omega^2 r-\frac{\omega^2}{c+c'}\cos\theta,$$

along the tangent and normal to the path of P, respectively, where c, c' are the curvatures of the two pole-curves at A, and θ is the angle which AP makes with the common normal at A. Deduce the curvature of the path of P.

10. A number of equal particles m are at the vertices of a regular polygon whose sides are formed by strings of length b. Prove that if the polygon revolves in its own plane about the centre the tension of each string is mv^2/b, where v is the velocity of each particle; and deduce the formula for the tension in a revolving circular chain.

11. A particle is subject to a force constant in magnitude and direction, and to a force normal to the path and proportional to the velocity. Prove that the path is a trochoid.

12. The point of suspension of a pendulum of length l yields horizontally through a small space X/μ, where X is the horizontal pull on it. Prove that the length of the equivalent simple pendulum is increased by mg/μ, where m is the mass of the bob.

13. The bob of a simple pendulum of length l makes complete revolutions, and the velocity v at the level of the point of suspension is large compared with $\sqrt{(gl)}$. Prove that the times of describing the upper and lower halves of the circle are approximately

$$\frac{\pi l}{v}\left(1\pm\frac{2gl}{\pi v^2}\right).$$

14. Two identical pendulums of length l swing in a vertical plane, being suspended from two points of a rigid horizontal bar which is not quite rigidly fixed, but yields horizontally to the pull of the pendulums by an amount $\beta l(\theta+\phi)$, where θ and ϕ are their inclinations (reckoned in the same sense) to the vertical, and β is small. Form the equations of motion, and solve them, determining the constants so that initially one bob is at rest in its mean position, and the other is started from its mean position with a given small velocity.

Prove that the two pendulums alternately will be nearly at rest in their mean positions, at intervals of about $1/2\beta$ periods.

15. Two particles m_1, m_2, moving with velocities v_1, v_2, in directions inclined to one another at an angle a, collide and coalesce. Prove that their subsequent velocity is

$$\frac{\sqrt{(m_1^2 v_1^2+2m_1 m_2 v_1 v_2\cos a+m_2^2 v_2^2)}}{m_1+m_2},$$

and that the loss of energy is

$$\frac{m_1 m_2}{2(m_1+m_2)}(v_1^2-2v_1 v_2\cos a+v_2^2).$$

16. A rigid body hangs from a horizontal axis; prove that the radius of gyration about this axis can be computed if we know the period of a small oscillation, and also the deflection produced by a small force acting in a given horizontal line below and at right angles to the axis.

Work out the result numerically having given that the period is 1·566 sec., and that a force equal to one-hundredth of the weight of the pendulum, acting in a line 3 ft. below the axis, displaces the point of application through ·072 of an inch. [5·48 ft.]

17. A cylinder rolls down a plane whose inclination to the horizontal is α, and its length makes an angle β with the lines of greatest slope. Prove that its centre has an acceleration

$$\frac{a^2}{a^2+\kappa^2} g \sin \alpha \sin \beta,$$

where a is the radius, and κ the radius of gyration about the axis.

18. A rigid body symmetrical with respect to a plane through its mass-centre G is moving in this plane with a uniform velocity of translation. If a point P in this plane, such that GP is perpendicular to the direction of motion, is suddenly fixed, prove that the fraction κ^2/k^2 of the original energy is lost, where κ and k are the radii of gyration about axes through G and P, respectively, perpendicular to the plane of symmetry.

19. Two equal rods AB, BC in a straight line, hinged together at B are moving at right angles to their length when the end A is suddenly fixed. Prove that the initial angular velocities are as $-3:1$.

20. Three equal uniform rods AB, BC, CD, hinged at B and C, are at rest in a straight line. Prove that if the rod AB is struck at right angles at any point P, the impulsive reaction at B is four times that at C.

Also that AB will not rotate initially if $AP:PB=11:7$.

21. A tube of mass M and length $2a$ lies on a smooth table, and contains a particle m close to its centre, which is also its centre of mass. Initially a rotation ω is given to the tube. Prove that when the particle leaves the tube the angular velocity is

$$\frac{(M+m)I\omega}{(M+m)I+Mma^2},$$

where I is the moment of inertia of the tube about its centre.

22. A spherical vessel of radius a initially at rest is free to rotate about a fixed axis. It is filled by liquid rotating about the same diameter with an angular velocity $\omega(r)$ which is a function only of the distance r from the centre. Prove that when all relative rotation has been destroyed by friction the angular velocity of the vessel will be

$$\frac{8}{3}\frac{\pi\rho}{I}\int_0^a \omega(r) r^4 dr,$$

where I is the moment of inertia of the whole mass and ρ the density of the fluid.

23. A man of mass m stands at A on a horizontal lamina which is free to rotate about a fixed point O. Initially both are at rest. The man then walks along the circle on OA $(=a)$ as diameter, returning to his starting point. Prove that the lamina will have turned through an angle

$$\pi\left\{1-\sqrt{\left(\frac{I}{I+ma^2}\right)}\right\},$$

where I is the moment of inertia of the lamina about O.

24. A massive globe is rotating about a fixed diameter with angular velocity ω. A man whose mass is m starts from either pole and travels with constant velocity v along a meridian. Prove that when he reaches the opposite pole the rotation of the globe will have been retarded by an angle

$$\frac{\pi\omega a}{v}\left\{1-\left(\frac{I}{I+ma^2}\right)^{\frac{1}{2}}\right\},$$

where I is the moment of inertia of the globe, and a its radius.

25. Two spheres of radius a and density ρ are in contact. Prove that the pressure between them due to their mutual gravitation is equal to the attraction of the earth on a mass

$$\frac{1}{3}\frac{\pi\rho^2a^4}{\rho_0 R},$$

where R is the earth's radius, and ρ_0 its mean density.

Assuming $\rho=8$, $\rho_0=5{\cdot}6$, $R=6{\cdot}38\times10^8$ C.G.S., find the radius a in order that the pressure may be a kilogramme. [4·8 metres.]

26. In a parabolic orbit about the focus the time of describing an angle θ from the apse is

$$\frac{l^{\frac{3}{2}}}{2\mu^{\frac{1}{2}}}\left(\tan\tfrac{1}{2}\theta+\tfrac{1}{3}\tan^3\tfrac{1}{2}\theta\right).$$

27. A particle is subject to a central acceleration μ/r^2, and is projected from a great distance with velocity v so that if there were no attraction it would pass the centre of force at a distance b. Prove that its direction of motion will ultimately be turned through an angle $\pi-2\alpha$, where

$$\alpha=\tan^{-1}(v^2b/\mu).$$

28. Prove that three particles m_1, m_2, m_3 situate at the corners of an equilateral triangle and attracting one another according to the law of the inverse square can describe circular orbits about their common centre of mass with the angular velocity

$$\left\{\frac{\gamma(m_1+m_2+m_3)}{a^3}\right\}^{\frac{1}{2}},$$

where a is the side of the triangle. (Laplace.)

29. A planet J and the sun S are describing circles about their common centre of mass G, with angular velocity n, and P is a satellite whose reaction on S and J may be neglected. If x, y be the coordinates of P relative to axes with S as origin, such that the axis of x passes always through G, prove that

$$\ddot{x} - 2n\dot{y} = \frac{\partial \Omega}{\partial x}, \quad \ddot{y} + 2n\dot{x} = \frac{\partial \Omega}{\partial y},$$

where $$\Omega = \frac{\gamma S}{r} + \frac{\gamma J}{\rho} + \tfrac{1}{2}n^2 \frac{Sr^2 + J\rho^2}{S+J},$$

and $$r = SP, \ \rho = JP.$$

30. A number of particles are projected simultaneously in various directions from the same point under a central acceleration $\mu \times$ (dist.). Prove that after a time $\frac{1}{2}\pi/\sqrt{\mu}$ their velocities will be parallel and equal.

31. Prove that in Ex. 4, Art. 86, the path of the particle on the table lies between two fixed circles unless the angular momentum is zero.

32. Two particles m_1, m_2 are connected by a string of length $a_1 + a_2$ which passes through a smooth ring on a smooth horizontal table, and are describing circles of radii a_1, a_2 with angular velocities ω_1, ω_2 respectively. Prove that

$$m_1 \omega_1^2 a_1 = m_2 \omega_2^2 a_2;$$

and shew that the period of a small oscillation about this state is

$$2\pi \sqrt{\left\{\frac{m_1 + m_2}{3(m_1 \omega_1^2 + m_2 \omega_2^2)}\right\}}.$$

33. If a particle be subject to a central acceleration

$$\frac{\mu}{r^2} e^{k/r},$$

a circular orbit will be stable only if its radius is greater than k.

34. Prove that in the case of a central force varying inversely as the cube of the distance

$$r^2 = At^2 + Bt + C,$$

where A, B, C are constants.

35. A particle subject to a repulsive force varying as $1/r^3$ is projected from infinity with a velocity which would carry it to a distance a from the centre of force if it were directed towards the latter. Actually it is projected along a line whose distance from the centre of force is b. Prove that its least distance from the centre will be $\sqrt{(a^2 + b^2)}$, and that the angle between the asymptotes of its path will be $\pi b/\sqrt{(a^2 + b^2)}$.

36. A particle subject to a repulsive force varying as $1/r^5$ is projected from infinity with a velocity V which would carry it to a distance a from the centre of force if it were directed towards the latter. Actually it is projected

so that it would pass the centre of force at a distance b if there were no repulsion. Prove that the least velocity of the particle is

$$V\frac{b^2}{a^2}\left\{\left(\frac{a^4}{b^4}+\frac{1}{4}\right)^{\frac{1}{2}}-\frac{1}{2}\right\}^{\frac{1}{2}}.$$

37. If the retardation of a train is $a+bv^2$, prove that if the engines are shut off when the speed is v_0 the train will come to rest in a time

$$\frac{1}{\sqrt{(ab)}}\tan^{-1}\left(\sqrt{\frac{b}{a}}\cdot v_0\right)$$

after travelling a distance

$$\frac{1}{2b}\log\left(1+\frac{bv_0^2}{a}\right).$$

38. Prove that the intrinsic equation of the path of a projectile under a retardation kv^2 is

$$e^{2ks}=1+k\rho_0\{\sec\psi\tan\psi+\log(\sec\psi+\tan\psi)\},$$

where s is the arc measured forward from the highest point, ρ_0 is the radius of curvature at this point, and ψ is the inclination of the path below the horizontal.

Prove that for small values of s

$$\psi=(s+ks^2)/\rho_0.$$

39. A circular tube rotates in its own plane with angular velocity ω about a point O on the circumference. Prove that the relative motion of a particle in it is subject to the equation

$$\frac{d^2\theta}{dt^2}+\omega^2\sin\theta=0,$$

where θ denotes angular distance from the point of the tube opposite to O.

40. A rigid tube in the form of a plane curve revolves about a point O in its plane with constant angular velocity ω. Prove that the small (relative) oscillations of a particle about a point P of the tube whose distance from O is stationary are determined by the equation

$$\frac{d^2s}{dt^2}+\omega^2\left(\frac{r}{\rho}-1\right)s=0,$$

where $r=OP$, and ρ is the radius of curvature at P.

Hence shew that the equilibrium of the particle when at P is stable or unstable according as P is a point of maximum or minimum distance from O.

41. A particle is constrained to move on a smooth spherical surface of radius a. Prove that if projected with velocity v along the horizontal great circle it will fall through a vertical height ae^{-u}, where u is defined by

$$\sinh u=\frac{v^2}{4ga}.$$

Prove that if v^2 is large compared with $4ga$, the result is $2ga^2/v^2$, approximately.

42. A horizontal rod AB rotates with constant angular velocity ω about its middle point O. A particle P is attached to it by equal strings AP, BP. If θ be the inclination of the plane APB to the vertical, prove that

$$\frac{d^2\theta}{dt^2} - \omega^2 \sin\theta \cos\theta = -\frac{g}{l} \sin\theta,$$

where $l = OP$. Deduce the condition that the vertical position of OP should be stable.

43. If a system of particles m, m', ... attract one another with forces varying as the nth power of the distance, and r, r', ... denote their distances from the centre of mass of the system, prove that

$$\frac{d^2}{dt^2}(\Sigma mr^2) = -(2n+6)\, V + \text{const.},$$

where V is the potential energy of the system.

Prove that if $n = -3$ a stable configuration is impossible.

APPENDIX

NOTE ON DYNAMICAL PRINCIPLES

ALTHOUGH the day is long past when there could be any serious difference of opinion as to whether a particular solution of a dynamical question is or is not correct, the proper formulation of the *principles* of Dynamics has been much debated, and especially in recent times.

To a student still at the threshold of the subject the most important thing is that he should acquire as rapidly as possible a system of rules which he can apply without hesitation, and, so far as his mathematical powers will allow, with success, to any dynamical question in which he may be interested. From this point of view it is legitimate, in expounding the subject, to take advantage of such prepossessions which he may have as are serviceable, whilst warning him against others which may be misleading. This is the course which has been attempted in the present work, as in most elementary accounts of the subject.

But if at a later stage the student, casting his glance backwards, proceeds to analyse more closely the fundamental principles as they have been delivered to him, he may become aware that there is something unsatisfactory about them from a formal, and even from a logical standpoint. For instance, it is asserted, as an induction from experience, that the velocity of a body not acted on by any force will be constant in magnitude and direction, whereas the only means of ascertaining whether a body is, or is not, free from the action of force is by observing whether its velocity is constant. Again, by velocity we mean, necessarily, velocity relative to some frame of reference, but what or where this frame is, is nowhere definitely specified.

If the student's intellectual history follows the normal course he may probably, after a few unsuccessful struggles, come to the conclusion that the principles which he is virtually, though not altogether expressly, employing must be essentially sound, since they lead invariably to correct results, but that they have somehow not found precise and consistent formulation in the text-books. If he chooses to rest content in this persuasion, deeming that form and presentation are after all secondary matters, he may perhaps find satisfaction in the reflection that he is in much the same case as the great masters of the science, Newton, Euler, d'Alembert, Lagrange, Maxwell, Thomson and Tait (to name only a few), whose expositions, whenever they do not glide hastily over preliminaries, are all open, more or less, to the kind of criticism which has been referred to.

The student, however, whose interest does not lie solely in the applications, may naturally ask whether some less assailable theoretical basis cannot be provided for a science which claims to be exact, and has a long record of verified deductions to its credit. The object of this Appendix is to indicate the kind of answer which may be given to this question.

The standpoint now generally adopted is a purely empirical one. The object of all science, it is held, is to give an account of the way things go on in the world*. Guided by experience we are able to frame rules which enable us to say with more or less accuracy what will be the consequences, or what were the antecedents, of a given state of things. These rules are sometimes dignified by the name of 'laws of nature,' but they have relation to our present state of knowledge, and to the degree of skill with which we have succeeded in giving more or less compact expression to this. They are therefore liable to be modified from time to time, or to be superseded by more convenient or more comprehensive modes of statement.

* As regards Mechanics this is expressed by Kirchhoff as follows: "Bei der Schärfe, welche die Schlüsse in der Mechanik sonst gestatten, scheint es mir wünschenswerth, solche Dunkelheiten aus ihr zu entfernen, auch wenn das nur möglich ist durch eine Einschränkung ihrer Aufgabe. Aus diesem Grunde stelle ich es als die Aufgabe der Mechanik hin, die in der Natur vor sich gehenden Bewegungen zu *beschreiben*." *Mechanik*, Leipzig, 1876.

Again, it is to be remembered that we do not aim at anything so hopeless, or indeed so useless, as a *complete* description of any phenomenon. Some features are naturally more important or more interesting to us than others; by their relative simplicity and evident constancy they have the first claim on our attention, whilst those which are apparently accidental and vary from one occasion to another are ignored, or postponed for later examination. It follows that for the purposes of such description as is possible some process of abstraction is inevitable if our statements are to be simple and definite. Thus in studying the flight of a stone through the air we replace the body in imagination by a mathematical point. The size and shape of the body, the complicated spinning motion which it is seen to execute, the internal strains and vibrations which doubtless take place, are all sacrificed in the mental picture in order that attention may be concentrated on those features of the phenomenon which are in the first place most interesting to us. At a later stage in our subject, the conception of the ideal rigid body is introduced; this enables us to fill in some details which were previously wanting, but others are still omitted. Again, the conception of a force as concentrated in a mathematical line is as artificial as that of a mass concentrated in a point, but it is a convenient fiction for our purpose, owing to the simplicity which it lends to our statements.

The laws which are to be imposed on these ideal representations are in the first instance largely at our choice, since we are dealing now with mental objects*. Any scheme of Abstract Dynamics constructed in this way, provided it be self-consistent, is mathematically legitimate; but from the physical point of view we require that it should help us to picture the sequence of phenomena as they actually occur. The success or failure in this respect can only be judged *à posteriori*, by comparison of the results to which it leads with the facts. It is to be noticed, moreover, that available tests apply only to the scheme as a whole,

* Cf. Maxwell: "The bodies which we deal with in abstract dynamics are just as completely known to us as the figures in Euclid. They have no properties whatever except those which we explicitly assign to them." *Scientific Papers*, vol. 2, p. 779. The Article quoted was written in 1879.

since, owing to the complexity of real phenomena, we cannot subject any one of its postulates to verification apart from the rest.

It is from this point of view that the question of relativity of motion, which is often felt to be a stumbling-block on the very threshold of the subject, is to be judged. By 'motion' we mean of necessity motion relative to some frame of reference which is conventionally spoken of as 'fixed.' In the earlier stages of our subject this may be any rigid, or apparently rigid, structure fixed relatively to the earth. When we meet with phenomena which do not fit easily into this view, we have the alternatives, either to modify our assumed laws of motion, or to call to our aid adventitious forces, or to examine whether the discrepancy can be reconciled by the simpler expedient of a new basis of reference. It is hardly necessary to say that the latter procedure has hitherto been found to be adequate. In the first instance we adopt a system of rectangular axes whose origin is fixed in the earth, but whose directions are fixed by relation to the stars; in the planetary theory the origin is transferred to the sun, and afterwards to the mass-centre of the solar system; and so on. At each step there is a gain in accuracy and comprehensiveness; and the conviction is cherished that *some* system* of rectangular axes exists with respect to which the Newtonian scheme holds with all necessary accuracy.

Similar remarks apply to the conception of time as a measurable quantity. From the purely kinematic point of view the t of our formulæ may be any continuous independent variable, suggested (it may be) by some physical process. But from the dynamical standpoint it is obvious that equations which represent the facts correctly on one system of time-measurement might become seriously inadequate on another. The Newtonian system postulates (virtually) not only an absolute geometrical basis of reference, but also some absolute time-scale, which (so far as observations go) is hardly distinguishable, if at all, from the practical scale based on the rotation of the earth.

* This system is of course not uniquely determinable. If a system A can be found answering the requirements, any other system B which has a motion of *translation* with constant velocity relative to A will serve equally well (Arts. 2, 22).

The obviously different degrees of inertia of actual bodies are imitated, in our ideal representation, by attributing to each particle a suitable mass- or inertia-coefficient m, and the product $m\mathbf{v}$ of the mass into the velocity is called the momentum. On the Newtonian system, change of velocity is attributed to the action of force; and if we agree to measure the force $\mathbf{P}$ by the rate of change of momentum which it produces, we have the vector equation

$$\frac{d}{dt}(m\mathbf{v}) = \mathbf{P}.$$

From this point of view the equation is a mere truism; its real importance rests on the fact that by attributing suitable values to the masses m, and by making simple assumptions as to the values of $\mathbf{P}$, we are able to obtain adequate representations of whole classes of phenomena, as they actually occur.

Similar remarks might be made with respect to the law of 'action and reaction,' and the further postulates which are introduced when we deal with 'rigid' bodies; but it is unnecessary to go through these in detail*.

In conclusion we would remark that the preceding considerations have been brought forward with a view of placating the conscience of the student, rather than with the intention of providing him with a new set of working ideas. It is well to recognize, occasionally, the artificial character, and the limitations, of our methods; but it is by no means desirable that the student should endeavour always to frame his dynamical thoughts and imaginations in terms of the somewhat shadowy images of abstract theory, rather than of the more familiar and concrete notions of force and inertia, which, however crude they may seem to the logician, have been the stock in trade of almost all the great workers in the science.

* The substance of this note is taken mainly from the article on 'Mechanics' in the eleventh edition of the *Encyclopædia Britannica*, by kind permission of the proprietors.

INDEX

[The numbers refer to the pages.]

Printed in the United States
By Bookmasters